P9-ELG-747

THE COMMERCE OF CARTOGRAPHY

THE KENNETH NEBENZAHL, JR., LECTURES IN THE HISTORY OF CARTOGRAPHY

Published in Association with the Hermon Dunlap Smith Center for the History of Cartography, The Newberry Library

Series Editor, James Akerman

The Commerce of Cartography

MAKING AND MARKETING MAPS
IN EIGHTEENTH-CENTURY FRANCE AND ENGLAND

MARY SPONBERG PEDLEY

THE UNIVERSITY OF CHICAGO PRESS
CHICAGO AND LONDON

MARY SPONBERG PEDLEY is adjunct assistant curator of maps at the William L. Clements Library at the University of Michigan, a Latin instructor in the Ann Arbor Public Schools, and an associate editor of *Imago Mundi: The International Journal for the History of Cartography.*

The University of Chicago Press, Chicago 60637
The University of Chicago Press, Ltd., London

Printed in the United States of America

14 13 12 11 10 09 08 07 06 05 1 2 3 4 5
ISBN: 0-226-65341-2 (cloth)

Library of Congress Cataloging-in-Publication Data

Pedley, Mary Sponberg.
The commerce of cartography : making and marketing maps in eighteenth-century France and England / Mary Sponberg Pedley.
p. cm. — (Kenneth Nebenzahl, Jr., lectures in the history of cartography)
Includes bibliographical references and index.
ISBN 0-226-65341-2 (alk. paper)
1. Map industry and trade—France—History—18th century. 2. Map industry and trade—England—History—18th century. 3. Cartography—France—History—18th century. 4. Cartography—England—History—18th century.
I. Title. II. Series.
Z286.M3P43 2005
381'.4591241'09033—dc22

2004015210

♾ The paper used in this publication meets the minimum requirements of the American National Standard for Information Sciences—Permanence of Paper for Printed Library Materials, ANSI Z39.48-1992.

For

Grace

Annie

&

Eila

CONTENTS

APPENDIXES

ILLUSTRATIONS

FIGURES

PLATES
(FOLLOWING PAGE 174)

PREFACE

This book took root twenty years ago when I was doing research on the Robert de Vaugondy family of geographers, who printed and published their own maps. I wanted to understand the economic background of their intellectual and critical success but commercial failure. Supported by a fellowship from the National Endowment for the Humanities in 1984, I trolled the Paris archives for documents relating to the map trade in the eighteenth century. Summer visits and a short-term research grant from the American Council of Learned Societies in 1990 allowed more archival time in France over the years. I did not fully appreciate the trade connections between Paris and London until I began editing for publication the business letters from Continental map dealers to the London firm of Jefferys and Faden.[1] When I was invited to give the Kenneth Nebenzahl, Jr., lectures at the Newberry Library on the subject of the map trade, I decided to stretch the French canvas to include England to compare these two markets in light of the close trade relations between their mapmakers and map sellers. A John Simon Guggenheim Memorial Foundation Fellowship in 2001 provided support to enlarge the scope of my lectures and this study.

Having come to the English map trade fairly recently, I have relied heavily and with gratitude on the work done by other researchers, who have studied individual maps and ferreted out the primary sources for English cartography. I benefited from Tony Campbell's excellent guide to the cartographic holdings in the Manuscripts Division of the British Library. There are rich veins still untapped in the Print Room of the British Museum in the Public Record Office (now the National Archives), and in the Admiralty and Colonial Office records, as well as in the records of regional archives, particularly for county mapping projects, as the work of J. B. Harley, Donald Hodson, Paul Laxton, Laurence Worms, and others shows.

The Archives nationales in Paris hold rich deposits for the study of the economics of the map trade. The records of the French hydrographic office, the Dépôt des Cartes, Plans, et Journaux de la Marine, contain the accounts and occasionally the daily notes describing its administration. In addition, the marine archives house the papers of several well-known cartographers of the eighteenth century: the Delisle family (especially Guillaume and Joseph-Nicolas), Philippe Buache, the marquis de Chabert, and the comte de Fleurieu. Because legal transactions were regularly notarized and copies deposited in the Minutier central, this archive is a valuable source of biographical and economic information for any study of the artisanal life in Paris. So too the records found in Série Y: Châtelet de Paris et Prévôté d'île-de-France, especially those of the Chambre de police, cover various lawsuits and complaints brought within the artisanal community of Paris. The Collection Anisson in the Département des manuscrits of the Bibliothèque nationale de France contains records concerning the book trade in mid-eighteenth-century Paris; they also reveal colorful disputes and litigation. The Département des Cartes et Plans of the Bibliothèque nationale de France holds the papers of various geographers (among whom are Buache, Buache de la Neuville, Dezauche, and Robert de Vaugondy), many of which concern their commercial dealings. These holdings may be supplemented with those in the Bibliothèque de l'Institut, where collections—literary and cartographic—formed by members of any of the five academies may be found.

This brief description of sources in part explains what might be perceived as an imbalance between the primary sources available to me for the French part of this book, and my heavy reliance on secondary material or primary material already published for the English section of the book. I would not dare suggest that this work is complete. What I have provided here is a beginning, a gathering of the data as I have found it and an attempt to piece together the jigsaw puzzle of the map trade.

ACKNOWLEDGMENTS

First and foremost, my warmest thanks to Mr. and Mrs. Kenneth Nebenzahl for their generous support of the Kenneth Nebenzahl, Jr., Lectures in the History of Cartography and for the opportunity afforded to me to share my work with a receptive audience in the autumn of 2001. The occasion was enriched by my fellow lecturers, David Woodward, Peter van der Krogt, and Markus Heinz, whom I thank for their enthusiastic and enlightening participation. These lectures would not have taken place at all without the invitation and encouragement of James Akerman, who saw possibilities in my research well before I did. As always, the staff at the Newberry Library stimulated a taste for maps by the exhibit especially for the occasion of the lectures, *The Cartographic Treasures of the Newberry Library*, curated by Jim Akerman and Robert Karrow. To Jim and Bob, who have been good friends for many years, I owe very deep thanks. Their good cheer and generosity have made the Newberry Library a welcome center for cartographic research, and their readiness to listen and answer questions and share their own work makes them valued colleagues. To fellow workers in the Newberry vineyard, Arthur Holzheimer and Pat Morris, I also give thanks for the insights they have shared and help they have given.

My work in the history of cartography finds its home in the Map Division of the William L. Clements Library of the University of Michigan. John C. Dann has actively supported research on the history of cartography throughout his tenure as director of the library. His keen appreciation of the importance of maps as historical documents has strengthened the map collection and encouraged his map curators to continue a Clements tradition of published research on the history of cartography. The staff of the

Clements Library counts scholarship and inquiry as an active component in the life of the library, and their interest and support have added much to this study. Former curator of maps David Bosse has long been an ally in studying the map trade, and I have benefited enormously both from nearly twenty years of ongoing conversations about shared research and from his willingness to read and comment on what I have written. Brian Dunnigan, current map curator at the Clements, now cheerfully suffers and shares many of my research dilemmas and has contributed to many of the discoveries. John Harriman has ever been a stalwart aide in matters practical, historical, and poetic. Barbara De Wolfe, Don Wilcox, and Clayton Lewis have listened to many of the ideas contained here. Volunteer and Michigan Map Society stalwart Carmen Miller has helped with a long-term search for map advertisements, aided by fellow volunteer Cheryl MacKrell. The Clements Library has also contributed the majority of illustrations for this work, for which I am very grateful; they have been skillfully and patiently prepared by Paul Jaronski of the University of Michigan Photo Services.

The Map Library of the University of Michigan's Graduate Library also houses a significant antiquarian collection of maps, looked after by Karl Longstreth, Tim Utter, and Jerry Thornton (now retired), who over the years have provided cheerful and generous assistance.

No work on French maps begins without enjoying time in the Département des Cartes et Plans of the Bibliothèque nationale de France. Many happy hours there have been spent in the company of Monique Pelletier, who has provided much valued support to many of my endeavors. Catherine Hofmann has been unstintingly generous in helping me unravel some of the complexities of the French map trade. Mireille Pastoureau, whom I first met in Cartes et Plans and whose hospitality and generosity of scholarship made a significant impression, is now in charge of the Bibliothèque de l'Institut; her friendship and aid in the early years of my research are gratefully remembered. Other friends in Paris have also added to my research and to the joy of visiting that capital: Isabelle Raynaud, with her knowledge and appreciation of eighteenth-century history, and Dudley Barnes, a walking encyclopedia of maps, who has suggested examples for many of the themes in this book. I also thank Pierre Casselle of the Bibliothèque de la Ville de Paris and Corinne Le Bitouzé of the Département des Estampes of the Bibliothèque nationale de France for allowing me access to their theses on the print trade in Paris.

In London, work in the cramped quarters of the old map room of the British Library was always brightened by Helen Wallis's welcome. Her successors, Tony Campbell and Peter Barber, continue that tradition in the more expansive surroundings of the new British Library. I appreciate Tony

Campbell's support and advice at crucial stages in this project, and Peter Barber has shared generously from his broad background for much of this research. Surekha Davis, Geoff Armitage, and Andrew Cook also have provided help and ideas for various aspects of this book. Sarah Tyacke and the map specialists Geraldine Beech and Rose Mitchell made working in the Public Record Office, now the National Archives, a particular pleasure. The cartographic researcher's special joy in London is visiting the map room of the Royal Geographic Society, where Francis Herbert answers nearly every question with an apt example from his prodigious memory. No work on the London engravers and cartographers of the eighteenth century goes far without help from Laurence Worms and Donald Hodson. Their authoritative work on the London map trade and their collegial sharing of their findings have enriched this study.

In the Geography and Map Division of the Library of Congress, Ron Grim and Jim Flatness have helpfully answered questions and provided examples. Susan Danforth of the John Carter Brown has read and commented on sections of this work, for which I am grateful. Ed Dahl, retired now as early map specialist from the Public Archives of Ottawa, continues to be an invaluable source of information and clarification, particularly on North American material. Margaret Pritchard of the Colonial Williamsburg Foundation generously helped with several illustrations. David Woodward and David Bosse read through the entire manuscript, and to them I am deeply indebted for their time, help, and advice. David Woodward's recent death has deprived map historians of a magnanimous colleague and cherished friend. Two anonymous readers and Josef Konvitz also read the manuscript for the University of Chicago Press and added significantly to the bibliography and clarification of many details, as well as suggesting an additional chapter. Old friends Betty Ingram, Ann Johnston, and Catherine Delano-Smith have listened to many parts of this book in its various permutations over many years and in various contexts. The book's editors at the University of Chicago Press, Christie Henry and Michael Koplow, have sorted out many problems with patience and flexibility. To them all I give credit for what is good in this book; the errors are mine alone.

My sister Ingrid has contributed to this work in more ways than she knows. Though she hears the word "cartography" only when I am around, she is one of my best audiences, and I thank her for her sympathetic ear and insightful comments. My husband John has contributed in all the ways he does know, for which my love and gratitude are boundless. This book is dedicated to the memory of three women whose lives are models for me of energy, enthusiasm, and commitment: Grace, Annie, and Eila.

MONETARY UNITS

In France, the *livre* was divided into twenty *sous* (or *sols*). Each sou was made up of twelve *deniers*. An *écu* was equal to three livres. Thus, 1 livre was worth 240 deniers, while 1 écu was worth 720 deniers. The French currency of écus, livres, sous, and deniers was converted to a decimal currency based on francs and centimes on 17 germinal an XI (7 April 1803).

In England, the *pound* was made up of twenty *shillings*. Each shilling was divided into twelve *pence* (abbreviated as "d"). English currency also used a denomination of one *guinea*, which was worth one pound and one shilling.

The rate of exchange between France and England fluctuated; in general, it was between twenty and twenty-five livres to one pound. For ease of conversion, a rough equivalent of one French livre to one English shilling might be used.

Further considerations of currency and exchange may be found in John J. McCusker, *Money and exchange in Europe and America, 1600–1775: a handbook* (Chapel Hill: University of North Carolina Press, 1978).

INTRODUCTION

This book sets out to study the commercial factors affecting the production and consumption of printed maps in eighteenth-century France and England. The world of printed maps in the eighteenth century included everything from maps used as illustrations in books and periodicals to large-scale estate and town plans, military maps, regional maps, small-scale geographical maps, wall maps, atlases, and globes. All were available in printed form during a period that saw a dramatic increase in the quantity and quality of consumer goods and an equally dramatic increase in the spending power not only of the wealthy, but also of the middling and even working classes.[1] Benefiting from these favorable conditions, the printed map became the most accessible form of geographic information for the largest number of people throughout Western Europe.

The centers of printed map production in Europe during this period were Paris, London, Amsterdam, and Nuremberg. These cities enjoyed mobile labor pools of copperplate printers and engravers and access to supplies of copperplate and paper. The map publishers in these urban centers were trade partners as well as commercial rivals. The map trade in Amsterdam and Nuremberg can claim its own historians,[2] and the print trade in Paris and London has been well studied.[3] The study of the map trade in Paris and London has been confined more to individual mapmakers and specific enterprises than to a study of the trade networks and labor resources that assured the primacy of these two capitals in printed map production.[4] Paris and London were the epicenters of Enlightenment science; their respective academies supported and promulgated geodetic and geographic studies, the dissemination of which relied on the production of maps.

The Enlightenment is a period of human endeavor when reason and logic superseded religious belief and superstition as explanations for natural phenomena. Its origins lie in the evolution of scientific thought in the seventeenth century. The emphasis of Francis Bacon on observation and hypothesis as the determinants of facts presaged a revolutionary shift in the way in which nature and its forms were described. In its effect on cartography, this revolution passed in two waves. The first ran for roughly one hundred years, beginning around 1650, when the publication and dissemination of the works of Galileo Galilei and the *Geografia Riformata* of G. B. Riccioli began to have a profound effect on thinking about the shape of the earth and its measurement. More accurate methods of triangulation and measurement of longitude were developed in order to answer questions of global shape and size, techniques that gradually were fully embraced by the greater mapmaking community. The second wave began c. 1750, as state agencies emerged to undertake large national surveys and as state-supported schools began to meet the growing demand for trained surveyors and engineers, all of whom created maps. This phase lasted to the first quarter of the nineteenth century, when the development of lithography as a medium of map printing, the establishment of a metric system, the growth of statistics, and the emergence of geography as an intellectual discipline refocused the efforts of cartographers. Equating the Enlightenment with the eighteenth century incorporates an idea of the "long eighteenth century," c. 1650 – 1820.

These broad developments are paralleled in the long eighteenth centuries of the political history of both France and England. In France, the period runs from the majority of Louis XIV (1661) to the French Revolution (1789); in England, from the Restoration of Charles II (1660) to the end of the reign of George III (1820). During this time both nations, though often at war with each other and others, enjoyed internal political stability, increasing prosperity, commercial expansion, and consumer growth. They had both recovered from a long period of religious wars followed by the civil wars of the mid seventeenth century.

The content of maps felt the effect of the seventeenth century's legacy of science. The study of the natural world made possible by the microscope and of the extraterrestrial world made possible by the telescope radically changed how people understood their place in the universe. That natural phenomena could be observed, repeated by experiment, and explained by hypothesis and repeated examination established a new understanding of "fact" and "truth." The seminal works of Galileo Galilei, Francis Bacon, and

René Descartes formed the foundations of the two institutions that would dominate scientific life in the eighteenth century: the Royal Society in London (est. 1660) and the Académie Royale des Sciences in Paris (est. 1666). Experimentation, analysis, and critical discussion flourished in both. Concurrent with the foundations of these institutions, the astronomical observatories built in Greenwich and Paris (figure 1) further stimulated inquiry, observation, and the collection of data. A growing understanding and use of trigonometry for triangulation in mapmaking and the use of the telescope's extraterrestrial observations were factors in the determination of longitude and the shape of the earth.

The importance of these institutions to cartography can be summed up in the three new ideas about the position, shape, and measurement of the earth. First, educated society was gradually accepting the sun-centered Copernican solar system. This led to a radical rewriting of texts concerning geography and the place of the earth in a solar system. Second, Isaac Newton's laws of gravity postulated that the earth was an oblate spheroid. This changed accepted notions about variations in the length of a degree of latitude arc as one moved towards or away from the earth's poles. Third, a movement grew to determine longitude accurately and with it the more precise location of places on the earth's surface. Proof of the shape of the earth by measurement required accurate timekeeping and measuring instruments that would be unaffected by extremes of temperature, weather, or other conditions. These three intertwined ideas were developed mathematically, experimented with, and refined over the course of the eighteenth century, using information collected and analyzed by members of the Royal Society and Académie Royale. Mapmakers, even if not themselves members of these two societies (and many of them were), relied on the results of their research, which were published regularly in the *Philosophical Transactions* in London and the *Journal des Sçavans* in Paris. The desire for accuracy based on observation created the rhetorical template for map titles in the eighteenth century: "A new and accurate / new and complete / new and correct / new and exact map"; "Nova et accuratissima tabula," "Nouvelle carte . . . assujettie aux observations astronomiques/ aux observations nouvelles/ dressée sur les relations les plus modernes / les plus authentiques les plus récentes / faites sur les lieux."

The steady energy of scientific inquiry drew workers in the print trades to the two capitals as surely as Newton's gravitational force, for there was much to be published and even more to be illustrated. Of the two available means of reproducing graphics—the woodcut and the copperplate engraving—intaglio engraving on copperplate had became the choice in Europe in

FIGURE 1. Sebastien Le Clerc, frontispiece to Claude Perrault, *Mémoires pour servir à l'histoire naturelle des animaux* (Paris, 1669). Louis XIV and minister Jean-Baptiste Colbert visit the Académie Royale des Sciences. In the middle of the scene, Colbert shows the king a plan of a fortified town. In the background, the Paris observatory, under construction, is visible through the window. (Private collection.)

the late sixteenth century and remained so.[5] The opportunities for copperplate engravers in both London and Paris were many. The shifting politics of the late seventeenth century in Europe created waves of emigration following the Wars of Religion. Catholic engravers from the Low Countries traveled to Paris, while French Huguenots fled to London. By 1700, Paris and London could become European centers of the map and book trade. But the trade itself was European and, in Western terms, global. Many participants in the creation of cartography were not immediately linked with one government or political entity, but were itinerant and went where the job market led.

A further stimulus to the market for maps can be found in the creation of a number of state-sponsored educational institutions, some of which were founded early in the eighteenth century, but many of which date from midcentury and mark a new phase in cartographic efflorescence. The teaching of cartographic skills was incorporated in the curriculum of the French École d'Artillerie, founded in 1720. Its renown was such that the Royal Military Academy at Woolwich, established in 1741, was modeled on it. Other institutions followed suit, placing a high priority on the skills required for surveying, topographical drawing, measurement, and the taking of elevations. The École des Ponts et Chaussées was created in 1747 as a "Bureau des dessinateurs," becoming an école after 1775. The École Royale Militaire, founded in 1751, was split into eleven provincial écoles in 1776. The École Royale du Génie de Mézières (1748) was established to train engineers and fortifications experts; César-François Cassini de Thury set up his own training program for his *ingénieurs-géographes* working on the *Carte de France* from 1750. After 1763, when the Treaty of Paris marked the end of Seven Years' War, the military engineers released from active duty formed the Bureau des ingénieurs géographes at Versailles, joined by the similarly relieved *gardes de la Marine.* Such institutions created a cadre of civil servants who had direct experience in surveying and mapmaking.[6]

In England, fewer institutions incorporated such strong state support for the teaching of techniques applicable to mapmaking. Nonetheless, training given at the Royal Military Academy at Woolwich and the Naval Academy at Portsmouth, as well as the work of the observatory at Greenwich and in the Ordnance Drawing Room in the Tower of London[7] demonstrate increasing map literacy and production in the military class. On the civil side, the Society for the Encouragement of Arts, Manufactures and Commerce from 1759 offered premiums for the best trigonometric survey of an English county. This has been seen as one of the main factors accounting for the increase in large-scale mapping in England, accompanied by increased attention on the part of landowners for the care and improvement of their estates.[8]

[THE EVIDENCE FOR PRINTED MAP CONSUMPTION]

In spite of war and political upheaval, or rather because of it, the market for maps was buoyant. By the eighteenth century, maps and globes had become a regular feature in the life of the literate. The number of maps found in personal libraries in both England and France increased throughout the century.[9] Maps came in all forms and sizes, from those bound in pocket-sized atlases to very large multisheet maps hung from specially designed rollers. As wall decorations, maps were described as "the most commodious ornament for everyman's House."[10] William Leybourne echoed these sentiments in his *Treatise on plots* (1653); maps were a "neat ornament for the lord of the manor."[11] In the private library of Samuel Pepys, prints were "commodiesly hung with an ordinary panell of our modern wainscotting, such as those of Paris and London in my study." [12] Similar interest in maps existed in France; the comte de Seignelay, son of the great minister Jean-Baptiste Colbert, had twenty maps hanging in his *grand cabinet.*[13] Royal mistresses understood the potent effect of maps. During the Seven Years' War, Madame de Pompadour bothered Louis XV's generals with maps adorned with her beauty spots to suggest troop movements.[14] Printed maps illustrated French and English books, particularly history and military memoirs, as well as novels. Maps featured regularly in English periodicals, such as the *Gentleman's Magazine* and the *London Magazine.*[15] Wall maps, atlases, and globes were an essential furnishing in the schoolroom, as contemporary illustration attests (figure 2).

One may see that maps formed part of the print trade in the work of Casselle and Le Bitouzé and from regular announcements of the publication of maps in such journals as the *Mercure de France* and the *Journal de Paris,* as well as the more detailed advertisements and lengthy reviews of maps in the *Journal des Sçavans* and the *Mémoires de Trévoux.*[16] Timothy Clayton's study of the print trade in England in the eighteenth century similarly emphasizes that maps formed a regular and integral part of the trade in graphic art, advertised in newspapers, weekly and monthly magazines, broadsides, and trade cards.[17]

Further evidence of the sale and collection of maps may be found in auction catalogues,[18] in subscriber lists for maps and atlases,[19] and in diaries and travel journals (this last remains an open and fertile field for study).[20] Jean Chatelus' study of the inventories of artisans and merchants in Paris in

FIGURE 2. Frontispiece from Nicolas Lenglet Du Fresnoy, *Méthode pour étudier la géographie,* 3rd ed. (Paris, 1741). WLCL, C2 1741, Le. (William L. Clements Library, University of Michigan.)

FIGURE 3. A watercolor of the interior of a shipping merchant's office, displaying maps on the wall and globes above the door. By A. C. Hauck, Rotterdam(?). 1783. (Colonial Williamsburg Foundation.)

the second quarter of the eighteenth century reveals the role of maps in the private home. Thirty-seven percent of merchants' listed works of art were prints; for the artisans, 40 percent (figure 3). Many of these prints were geographical maps, frequently found hanging in the studies, corridors, and antechambers of their Paris apartments. Chatelus cites specific examples including two large maps: one, a mappemonde (a map of the world in two hemispheres) in the corridor of the wife of a merchant jeweler; the second, a map of Europe under the stairway at the house of a merchant in Passy. He points out a Paris general merchant who owned a collection of seventy-five maps, though such a case seemed to be exceptional.[21]

Maps permeated decorative schemes beyond the mere hanging on a wall or livening up a book. They were pasted on screens, used on fans, "printed on Silk of a Sarsh window, or to carry in the pocket in a little room, as an Handkerchief."[22] A "curious map of Ireland for a watch case" was created for one discerning owner.[23] The novelist Laurence Sterne evoked the map's appeal in this description Uncle Toby looking at a map of Namur in *Tristram Shandy:*

> The more my uncle Toby pored over his map, the more he took a liking to it;—by the same process and electrical assimilation, . . . thro' which I ween the souls of connoisseurs themselves, by long friction and incumbition, have the happiness, at length, to get all be-virtu'd,—be pictur'd,—be-butterflied, and be-fiddled.
>
> The more my uncle Toby drank of this sweet fountain of science, the greater was the heat and impatience of his thirst.[24]

"This sweet fountain of science" flowed for everyone with eyes to see and money to buy the printed maps that constituted commercial cartography.

[COMMERCIAL CARTOGRAPHY AND THE MAP TRADE]

The phrase "commercial cartography" hasn't been precisely defined. It has been used loosely to describe printed maps made for sale.[25] It occasionally assumes a faintly pejorative air, tending to denote maps of a lesser value, whether aesthetic or geographic or intellectual.[26] The phrase also attempts to distinguish maps produced by what we would currently call the private sector as opposed to "state" cartography produced by the public sector.[27] Such tidy divisions between public and private do not serve the complex overlap between those mapmakers who drew maps for the government and who published them for themselves. Nor does it incorporate the notion that maps were sometimes sold in order to fund the making of more maps, and not necessarily to meet the modern idea of market demand.

All printed maps were in some way commercial, whether they were created for sale or not. Every aspect of map production from survey and compilation to printing and distribution involved specialized labor and materials to which a value was attached. The exchange of money or goods that paid for each stage of printed map production created its own commerce. Simply put: behind every printed map was someone waiting to be paid. This queue of debt and credit exerted its own force on the content and production or nonproduction of maps. In most cases, maps were printed in order to be sold; their commercial value derived from their potential sales value. Commercial cartography, then, is the production of printed maps to be sold to a public market.

The marketplace of the eighteenth century incorporates not only economic questions but legal, intellectual, and aesthetic ones as well. The market for printed maps was a competitive one that demanded attention from the guilds, the courts, and the critics. Laws and customs concerning who could produce and sell maps and how maps were judged occupy much of this book. What was meant in the eighteenth century by a "good" map? The contemporary criticism of maps tells us what was desired, what was regarded as important for a map to sell. This discussion addresses not only the issues of accuracy and beauty but also those of authenticity. Whatever the commodity, the marketplace is ever a breeding ground for copies and fakes. Whether geographic ideas can be "owned," whether a location can be patented, what makes a map "good" are questions still with us.

[THE MAP TRADE AND THE HISTORY OF CARTOGRAPHY]

Maps are "graphic representations that facilitate a spatial understanding of things, concepts, conditions, processes, or events in the human world."[28] Because of their capacity to simplify the complex, maps can be seen as "simple iconic devices" wielding "extraordinary authority."[29] Thus, they may be perceived to contain hidden messages that can exert a powerful propaganda force, as has been argued by recent scholars.[30] They have taught us that a map is not a neutral document displaying an objective view of reality but is rather an image laden with meaning, open to dissection, analysis, and interpretation. We consider the choices made by the map's author and how the map was used. We now understand maps as artifacts that share with the world of art the translation of perceptions into graphic form and with the world of mathematics the means to translate three-dimensional space onto a two-dimensional plane. Of singular importance in understanding the meaning of maps and their influence is the work of placing maps in their political, cultural, intellectual, and social contexts.

The market provides one aspect of the social context of maps. Mapmakers sold printed maps to support themselves and to support the process of making maps. As with other consumer goods, the knotty question remains to what extent the producer created the market to which a consumer responded. To understand more clearly the market for maps is to work towards understanding whether a printed map satisfied a perceived demand or attempted to encourage new thinking. Johannes Blaeu, the seventeenth-century Dutch mapmaker and publisher, captured the question when he stated that

> the Earth does not owe gratitude only to those who create books, unite maps to the art of Geography, and fit lands to sky and sky to lands as with a plumbline; but also to those who *persuade, urge, correct, and increase these works, promote them with money and expense,* strengthen the feeble, recover the lost, and give

shape and polish to the deformed, so that the illustrious offspring can be born and reborn and appear with beauty into the light and faces of men.[31]

Printing maps required promotion, money, and expense. Maps could be used to "persuade, urge, correct, and increase" the works of geography already known. They could instigate intellectual change or merely reflect contemporary mind-sets. A study of the map trade helps to illuminate some of the impulses for creating maps for sale. By defining more clearly the putative and actual consumers of maps, we can gauge how the content of a map might reflect consumer demand. By understanding the sources of money for map production, we can test assumptions about pressure from institutions or social groups affecting the content of maps. By knowing what economic considerations were inherent in map production, we can assess questions of originality and falsehood, of truth and error in a map's detail.

Much recent research and speculation have revolved around issues of social power as reflected in the content of maps.[32] The influence of powerful social groups on mapmaking has served as the starting point for numerous map studies, which ask whether such groups overtly or covertly suppress, alter, or determine what information is included in a map. When we piece together the story of an individual map, however, we discover that there are many factors involved with a map's creation and reproduction. This book is less concerned with the power structures inherent in the map trade than in what was economically possible and economically profitable for map producers. While individuals, institutions, or agencies at one end of the social spectrum—landowners, ministries, military and educational establishments, scientific academies—often provided the impetus for mapmaking endeavors, the process of printing and distributing their efforts often fell to private individuals in the lower echelons of society. Similarly, some map producers created maps to supply a demand created by their own advertising. The overlapping interests, energies, and economies of state and private enterprises in the market for eighteenth-century printed cartography blur many of the distinctions between public and private. Economic considerations affected the way in which maps became printed objects and entered the public sphere, subject to use, interpretation, and contribution to a discourse occasionally far removed from their original purpose. The process of compilation, publication, and distribution was subject to the availability of resources—money, labor, and materials. These market conditions could affect the content and form of the final printed map as much as the availability of the raw survey data that comprised the map's content. The same conditions could also prevent a manuscript map from ever taking printed form, limiting its distribution to a wider audience.

The role played by the economics of mapmaking and map publishing has not always been fully developed in the scholarship of the history of cartography. The costs of production and distribution of maps are often touched upon in studies of individual maps or groups of mapmakers,[33] but a more global view that might provide comparanda for costs and a picture of trends has been lacking. This book attempts to provide a wider lens for surveying these costs. In addition to the economic constraints on map production is the role of the consumer and consumer demand as perceived by the map producer. Advertisements for and criticism of maps in the popular press form one group of sources for understanding demand; the discussion of maps in private correspondence, journals, and memoirs form another. Here, too, I have tried to draw together disparate material to form a more cohesive sense of how maps were analyzed and assessed in their own time. To accomplish these tasks, I have focused on map production and consumption in eighteenth-century France and England.

[WHY FRANCE AND ENGLAND]

Maps were a social and cultural phenomenon throughout Europe; the trade in maps crossed borders with ease. As we shall see, the appeal of printed maps lay in their attractiveness, their legibility, their size, and the reputation of their author, all attributes external to the content of the map. These features were international, even though national styles and tastes may be discerned (preferences for more or less decoration, for a particular language, for a certain cartographer). Consequently, the market for printed maps, like the market for prints, was international.

The role of France and England in the international market derived from their political and military rivalry throughout much of the eighteenth century. Their maps preserved their national perceptions of empire and illustrated their respective histories. Though rivals for global hegemony, they were commercially intertwined, providing each other with raw materials and manufactured products, including printed materials, in trade that survived all wars. Socially, their respective populations fascinated each other. One only needs remember the English "milords" who began their Grand Tours in France en route to Italy or note the French who visited Britain seeking commercial and industrial expertise.

The map trade also exemplified the close relationship between the two countries. Like Holland, England provided a home and work for French

Huguenots in the last decade of the seventeenth century, after the revocation of the Edict of Nantes. Huguenot copperplate engravers became a significant part of that workforce in London. The Huguenot settlement of the area around Soho in London welcomed later French engravers brought over by English publishers to work on large projects from the 1720s through the 1750s.[34] These engravers brought with them design ideas formed by the rococo taste advocated by the court of the regent, the duc d'Orléans, in France. Such styles began to influence map design, as may be seen in the work of the garden designer and surveyor Jean (John) Rocque, who established what one scholar has named the French school of surveyors.[35] Thus a strong French influence in design and engraving may be discerned in maps printed in London in the second quarter of the eighteenth century.

It was not only in design but also in content that France exerted great influence over its cross-Channel rival. English geographers admired their French counterparts, the *géographes de cabinet,* whose methods of compilation and analysis of sources were not so much imitated by the English mapmakers as their maps were copied. The works of Guillaume Delisle, Jean-Baptiste Bourguignon d'Anville, Jacques-Nicolas Bellin, and Gilles and Didier Robert de Vaugondy were repeatedly copied, published, and criticized in England. Even when the English commentator found the French map to be chauvinistic and imperialistic, he still used the French cartography as the basis for maps of areas outside the British Isles.[36] Both countries created royal titles for outstanding geographers, but there were numerous French *géographes du roi* and only one English geographer to the king; the *géographes* were also better funded. The French could claim a government that supported map creation to a greater degree than the English. By the middle of the eighteenth century France had embarked on a national survey, boasted a long-established corps of military engineers devoted to mapping and survey, and was committed to producing and revising charts of its coastal waters. The English government approached these cartographic activities in a more ad hoc fashion. The work of the Académie Royale des Sciences, supported by the French royal purse, investigated the means to establish longitude not for practical navigational purposes but also to answer the larger question of the shape of the earth. England's Royal Society was similarly engaged in the intellectual question of longitude, while its Board of Admiralty pursued the practical aspects. Many French and English scientists were corresponding members of these two scientific societies. France also was home to a scholarly tradition of historical research grounded in the search for ancient documents that would reveal local customs, measures, mores, place names, and geographical and political change as well as social and economic detail. Such an approach, as exhibited by the Benedictines in their research and teaching, was supported by the state in the form of subventions towards research and publication.[37] Secular antiquarians engaged in similar efforts and approaches to history writing in England, which resulted in some large cartographic enterprises. By contrast with the state-sponsored work of the Benedictines, the English antiquarians undertook their projects at their own expense or through private fund-raising in the form of subscriptions.[38]

Both countries shared an enthusiasm for and interest in maps; they exchanged maps as a regular part of the traffic in printed materials. Yet the differences in their national support for mapmaking and map publishing encourage us to think we may profitably study both of them in order to provide a largely European picture of the process of the getting maps to print.

[THE ORGANIZATION OF THIS BOOK]

This book follows the process of mapmaking from the inception of a map to its arrival in the hands of a consumer. Chapter 1 introduces the participants in the process of gathering data and compiling and producing printed maps. Chapter 2 focuses on the costs of each of these stages of production. Chapter 3 studies prices, distribution, the various means of financing map production, the obstructions to cash flow, and methods of cutting costs. Chapter 4 looks at two protective mechanisms set up to ensure some degree of authorial control over the content of maps: privilege and copyright. In chapter 5, the story of a specific series of maps exemplifies each step of the process from the gathering of data in a survey through printing by two separate publishers, and the copying of the map by two foreign publishers. Chapter 6 analyzes the range of definitions of a "good map" from the points of view of mapmakers and their audiences and sets out the programs proposed by mapmakers for improving maps.

Within these broad divisions, several themes are played. One that stands out among them is the individuality and idiosyncrasy of map production in this period. The world of the map trade did not include large businesses or the complete institutionalization of any form of mapmaking. Rather, it was comprised of a relatively small number of individuals, many of whom were known to each other, worked with each other, cooperated with and criticized each other. The ambitious surveyor, the map-compiling engraver, the entrepreneurial bookseller, the government minister "who never traveled without a compass and small level for taking elevations,"[39] each played a significant role in any map's production. Their mutual dependency and shared economic interests blurred the division between the public cartographer employed by the state and the private publisher. The reliance of French and English governments on the private sector for maps and the reluctance of these governments to enter the map trade becomes readily apparent.[40]

Yet the private enterprise, in response to the high costs of data gathering and compilation, was often content to reproduce antiquated and outdated material under new titles with claims to authority propagated by the growing scientific awareness of its public. The addition of such phrases as "corrected by the latest observations" or "made from surveys on the spot" attests to the growing influence of Baconian ideals of verifiability: that facts are based on observation and testing. However, observation and testing, when translated into mapmaking as survey and compilation, were high-cost activities, and their results were not always challenged when transformed into a map. The map producer could be easily seduced into copying material from others without critical judgment. For who was to judge a map? The range of publications and the subject of maps in private discourse reveal the rhetorical vocabulary used about maps and their qualities. In a competitive market, mapmakers judged other mapmakers through the publication of mémoires, through articles in the journals of learned societies, through presentations of papers, and in private correspondence. The debate over accuracy and truth of representation was sharp and often personal.

The effect of these debates on the buyer is difficult to ascertain, for many of them were more concerned with a map's aesthetic qualities: legibility, coloring, size, the quality of the engraving. A few mapmakers expressed a desire for some kind of quality control of maps before they entered the marketplace; they well understood the map's inherently deceptive qualities and knew the ease with which maps could be made. The failure of any such institutionalization to develop is one of the hallmarks of cartography that we bear in mind even today.[41]

Maps are one of the most porous sources of information in the eighteenth century; apparently easy to apprehend, yet often difficult to read in their many subtexts. As with the print, the questions of whether one is dealing with a reproduction or an original work and what the nature of the representation of reality is should inform our research. In drawing together a wide range of sources, from both published and unpublished material, I have tried to compile a different map of the cartographic landscape of the eighteenth century. It represents the map trade at a very large scale, from the cost of a copperplate to sharp criticism of a nautical chart. It is not a global view of cartography in all its manifestations, but a route map through the account books, ill kept though they were. It seeks to illuminate and explain the growing taste for maps in eighteenth-century France and England.[42]

PART ONE

MAKING MAPS

CHAPTER ONE

GETTING TO MARKET

[ESTABLISHING AUTHORITY]

To be a mapmaker in eighteenth-century London or Paris required little more than an interest and the will to set pen to paper. No special skills or training were required. Anyone who was so inclined could make a map. The eighteenth-century hydrographer Jacques-Nicolas Bellin recognized this when he wrote

> Nothing is more commonplace or easier than making maps. Nothing is as difficult as making them fairly good. A good geographer is all the more rare for needing *nature* and *art* to be united in his training.[1]

Bellin's description of a "good geographer" (to be understood as meaning a good cartographer, since the latter term was not used until the nineteenth century)[2] implies two categories: the makers of maps and the makers of pretty good maps. These blurred categories signify the close relationship between the geographer possessed of sophisticated interests and skills and the mere map compiler whose access to the marketplace depended only on sufficient capital.

[THE SURVEY]

There are, in the end, only two ways to make a map: by going outside, and by staying inside; that is, either by one's own direct observation or by the compilation of the work of others. Maps originate with direct observation. They often begin with measurements performed on the spot by a surveyor (figure 4) or a verbal description of land by an observer. The surveyor not only mea-

FIGURE 4. Frontispiece, Dupain de Montesson, *L'Art de Lever des Plans* (Paris, 1763). L 99.247. (Newberry Library, Chicago.)

sured and drew but assessed value and planned improvements, for a variety of purposes, such as taxes, rents, sales, improvements, or lawsuits concerning boundaries.[3] Surveys were also made for noneconomic reasons, such as to establish longitude and latitude in order to determine the length of a degree, a measurement important in ascertaining the shape of the earth. Such surveys were not always graphic in nature; they could also be textual. While the information included on a survey varied according to the desires of the landowner, the raw data of the survey provided two important features of a map: size and shape of an area, with aspects of its terrain or features.

Surveyors themselves sometimes became the compilers and engravers and publishers of printed maps. In addition, they were an integral part of local communities, often in roles beyond that of land measurer. Sarah Bendall's work on surveyors in Cambridgeshire describes their other functions as mapmakers, land agents, tithe surveyors, valuers, auctioneers, attorneys, stewards to public institutions such as hospitals and schools, and even portrait copier.[4] R. Campbell hints at their position on the social ladder of trades in his guide to parents on the "instruction of youth in their choice of business."

> The Land-Surveyor is employed in measuring Land, and laying it out in Gardens and other kinds of Policy about Gentleman's Seats. To have a good Taste this Way he ought to travel to France and Italy, and to have a Liberal Education, but especially a thorough Knowledge of Geometry and Designing. They may earn a Guinea a Day when employed in laying out, and are always esteemed above a Mechanic.[5]

These sentiments are echoed by Peter Bernard Scalé's call for apprentices to "one of the most profitable and genteel professions."[6] The multifaceted occupation of surveyor was not limited to the English side of the Channel. Roger Desreumaux' research into the lives of local surveyors in northern France has found a similar range of roles: surveyor, mapmaker, lawyer, steward, agent, valuer.[7]

All surveyors required the abilities to measure, to draw to scale, and to draw well. In addition, at the largest scale, local land surveyors who worked for land owners—drawing estate plans, measuring tracts of forests or other lands, drawing up land registers—had to be able to value and to evaluate the worth of the land itself. These skills made them indispensable to their employers, the landowners. To the tenant, the surveyor had ever been associated with increased rates, rents, taxes, and tithes. However well respected and local the surveyor, he was not always a welcome presence on the land, and local resistance, if not outright obstruction, often marred the surveyor's work. In spite of his permits from the Crown to survey Brittany, the French astronomer and surveyor César-François Cassini de Thury found local inhabitants to be suspicious and obtuse. They threw stones at the surveyors, stole their equipment, refused to furnish horses or guides, accused the surveyors of being spies, and threatened to burn the manuscript plans, because of their fear of taxes and distrust of the taxman ("méfiance fiscale").[8] In Scotland in 1740, the Rev. Alexander Bryce met opposition along the shore of Caithness as he attempted to make astronomical observations to fix points along the coast. The locals "were not anxious that navigation should be made safer lest it should deprive them of the spoils of the numerous wrecks along that coast."[9]

Other types of surveyors added more skills to those of the estate surveyor. Military surveyors needed the ability to understand, measure, and draw topographical features that might influence the disposition of troops or the conduct of a campaign. Civil surveyors and engineers required the mathematical skills to measure angles and determine distances needed for the triangulation of large areas of land and coastline; they also drew upon astronomical data for the determination of longitude and latitude. Their work demanded instruments of measurement whose improvement over the course of the century allowed them to refine and correct their work to an ever-increasing degree of accuracy.

A central interest of the Enlightenment caused the surveyor also to assume the role of scientist, measuring land and water not just for its valuation or location but for determining the shape of the earth. The great national survey of France, initiated by the Cassini family, began not by measuring the realm but by measuring the meridian of Paris, the line of longitude running through the capital. The execution of this measurement and the subsequent regional surveys made to complete the map of France required training different from that of the local land surveyor. By accurately measuring angles and baselines across France, the Cassini geographical engineers created a criss-cross web of triangulated measurements between places whose location on the earth's surface had been determined by astronomical observation. The *Carte de France* was one of the first national surveys to be executed through the combined techniques of astronomical observation and triangulation. The resulting map did not pretend to be an accurate topographical survey, but rather an accurate locational diagram of significant places.

Cassini de Thury described the attributes of this new breed of geographical engineer—he should be young and eager, ready to learn new skills. He would have to learn "new methods, very different from those of surveying," which were practiced through drill and exercises, aimed at achieving uniformity of characters used and style of representation.[10] The type of map, the *carte d'astronome,* was not surveyed in the same detail as a land survey, as one surveyor relates.

> We didn't go off into each village, into each hamlet in order to survey the plan. We haven't visited each farm or followed and measured the course of every river . . . such detail is required only of plans for some seignorial land; the reasonable size at which one ought to fix the map of a country doesn't allow one to be able to mark so many things without great confusion.
>
> We have preferred to use the summit of a mountain or high church towers for discovering the countryside. There, with the help of many guides from the nearest villages, we were shown and had named all the places that were in our view; we even took the alignments of places buried in the valleys, . . . right on the very spot, we prepared the map of this canton by means of angles that we had taken and on the estimate of the respective distances, joining to these the course of rivers, roads, etc., of which we had taken note with designs of all the places where we could.[11]

Whatever the differences between the methods of Cassini de Thury's geographical engineers and those of a land surveyor, both jobs required an understanding of math and of how to measure space. What the surveyor provided was a first-hand report that bore witness to the space measured; as a witness, the survey had the hallmark of truth. The survey was the closest the eighteenth century could come to the aerial view. Technically, it involved the use of measurement and mathematics to reconstruct a view of space *as if* from above. The continuous improvements in the instruments used—for example, more powerful lenses in telescopes allowing vision from greater distances, the increasingly fine gradations on angular measurement devices—meant an ever improving picture of terrain or coast, a closer and closer approximation of the "truth." The phrase "based on observations taken on the spot" appears in many eighteenth-century map titles, attesting to this claimed authority.[12]

In France, up until midcentury, surveyors were trained in the manner of other trades and crafts, on the spot, either through family tradition or in the artisan manner, from master to apprentice. By 1750, the state established institutional training for surveyors, first with the foundation of the École des Ponts et Chaussées in 1747 for civil engineers and the École du Génie in 1748, which included map surveying in its curriculum for military engineers. The *ingénieurs-géographes* of the army had been trained in an ad hoc way from the early seventeenth century, but after 1763, they had at Versailles a "bureau" where their work was vetted and improved, and where classes in foreign languages and mathematics were taught. César-François Cassini de Thury ran his own training program on terrain between Paris and Versailles to equip his men with the skills necessary for creating the *Carte de France*.[13]

From these institutions sprang a network of alumni who established schools or bureaus or trained yet more surveyors who worked both in the public domain and privately.

In England, fewer institutions enjoyed such support from the state. The Royal Military Academy at Woolwich (founded 1741) trained engineers for the army's fortifications and artillery. A study of the map training received by young draftsman working for the Ordnance in the Tower of London shows that the number of draftsmen (*dessinateurs*) was very small until 1752, when the Drawing Room was recognized as a department of the Ordnance. Even after this date the numbers remained small until the end of the century, never exceeding fifty.[14] In the private sphere, surveyors, like other tradesmen, trained other surveyors, either their own sons or apprentices. In addition, many of them taught mathematics; some established their own private schools of mathematics where a young person could obtain the rudiments necessary for a career either in business or in a trade like surveying.[15] Some offered the full range of mathematics and engineering principles, like Benjamin Donn in his Mathematical Academy in Bristol:

> At the Mathematical Academy in Library House, King Street Bristol, Young gentlemen are boarded and taught Writing, Arithmetic, book-keeping, navigation, and geography—elements of Algebra, Altimetry, architecture, astronomy, chances, conics, decimals, dialling, fluxions, fortifications, Gauging, geometry, gunnery, hydraulics, hydrostatics, levelling, mechanics, mensuration, optics, perspective, pneumatics, shipbuilding, surveying, trigonometry, plane and spherical; with the use of mathematical and philosophical instruments in a rational and expeditious manner, according to new improvements—Also courses of Experimental philosophy read on reasonable terms.[16]

Not all observation is pictorial. Vying for attention with graphic records were textual descriptions of land and landscape: deeds and cadasters, official reports and other legal documents, the diaries of travelers, the accounts of explorers, the relations of missionaries. All these provided sources for maps. Endowed with the authority of observation on the spot, they also acted as witnesses to a truth. For the user, the antiquity and language of written accounts could complicate their interpretation, but they nonetheless remained one of the sources for a map.

The manuscript map delivered by a survey could well stay in manuscript form and be copied as such for any users who wished it.[17] This was the fate of many surveys, particularly those large-scale plans of private estates or plats made for legal records; they were not intended for wide distribution. For a trade in maps to be possible, a crucial decision was required: whether to commit the survey to print.

[COMPILATION: THE SURVEY TRANSFORMED]

The second form of manuscript map that could become a printed map was created without going outside. This map was compiled from the literary as well as the graphic sources; the mapmaker looked for points of agreement or discord among the sources, synthesizing the similarities and determining a mean from among the differences. The French hydrographer, Jacques-Nicolas Bellin, described the work of the "good geographer," who was just such a compiler.

> From the start he needs to have memory, a love of work, patience, and a flair for order and arrangement. Then he needs a sufficient command of geometry and astronomy, after which comes the long and sterile study of voyages and the critical discussion of their reports and journals, sources continually filled with uncertainties and errors that even the most assiduous labor sometimes cannot conquer. Add to this some knowledge of foreign languages . . .[18]

The process of reading these literary accounts, taking various manuscript plans of the same place and coordinating them by determining their scale, analyzing the measurements used by the surveyor, and redesigning the maps on a projection that could accommodate the scale of the map is the process of compilation. It was most time-consuming and tedious, or as Bellin put it, "longue, ingrate, et dure."[19] His contemporary, Jean-Baptiste Bourguignon d'Anville agreed, describing the task of preparing maps as "astringeante et tyrannique," complaining that the mapmaker cannot slide (*glisser*) over anything as one might in writing, for scale and orientation always require great precision.[20]

A cartographer or hydrographer such as Bellin and the map draftsmen who worked with him compiled charts and maps from a vast array of sources. They used manuscript surveys made by engineers, manuscript reports and journals turned in by boat pilots, published travelers' reports, printed charts published in foreign cities, and old printed charts in the archives. The remarkable aspect of this process was that the mapmaker himself never went to sea or surveyed a single coastline. His work was based solely on his skills with math, with languages, with logic, combined with his critical sensibilities. He described himself as a *géographe de cabinet.* This method of mapmaking was the hallmark of French cartographers throughout the eighteenth century: the Delisles (Guillaume and Joseph-Nicolas), Jean-Baptiste Bourguignon d'Anville, Philippe Buache, and Gilles and Didier Robert de Vaugondy. With the exception of J.-N. Delisle, none of these geographers ever left Paris or traveled far afield, relying instead on a network of correspondents and their own reading and gathering of materials to create their maps.

After surveying, the compilation of material for a map required the greatest amount of capital, both intellectual and economic. Its most demanding requirement was the knowledge of mathematics needed to create a projection and scale for the map. Mapmaking also demanded knowledge of foreign languages to read travelers' accounts. Most of all, compiling a map required time. "You have to spend a considerable time preparing yourself and assembling the necessary information and often with the most assiduous work you can scarcely flatter yourself on being able to conquer the difficulties that present themselves."[21] The cost of compilation was high because of the time, materials, staff, and equipment required (figure 5).

[FRANCE]

Map compilers such as those described by Bellin were at the core of the printed map trade in France. Some, like Georges-Louis Le Rouge or Roch-Joseph Julien, had learned mapmaking skills in the army, as *ingénieurs-géographes* (military surveyors). Others, such as the members of the Cassini family and the Delisles, had been trained in astronomy. Philippe Buache had begun his career as an architect. Geographers like these emerged from an educational system that included math and science as part of its basic curriculum, a system that the geographers in turn served as teachers of history, geography, and mathematics. The educational opportunities in France during the long eighteenth century were predominantly organized by religious orders. Among these, the Jesuits stand out because of their emphasis on math and science. In their curriculum, geography was inextricably linked to the study of history; maps, atlases, and globes played an important role in teaching the young to understand the natural and political world.[22]

The formation of French geographers by their Jesuit teachers powerfully influenced mapmaking. The number of Jesuit schools grew steadily throughout French cities during the seventeenth century as they developed a curriculum that satisfied both the humanistic aspirations of the gentry and the utilitarian desires of a merchant class. The Jesuits incorporated into a

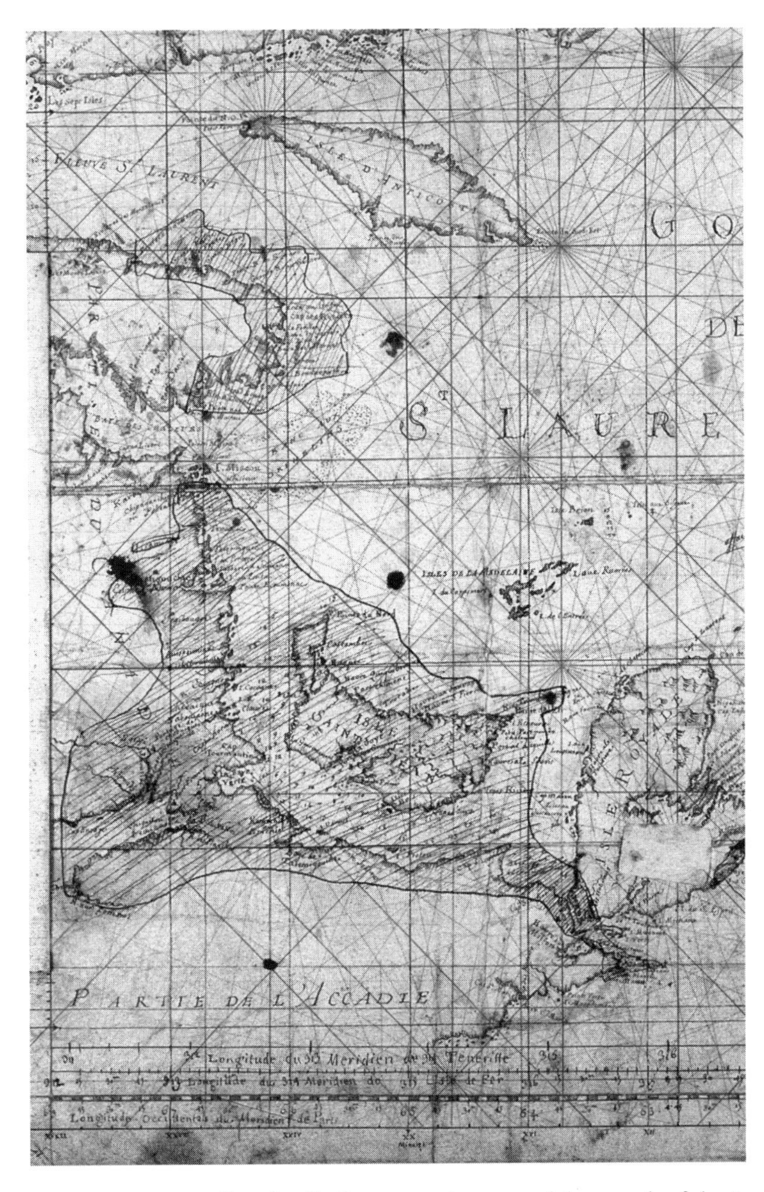

FIGURE 5. J.-N. Bellin, detail of a manuscript map of the mouth of the St Lawrence, showing corrections to the Ile de St Jean (Prince Edward Island) with part of the Ile Royale (Cape Breton Island). The area scratched out was redrawn on a different flap of paper, then attached to the larger map. Compare to full map shown in color plate 3. CP MAR 6/JJ/76 C ou III, no. 247A. (Archives nationales, Paris.)

traditional classical curriculum of ancient languages and literature the new and practical subjects of mathematical science; not just arithmetic, geometry, and algebra, but related subjects that required mathematical prowess, such as astronomy, optics and perspective, music, mechanics and hydraulics, applied geometry (for surveying or topographical drawing), and fortifications. To these they added the study of modern languages. As the purveyors of such "useful" knowledge, the Jesuits worked hard to win over the mistrustful denizens of the great commercial cities of France, whose city fathers saw no use in supporting schools that offered only the classical curriculum. In their eyes, schools teaching such useless literary subjects as ancient languages and texts only created dissatisfied youth who would not want to follow their parents into trade but would rather aspire to the unproductive life of a scholar.[23] The merchant class felt that such a dissolute pattern could only cause the demise of trade and economic ruin for cities such as Bayonne, Lyon, and Nantes. But the Jesuits could offer the mathematics and science necessary for the curricula offered by the new schools of hydrography created in the late seventeenth century in the major port cities by Colbert. Their expertise in mathematics and their interest in understanding ideas related to astronomy as promulgated in the works of Copernicus and Galileo made their curriculum one in which science sat easily alongside humanities. Theirs was the order capable of teaching navigation and chart making to future pilots and sailors.

If parents in the commercial cities were suspicious of the classical curriculum, the teachers of classical subjects were equally distrustful of subjects such as modern languages, science, and mathematics. "Men were not born to measure lines, to examine the relationship of angles and to spend their time considering the diverse movement of matter. Their spirit is too great, life too short, and time too precious to be occupied with such little things." objected the eminent philologist and jurist Bouhier, defending the study of literature against math and science.[24]

In the Jesuit curriculum, students were exposed early to maps and globes and encouraged to make use of them. A core of the science curriculum explored various explanations for the description of the universe, be it a Copernican sun-centered model or a Ptolemaic earth-centered one. The mathematics of a heliocentric system occupied astronomers and mathematicians for much of the eighteenth century; descriptions of various arrangements of the universe are to be found in many geographical textbooks. As more refined lenses for telescopes were developed throughout the century, the number and accuracy of the observations increased to support theories about the shape and structure of the solar system. The Jesuit fathers eagerly seized opportunities to make observations in their far-flung missions in the Far East and the Americas, thus becoming active participants in the continuing debate about the structure of the solar system and, ultimately, the shape of the earth. Students in Jesuit schools had immediate access to the Jesuit Relations, accounts written and published by the Jesuits about the establishment of their missions. Because of the Jesuit policy of priests not staying in any one mission for too long, fathers who had visited the distant and often exotic places described in the Relations returned to teach in the schools in France. The Jesuit curriculum encouraged active student participation; the display and occasional wearing of native costumes from the Americas or from China would not be unusual. A Jesuit college might have in residence a "savage" student brought back from Canada. The proximity of the exotic to the students in these colleges lent to geography an air of reality by endowing the names and places with a substance and import they might not have acquired if derived from texts and memorization alone.[25]

The legacy of the Jesuit approach was the encouragement of an interrogatory style and rational deliberation of possible answers. The early eighteenth-century Benedictine historians of St Maur adopted a similar approach as they compiled all relevant material from diverse sources to write ecclesiastical histories of different regions in France. Like the geographers, the monastic historians engaged in the "search for the truth . . . which consists of the certitude of facts."[26] Their goals evolved over the course of the century, from merely collecting and elucidating the works of St Benedict for the "enrichment for the Republic of Letters" to writing a more general (and secular) history of church and state in a desire to be "useful to the Nation."[27]

As a result of their early Jesuit training, *géographes du cabinet* like Delisle and d'Anville were imbued with a Cartesian ability to analyze and sift conflicting geographical evidence while preparing maps, and a readiness to take an all-inclusive approach to geographical sources. The defining feature of a French map from the first half of the eighteenth century was the printed mémoire, explaining how the map was made. The writings of the *géographes* show that they concentrated on the twin themes of name and place—nomenclature and location—, cartographic problems permeating the craft until the end of the eighteenth century. Their mémoires are packed with the descriptions of how they worked: how they analyzed the various names given to the same place, how they determined precisely the locations of these places, how they compared writers in various languages from various periods, both modern and ancient. No source was left unconsidered or undescribed.

Besides a common educational background, French geographers also were often associated with the king and his ministers of state in Paris. From the time of Louis XIII, the king appointed many *géographes du roi.* Upon presentation of a masterpiece map, the geographer would receive the reputation-enhancing title and very often, though not always, a pension, anything from 100 to 400 livres per year. The *premier géographe du roi* received a larger amount, 1,200 livres, and might be lodged in the Galeries du Louvre, residences reserved in the royal palace for state-supported artisans. What was expected of the *géographes du roi* varied from monarch to monarch, the demands being made not directly by the king but by his ministers.

Not all geographers and geographical publishers were appointed to the king of France. The Paris instrument maker and map publisher Louis-Charles Desnos held his royal brevet from Christian VII, king of Denmark, who granted him an annual gratification of 500 livres. In return, Desnos sent the king maps and books every year, amounting in some years to 200 livres worth of merchandise, thereby reducing the value of his pension, which Desnos complained had already shrunk to 433 livres by the costs of exchange.[28]

There were other ways in which French geographers received support from the state, whose purse strings were controlled by the Crown in the guise of the chancellor, not by parliament. Geographers taught in schools that were supported by local and central government. The army trained *ingénieurs-géographes,* many of whom became commercial mapmakers. The navy, through the Dépôt de la Marine, paid annual salaries to its hydrographers: Philippe Buache, followed by J.-N. Bellin. The Académie Royale des Sciences received a royal subsidy from which it was able to support the expenses and provide salaried posts for some of its members, among whom was the adjunct geographer.[29] The astronomer J.-N. Delisle held appointments both at the Royal Observatory and at the Collège Royale as a professor of mathematics.

Imbued then with geography and history as well as an understanding of mathematics, the geographers of early eighteenth-century France were well prepared to become the compilers of maps and thus are called *géographes de cabinet* (geographers of the study) for good reason. They came from educated families of the middle class. Sanson was the son of a merchant and city alderman in the town of Abbeville, in northern France; his father was known for his strong interest in geography. Claude Delisle, paterfamilias of another generation of mapmakers and scientists, was the son of a doctor; he abandoned an early career in the law when his love of history and geography took him to Paris to teach these subjects. Philippe Buache trained as an architect, winning first prize to the École Française in Rome before changing paths to study geography with Guillaume Delisle, and Didier Robert de Vaugondy's father Gilles was a mathematician.

The second group of map compilers was made up of military engineers. Having learned survey techniques in the army, men like Georges-Louis Le Rouge and Roch-Joseph Julien returned to civilian life to become map publishers. Surveyors like Giovanni Antonio Rizzi-Zannoni and instrument makers such as Louis-Charles Desnos also knew the techniques of compilation. Besides the reputations that they acquired through their teaching, these geographers also wrote justifying mémoires, published as brochures to accompany their maps or as articles in the *Journal des Sçavans.* These mémoires listed and analyzed the sources used for their maps, carefully explaining how information was accepted or rejected.

Once authority was established, a geographer's family name might retain enough value to support two or three generations of mapmakers. The continuation of map publication by a small number of families is a significant feature of French cartography. The stock of Nicolas Sanson, preeminent cartographer of the seventeenth century, whose protectors included Cardinal Richelieu and abbé Séguier, passed through his sons, Guillaume and Adrien and his grandson, Pierre Moullart-Sanson. Gilles Robert de Vaugondy inherited much of the Sanson stock and with his son, Didier, built upon the Sanson oeuvre with improvements and corrections; they used the notes and manuscripts collected by the Sansons for much of their own cartographic work.[30]

The Delisle family issued from Claude Delisle, who taught history and geography to the duc d'Orléans and other aristocrats of early eighteenth-century Paris and ultimately to the young Louis XV. His four sons, Guillaume, Joseph-Nicolas, Louis de la Croyère, and Simon Claude, formed a veritable cartographic workshop in the early years of the eighteenth century. The eminent geographer among the four was Guillaume, while Joseph-Nicolas became an important astronomer. The Delisle stock-in-trade was eventually passed on to Guillaume's son-in-law, Philippe Buache (though not without a legal struggle), and from him to his nephew Jean-Nicolas Buache de la Neuville, who carried on publishing Delisle-Buache maps until 1780, when he sold the *fonds* to Jean-Claude Dezauche.[31]

D'Anville is the only one of this group of dominant geographers who had no sons to whom he passed on the stock; his student Barbié du Bocage undertook the cataloguing and preservation of his work after it was acquired by the government before the geographer's death.

The contrast to the situation in England in the same period is striking. We find very few, if any, geographers who come from an educational tradition of humanistic study combined with sciences. The pattern of formal education in England in the long eighteenth century was markedly different from that in France. The state played little or no role in education or formal training at the most basic level except through the training of military officers and engineers. In terms of curriculum, the same tension existed in England as in France: a classical education was the token of gentility. Tradesmen and mechanics studied the modern languages, modern history, mathematics, and geography; gentlemen studied the ancients. Private academies, charity schools, or tutors were the only resort for those interested in studying subjects that would equip one to become a geographer: languages, mathematics, geography, and history. Such "mathematical schools" were aimed at young people wishing to enter commerce and the trades.[32]

Thus, the *London Tradesman* gives "advice to parents" in this summary of what "parts of Education are universally useful in almost all Trades."

> Reading and Writing . . . a tolerable Notion of Figures [as found in] common Arithmetic. Drawing or Designing is another Branch of Education that ought to be acquired early, and is of general Use in the lowest mechanic Arts.

Campbell mourns the lack of drawing and design practice in England, which accounts, in his view, for the superiority of foreign workmen, especially French, noting that

> this is the best Reason . . . why English Men are better at improving than finding out new Inventions. The French King is so sensible of the great Advantage of Drawing, that he has, at the public Expence, erected Academies for teaching it in all the great Cities in his Dominions; where the Youth are not only taught gratis, but the Parents are obliged by the Magistrates to send their Children to these Schools. . . .
>
> The sooner the Child is put to this study, the greater and easier will be his Proficiency . . . especially to be able to delineate on Paper a Plan of every Piece of Work he intends to execute: This much the meerest Dunce in Nature can acquire, much sooner than he can learn to write.[33]

Many producers of printed maps in England called themselves geographers. Some were tradesmen and mechanics, emerging from the print trade or the manufacture of scientific instruments, having been trained as apprentices to engravers or instrument makers. Their education was rudimentary in that they received neither a strong training in the classics nor a broad background in the emerging sciences. They learned about maps on the trot, especially print engravers and sellers, relying heavily on the import of geographic ideas from the Continent, where French, German, and Italian geographers were setting the pace.[34] For material gleaned from surveys done at home in the British Isles, they engraved and printed the work of local surveyors. They produced their maps without a justificatory memoir, though some, like Herman Moll, might annotate their maps with text printed on the map itself containing information about sources. With few exceptions, geographers and mapmakers were not members of the Royal Society, though their work was often presented there, such as the plan of London executed by Jean Rocque and John Pine in the early 1750s.[35] While the Royal Society concerned itself with issues and projects surrounding the search for an accurate means to determine longitude and the shape of the earth, it had neither funds for nor interest in strictly cartographic enterprises. The society's *Philosophical Transactions* published only two articles concerning maps during the whole of the century, as opposed to the almost monthly article or review of maps found in the *Journal des Sçavans,* organ of the French Académie Royale des Sciences.[36]

We can ascertain the important role of engravers and printers in the map trade from studying the list of holders of royal warrants. The titles of geographer to the king, hydrographer to the king, and chorographer to the king were bestowed on engravers and publishers, not on men who wrote about geography or worked on the compilation of maps in the French style. Except for John Cowley, who wrote on geographical and mathematical themes, the geographers to the king of the eighteenth century—John Senex, Herman Moll, Emanuel Bowen, Thomas Jefferys, William Faden—were engravers and map publishers, not scholar-geographers like the French *géographes de cabinet.* As geographers to the king, their role was commercial: to supply maps to the Crown. Their titles do not seem to have been accompanied by any annual stipend (despite the previous century's precedent of some monetary support granted to John Ogilby and his relative William Morgan). Neither the Crown nor Parliament was disposed to providing funding for the survey and measure of the realm or to establishing any mapmaking agency.

This is not to say that monarchs were without interest in maps: Charles II, William of Orange, Prince George of Denmark (consort of Queen Anne), George II and George III, and William, duke of Cumberland were all map aficionados who collected and used maps in their private and public capacities.[37] But their fiscal capacity to support mapmaking was limited by Parliament, and Parliament has left no trail of money for maps. The English makers of printed maps had their roots in the world of trade and looked to the marketplace to support cartographic endeavors, whether through subscriptions, sales of other products (instruments, prints, trade cards, design books), or the patronage of the wealthy.

Because English mapmakers were largely self-taught, they felt no compulsion to justify or explain their work, creating a map culture in which the critical faculties required for mapmaking were little emphasized. Bradock Mead (c. 1680–1757), who made maps using the name John Green, recognized the problem posed for cartography by such lack of intellectual training.[38] He describes the situation in the dedicatory preface to his book, *The Construction of Maps and Globes.*

> The Abuses of negligent and unskilful Geographers had long since made something of this Kind [of book] necessary, in order to put a Stop to those spurious Maps and incorrect Books which were daily publish'd by them and continu'd more and more to involve Geography in Error and Contempt. . . . The best Accounts of Travellers, are not free from Errors; their many irreconcilable Differences perplex and mislead us, and much of countries remain undiscover'd. . . . Is it not because proper Persons were never purposely employ'd to make discoveries in foreign Parts?[39]

Mapmaking from observation to compilation was a labor-intensive process, but it was not lonely. Surveyors and compilers were surrounded by assistants who helped with the more tedious aspects of the job, from carrying equipment to working out long mathematical operations. At the stage of printing a map, the process continued to be dependent on the skilled labor of the engravers and the printers. Only three trades represented here—surveying, engraving, and printing—restricted entry, for these required training, supervision, and some form of mastership. The job of the map compiler, perhaps the greatest source of printed map material, was something anyone could do. The variations that we will see in the cost of labor for these activities may be accounted for by restrictions on the source of labor where it mattered most: in the survey, where a printed map begins, and in its engraving and printing, where it ends.

CHAPTER TWO

THE COSTS OF MAP PRODUCTION

[COSTS OF SURVEYING]

Evidence for the costs of survey is more accessible in France than in England since many medium- to large-scale surveys of regions, provinces, and the kingdom were initiated by local and central government, whose archives still retain details of proposals, budgets, and final bills for these works. François de Dainville explored many of these archives for his works on the history of seventeenth- and eighteenth-century cartography in France. The work of surveyors for local landowners is less accessible except where family records have been deposited in local archives. See appendix 1 for the costs of map production in France, and appendix 2 for the costs in England.

The survey of land, a coast, a town naturally varied in cost according to the gross area to be covered. The costs of survey were not limited to the surveyor's wage alone. They also included expenses for instruments for measuring distances and angles and for drawing, such as the circumferentor, quadrant, and plane table. Surveys required horses for people, mules for equipment, food and fodder, servants to help carry, set up, and cook, surveyors' assistants, and guides. Thus, the breakdown of survey costs could vary widely, depending on local prices and customs. Total estimated costs could sound frighteningly large. In 1721, an estimate for a survey of Brittany was put at 12,000 livres per year.[1] In midcentury, the director of the *Carte de France,* Cassini de Thury, told the king of France that the cost of the survey of the kingdom for a national map would be 40,000 livres per year. This figure included not just the survey work, but also the engraving of the ten plates a year. He would pay his surveyors by the plate, but this amount would vary according to the difficulty of the terrain surveyed.[2]

Animals could be more costly than people. The hire of mules for the

Cassini survey of France was estimated at 10 livres per day. Costs naturally rose with the number of animals required: fodder for six horses and two mules totaled 26 livres per day for the work of the Cassini survey in the Languedoc.[3] Compared to beasts of burden, the guides or *indicateurs* seemed good value at 30 sous per day.[4] The importance of the guide was not to be underestimated. J.-N. Delisle was willing to pay his *indicateurs* 600 livres per year, double the military rate, for his projected survey of Brittany in 1720. The guide's service was indispensable for survey work, for he helped with routes and place names, especially in interpreting the local pronunciation and dialects.[5] In addition to guides, a survey also needed someone to look after the animals, someone to help with the equipment, such as erecting the tent that protected the surveyors and designers, and the draftsman or *dessinateur,* who helped create the manuscript map.

Even more costly than people were the instruments required for surveying. The accounts for a 1728 survey of the province of Languedoc spelled out such expenses in detail: from a small (4 inch) quarter-circle compass at 20 livres to a very large (3½ foot) quadrant at 1,600 livres. Other equipment included a box of colors, 20 livres; six dozen pencils, 16 livres; two quires of large paper, 18 livres; and six large sheets of vellum, 12 livres. Instruments and other materials for the survey totaled 6,000 livres. This sum did not include equipment needed for hauling, such as two wheelchairs pulled by two horses, a cart pulled by two mules for the instruments, and four saddle horses for the surveyors.[6]

There is less evidence for surveying expenses in England, but large sums prevailed. John Seller proposed a survey of all England and Wales for his *Atlas Anglicanus* in 1679 for an estimated £1,441 5s.[7] One hundred years later, Benjamin Donn remarked that his survey of Devon alone had cost £2,000.[8] William Roy estimated in 1766 that one year of surveying for a general survey of England would cost £2,778 12s.[9] An occasional document might break the cost into its constituent parts. A proposed coastal survey of Wales came in at £450 per year, including a surveyor, five assistants, a ship's mate, and yearly repairs to the ship.[10] A further indication of the cost of survey may be found in the reaction to the premium for the "best survey" offered by the Society for the Encouragement of Arts, Manufactures and Commerce from 1759. The £100 incentive caused at least one surveyor to scoff in 1760 that such a prize was "too little for a man to execute a survey," though three hundred subscribers at a half guinea each (£172 10s) would be sufficient to bring him south of the border.[11]

As in France, costs could vary with terrain. In 1722, the surveyor Edward Laurence offered different rates for cleared, common, and enclosed lands.[12]

Bernard Romans, a British surveyor in North America in the last part of the eighteenth century, determined the cost of survey by how far the surveyor would have to travel from his home as well as by the nature of the acreage (appendix 3). If the acres were nearby, he would charge 2s 3d for 100 cleared acres and 4s 6d for 100 uncleared acres. (These amounts average to about a farthing per cleared acre, or a half-penny per uncleared acre.) The cost tripled to 6s 9d per 100 cleared acres if he had to travel more than twenty miles. Such a distance meant he would need men to row the boats, more to carry provisions and instruments, two more to carry chains, and two to act as trailblazers. Romans included in his calculations the legal costs of petitions, warrants, precepts, plats, certificates signed by the deputy surveyor, and fiats signed by the attorney general, as well as the governor's and secretary's fees. All these charges could add £35 4s 8d to the final bill, a stiff sum when set against the cost of beef—£3 for a barrel of choice—or the price of a fifty-pound ham at 1s per pound.[13] The duke of Richmond also chose to pay for the survey of his estates in Sussex by total acreage rather than by the job. His surveyors were paid 2d per acre, about four to eight times as much as Romans's prices for North America, for which they were also expected to produce a fair copy of the plan at a scale of six inches to one mile.[14]

[THE COSTS OF COMPILATION]

Surveying the land was a costly endeavor, and it was not the final cost prior to printing. As we have observed, the survey provided only one source of information to map compilers, who worked with a variety of materials. Determining the cost of compilation is less straightforward than that of the survey, for in the documents, the activity of compilation was often folded into other costs. For geographers in France, like Delisle and d'Anville, it was what they did for their livelihood along with teaching. Their earnings were derived from the intellectual effort of making and explaining maps.

Nonetheless, some evidence reveals the cost of compiling manuscript maps in the late seventeenth and early eighteenth centuries. In 1673, the publisher A. H. Jaillot paid Guillaume Sanson 176 livres for a two-sheet map. Yet in 1700, Jaillot paid only 150 livres for *all* the manuscripts belonging to the engineer Jean-Baptiste Louis Franquelin. These manuscripts included Franquelin's surveys of various places in Canada and an untitled large map, probably of Canada or North America, graduated with scale and positions marked, ready to traced and engraved on copper. To the 150 livres, Jaillot added the promise of one hundred printed copies of Franquelin's map.[15]

By 1728, d'Anville could charge Father Du Halde 600 livres for preparing three large and fifteen small maps to accompany Du Halde's *Histoire de la Chine,* a sum increased first to 800 and then to 1,000 livres as the project dragged on longer than anticipated by Du Halde.[16] This works out to 100 to 150 livres per large map—significantly less than Sanson received. Midcentury, Philippe Buache charged the Crown 1,200 livres for preparing thirty quarto-sized maps to tutor the dauphin (an average forty livres per map), many of them prepared in blank for the dauphin to fill in with geographical information.[17]

The role of a draftsman or *dessinateur* was significant in the preparation of a fair copy of the manuscript map for printing. D'Anville reckoned that a competent draftsman doing work requiring neither creativity nor ingenuity was worth four livres per day. Such a *dessinateur* could prepare about a square foot of map in two days—three days if the map was more detailed.[18] In the last quarter of the century, Giovanni Antonio Rizzi-Zannoni asked for over 2,100 livres to prepare a map of the Mediterranean Sea for the Dépôt de la Marine. His breakdown of costs serves as a model for the steps required and the charges for compiling a manuscript:

—600 livres for 2 *dessinateurs,* at 3 livres per day each, to redraw a large plan of the Aegean from the Département des Affaires Étrangères onto a Mercator projection, working night and day

—400 livres for the preparation of the longitude and latitude grid by Brossier

—400 livres for preparing other materials and sketches to cover the rest of the Mediterranean, using other recent maps,

—570 livres for placing the sketches onto the completed grid

—168 livres for a cartouche by M. Du Gourre

—2,138 livres total.[19]

The compiler of a map was always under pressure to give his map something to distinguish it from other maps; if not a more decorative allure then some additional bit of information. Some geographers were in a better position to do this than others. The maps created in the early part of the eighteenth century by Claude and Guillaume Delisle, father and son, have long been singled out as fostering a cartographic revolution because of their accuracy.[20] The Delisles benefited from their access to a wide range of people whose travels and writings added to the store of geographic information required for up-to-date maps. They were connected to the court by virtue of tutoring the royal children. Their regular visits to Versailles and the Palais Royal, home of the regent, Philippe, duc d'Orléans, allowed them to meet travelers returning from abroad who were reporting their findings to the king or to the regent. They interviewed explorers and recorded their descriptions of the New World and Far East for use on their maps. Even their home was well located. They lived on the rue des Canettes, around the corner from the church of St Sulpice.[21] The Sulpiciens were among the most active missionaries in New France, from whom the Delisles gleaned much information for their cartographic descriptions of the Mississippi. Claude Delisle had even sent specific instructions with three members of the LaSalle expedition for clarifications of certain locations. These questionnaires resulted in three manuscript maps being returned to Delisle as he and his son prepared their maps of North America.[22] Two of the Delisles worked abroad. Louis de La Croyère served first in Canada, then joined Jacques-Nicolas in Russia. Guillaume Delisle benefited from their reports to create accurate and up-to-date maps until his death in 1726.

Though not as intimately connected at court as a family like the Delisles, English cartographers similarly seized opportunities for acquiring geographic information. Dennis Reinhartz has well described Herman Moll's circle of friends, from the pirate adventurer William Dampier to the publicist and novelist Daniel Defoe and the antiquarian William Stukeley.[23] Thomas Jefferys, engraver and map publisher, continuously sought recent information for new maps. He made much use of Spanish material captured by Admiral Frankland in 1740 for the preparation of the maps accompanying the *Description of the Spanish Islands* (London, 1762), and he was not hesitant to solicit information in what might have been regarded as hostile territory.

In 1774, a curious report was filed in the archives the Dépôt de la Marine by one M. Cambernon de Bréville, who had surveyed the coasts of Peru while sailing under the Spanish flag from 1750 to 1755. From his survey work he had prepared a map of the coast from Panama to Cape Horn. He estimated the cost of his four years of work (which, he assured the reader, had run him many risks) at around 5,000 livres per year, a large sum compared to the salary of the navy's chief hydrographer at 4,000 livres per year.[24] He had presented his work to the ministers of the navy and sent the results of his research to the Dépôt de la Marine in 1766. He was told by the ministers not to communicate any of his results to the English, but he was not paid for his efforts; indeed, he claimed he retrieved his manuscripts from Bellin only with great difficulty. In an effort to pry some reimbursement out of the ministry, de Bréville reported that while at Versailles on 6 January 1768, he had been offered 500 guineas by an Englishman named Jefferys, "who came to France for this purpose and to offer me work in England." De Bréville refused and hoped that his patriotic resistance to Jefferys's rich blandishments would be translated into a lucrative reward.[25] There the story in the Marine archives ends, without any hint of whether de Bréville was ever paid. But police archives in Paris continue the tale, for Thomas Jefferys was in Paris later in 1768 with his business partner, the print seller Robert Sayer. Purportedly on a print-selling trip, they sold maps and prints from their rooms in the Hôtel d'Entragues on the rue de Tournon near the Jardin du Luxembourg. A police raid on their hotel revealed the sale of "a considerable quantity of the most indecent prints," fourteen of them, of which some could be seen only by the light of a candle. They were invited to leave town quickly, and we may surmise that Jefferys's activities in soliciting maps at Versailles earlier in the year may have been a factor in their deportation.[26]

D'Anville too used the most up-to-date sources possible. His maps of China to accompany the Reverend Father Du Halde's *Histoire de la Chine* were based on both Du Halde's material and other sources, as admitted by Du Halde in the preface to his work. The geographer's correspondence with Cardinal Passionei in Rome is filled with requests for up-to-date material, as were his letters to the diplomat Jean Michel Hennin.[27]

Access to a good survey also provided opportunities for the compiler. One of Cassini de Thury's engineers, Joseph-Dominique Seguin, was invited by the provincial government of Burgundy to prepare a fifteen-sheet map of the province. He reduced the fifteen-sheet map to a one-sheet map and a road map that he promised would sell as well as if not better than the large-scale map.[28]

Both geographers and their publishers sometimes found recent and even local sources difficult to come by. In the "Préface historique" to the *Atlas universel* (1757), in which he cites the sources for the atlas maps, Didier Robert de Vaugondy complained that a recently announced map of the western part of Provence was not available anywhere, even in Avignon, where it was to be published. The atlas map of Indochina was based on a map published by Homann Heirs of Nuremberg, which was itself a copy of a map in the *Neptune Orientale,* the collection of charts compiled for the French East India Company.[29]

Compilation was a process so costly and so demanding that it was the step, along with surveying, that a producer or publisher would like to avoid if he could. Nonetheless, it became a necessary expense if one wanted to bring new information to public notice or to say honestly that a map was adjusted to the most recent observations. New accurate information was expensive.

In England, we have a much dimmer sense of the costs of map compilation. Many printed maps were published by engravers who either simply copied maps imported from the Continent or those previously published in England, adding minor changes, or merely translating the maps into English. Thus we have very few records that tell us the costs of such compilation, for the engraver was working only for himself. One or two documents from the last quarter of the seventeenth century give widely varying figures for preparing a manuscript map. The accounts of the Committee for Trade and Plantations show £10 6s 10d spent in 1678 for a map of New England.[30] In 1679, John Seller and colleagues charged £850 to prepare the maps for the *Atlas Anglicanus* by using old surveys, and another £100 to add the topographical detail.[31] In a collection of documents for 1706 from the household of Prince George of Denmark, consort of Queen Anne, appear receipts for £10 15s for the draft of a battle plan of Ramillies by Gabrill de la Hay and £32 5s for a plan of Kensington House and gardens.[32]

[THE DECISION TO PRINT]

By the time a map, whether the result of a topographic survey or a compilation effort, reached its final manuscript form, it represented a considerable investment of capital. Its economic life could well stop there. It could be copied in manuscript and distributed in that form to a relatively small audience. Sources for the costs of copying maps are few. The seventeenth-century Down Survey of Ireland provides one example: the charge for copying of a particular section or complete map from that survey was 5s 6d in 1702, plus the charge for the search of half a crown (2s 6d). By 1718 the search fee had risen to 5s, with the map copying fee negotiable.[33] At this rate, it would require four hundred manuscript copies to pay for the salary of one of the surveyors for a year.[34]

On the other hand, the person or agency might choose to reproduce maps in manuscript not just to save the costs of engraving and printing, but also to save man hours. As a member of the staff of the Dépôt de la Marine put it, "Two people can copy fifty plans or designs while six will not [be able to] make four maps that demand a great deal of work, discussion, attempts, and arrangements."[35]

The decision to print a map was motivated by a number of factors, not all of them economic. In general, we might suggest that the closer the map was to its point of origin and the larger its scale, e.g., the land survey, the less likely it was to be printed; it was far more likely to be reproduced in manuscript. The likely candidates for printing were medium- to small-scale maps of large areas compiled by geographers and printed either by the geographer or a publisher planning distribution to a select public or for general sale. Much depended on the publisher's perception of popular demand, and this perception could often miss the mark. Large-scale topographical maps required large-scale investment in their production. When they were printed, they found only a small market, buyers who lived in the area mapped.[36] Other factors could prompt printing besides recouping compilation costs and desire for profit. One might be administrative efficiency, in order to supply a quantity of information for specific purposes, as in the building of roads and canals or planning and executing military strategy. Another would be the promotion of a specific enterprise such as a voyage of exploration or economic endeavor, as in the case of the South Sea Company. Finally, printed maps were well suited for the dissemination of scientific information, as found in maps presented to the Académie Royale des Sciences or to the Royal Society for distribution and discussion or maps resulting from the work of the members of these academies.[37]

But even the most dedicated scholars appreciated the need to sell their work. D'Anville noted that his map of ancient Gaul, which had "cost him many years, had not been undertaken for any pecuniary motive, although it would have been unreasonable to expect that he created it as a pure gift."[38] J.-N. Delisle, not only desiring to keep his network of fellow astronomers informed and apprised of his cartographic work, also felt the need to sell his maps "to compensate me for the expenses of the engraving and the printing, and to allow me to continue to publish such works."[39]

As maps increased in popularity during the eighteenth century, certain types of cartographic productions that previously had not been printed became grist for the printsellers' mill. For example, in England large-scale estate plans were not usually engraved and printed until the mid eighteenth century witnessed a significant increase in demand for these printed plans. The development of a market for highly decorative printed estate plans can be attributed to the work of John Rocque and his followers, beginning in the 1730s.[40] In the second half of the century, the number of surveys of English counties increased. Most of them were the work of a single man who also supervised or even performed the engraving and oversaw the printing and publication.

Once the decision was taken to print a map, an entirely new set of costs

FIGURE 6. The preparation of copper for engraving. Diderot and d'Alembert, *Recueil de Planches sur les Sciences, les Arts Libéraux, et les Arts Méchaniques,* supplement to the *Encyclopédie* (1762), vol. 3, plate 3. (Harlan Hatcher Graduate Library, University of Michigan.)

was faced. With very few exceptions—for example, small woodcut maps that occasionally appear in the periodical press in England—maps in the eighteenth century were printed from engraved copperplates (figure 6).[41] The prices of copper, the labor of engraving, the cost of paper, ink, and printing from a roller press were all elements that could be controlled in some detail by a map publisher, who was the person responsible for organizing and paying for the engraving and printing. The map engraving process itself is very well documented.[42] My purpose will not be to explain the process of engraving and printing, but rather to analyze the various costs from the sources available to us.

[COPPER]

Copper was sold by color. Red copper was best for its adherence to the cutting stroke of the engraver's tool, the burin. Yellow or brassy copper, with its higher proportion of the hardening agent zinc, was regarded as too hard, with a tendency to break the point of the burin. Copper was also sold by weight and sometimes by plate. Size and thickness of the plate, naturally, affected weight. A plate measuring 8 × 11 inches weighed eleven pounds; a folio plate of 24 × 36 inches, thirty-five pounds; a large or elephant folio of 28 × 41 inches, forty pounds.[43]

François de Dainville, taking his information from François Courboin, estimated that the price of polished, burnished copper fluctuated through the eighteenth century between 55 sous and 1 écu (3 livres) per pound; this is borne out by much archival evidence detailed in the notes to appendix 1.[44] Copper sold by the plate varied in size, usually between quarto and folio. A copperplate that had already been engraved could be pounded out, reburnished, and polished, or it could be sold as old plate. Appendix 1 shows a leap in the price of copper after the French Revolution, driven up by the use of copper for bottoming ships as a defensive measure.[45] English prices showed a similar range. Quarto plates varied from 3s 3d to 11s per plate, that is, from three or four pence to four shillings per pound. Folio plates cost between £2 to 3, or one to three shillings per pound.

A factor further obfuscating the price of copper is that very often when engravers took the responsibility of supplying the copper for the work, copper and engraving were included in the same price.

[ENGRAVING AND ETCHING]

In France, the map compiler, the *géographe de cabinet,* was usually not an engraver. Though geographers such as Delisle, Buache, Gilles or Didier Robert de Vaugondy, and d'Anville might appreciate and understand the work of the engraver, they rarely attempted to engrave their own work. By contrast, at least two of the surveyors of the *Carte de France* offered their services privately as surveyors, compilers, and engravers. The surveyors Seguin and Belleyme contracted to produce maps based on the Cassini survey for the provinces of Languedoc, Guyenne, and Burgundy and to engrave them or to supervise their engraving. Seguin advertised himself as one who could "survey, draw, and engrave with taste."[46] The comte de Fleurieu, as head of the Dépôt de la Marine, encouraged hydrographers to learn engraving skills so that they could set down the projection and the geographic outline themselves, to ensure accuracy.[47] The combination of surveying and engraving skill is found in England as well, most strikingly in the case of John (Jean) Rocque, a Huguenot garden designer and surveyor resident in London, who engraved and published his own surveys of estates around the city beginning in the 1730s.

The participation of surveyors and hydrographers in the engraving business caused little problem in Paris, for copperplate engravers constituted a free trade. They had no guild for protection or for regulation, though most were trained in an apprenticeship system governed informally by the guild of St Luke. Their location in the capital was not restricted, though they tended to keep shop near the book trade, where they had access both to buyers and to a variety of employers. In London, by contrast, map compilers were very often engravers; at least it is the engraver's name that often claims authorship of a map for lack of any other. Engravers belonged to city companies or guilds, organizations that regulated entry to the profession, working conditions, and rules regarding the conditions of the apprentice-master relationship.

In both countries, engraving was a family business, and we find shops and materials passed from father to son (Jaillot, Kitchin, Jefferys, Bowen), from father to daughter (Delahaye, Haussard), from husband to wife (Rocque, Lattré, Delahaye). In England the engraving community was strongly influenced by members of the French Huguenot families who resided in London (John Rocque, Bernard Scalé, Pierre André) and by French visitors, such as Hubert Gravelot, who taught and worked in London until the mid 1740s. Laurence Worms has also detected Nonconformist connections, as in the case of Thomas Kitchin.[48]

A single engraver's name on a map does not begin to reveal the number of workers who contributed to the finished product. An engraver's workshop consisted of assistants and apprentices, each of whom had his or her own specialty. One person would engrave with a burin the outlines of the map, while another would etch the hachures and undulating lines for water. A third would be responsible for etching topographic detail such as mountains or fields, while a final engraver would use the burin for place name lettering.[49] (Engraving, which involves cutting with the burin, produces only lines; etching, in which a design is burned into the plate with caustic chemicals, produces a more textured print.) Lettering itself was a specialty divided between engravers of roman and of italic. Typical of the engraving of the *Carte de France* and other eighteenth-century French maps were the two groups working on the map of Guyenne in southwest France: the engravers of the plan and the engravers of the lettering. Among the plan engravers was Joseph Perrier, whose talents were reserved for the "undulations of the sea and rivers and for retouching the plate with the burin"[50] (figure 7).

[EXPERIMENTS WITH OTHER ENGRAVING TECHNIQUES]

Engraving and etching remained the primary techniques for map reproduction from copperplate throughout the eighteenth century. However, some isolated experiments with map engraving were made during the latter part of the century using techniques developed to achieve the tonal effects of pencil, chalk, pen, and color on paper.[51] The simplest experimental method

FIGURE 7. The process of etching on copperplate. Diderot and d'Alembert, *Recueil de Planches sur les Sciences, les Arts Libéraux, et les Arts Méchaniques,* supplement to the *Encyclopédie* (1762), vol. 5, plate 1. (Harlan Hatcher Graduate Library, University of Michigan.)

was the printing in two colors from two separate plates, each inked with a different appropriate color. This method was used by the Dépôt de la Marine to print the rhumb lines for charts in one color (usually green or red) (plates 2 and 3) and the hydrographic and geographic information in black. The same technique was used to print one set of information over another, as in the two maps of Europe showing in red or bister the visibility of the eclipse of the moon, published separately by Lattré and Desnos (plates 4 and 5).[52]

Other publishers experimented with printing using sepia ink for effect. Roch-Joseph Julien used two plates for his *Tableau topographique qui comprend la partie septentrionale du Landgraviate de Hesse, Cassel, et de la Principauté de Waldeck* (Paris, 1762). One plate, with the geographical outlines, was printed in black; a second plate with the symbols for woodlands and the hachures for hills was printed in sepia. Georges-Louis Le Rouge printed at least one copy of the plans for his small *Recueil des fortifications, forts, et ports de mer de France* (Paris, c. 1750) entirely in sepia, making the plans look deceptively like manuscripts, especially when hand coloring was added.[53]

Printing from two plates of course doubled the cost of printing and required care in registration of the plates. Using a single plate and coloring the plate itself, *à la poupée,* or with a rag daubing on the colors, was also time-consuming but could produce effective results. Louis Marin Bonnet engraved and colored the plates for a book of military plans by Charles Fossé, employing techniques that were to become his specialty in color printing: the crayon, or chalk, manner and pastel manner of engraving (plate 1).

The chalk manner of engraving imitated the appearance of chalk on paper, giving the line a soft, crumbly tone. It involved the use of roulettes and *mattoirs* (tools with flat, roughened ends for scraping the surface of the plate) on the etching ground, and when printed in color from multiple plates the technique reproduced the uneven chalked surface of pastel drawings. This technique was closely allied with mezzotint, by which rollers and scrapers produce a highly textured plate resulting in prints with deeply shaded tones. Similarly, the stipple technique, which produced images through the massing of dots, and aquatint, which produced an etched image similar to watercolor, provided tonal qualities that replicated different drawing surfaces and materials.

Few of these techniques were used in map engraving, but certain examples stand out. Henry Pelham's map of Boston (*A plan of Boston in New England with its Environs* [London, 1777]) was etched in the aquatint manner by Francis Jukes, who specialized in using the technique for topographical views.[54] It is worth noting that the method was first presented in England by the artist and amateur scientist Peter Perez Burdett in 1772; Burdett, who also worked as a county surveyor, engraved part of his map of Cheshire.[55] Another surveyor, J. F. W. Des Barres, experimented with different engraving techniques on the charts and views printed in his *Atlantic Neptune.* Des Barres rendered the topography on many of his charts with what appears to be a technique similar to mezzotint, wherein the plate is roughened to produce a black effect, then the rough burrs smoothed away to achieve a velvety rendering of hills and landscape. Additional line work with the burin highlights the soft etched details of hill and coastline. Des Barres may have employed engravers like those working for Alexander Dalrymple, the hydrographer to the East India Company, in producing his *Collection of Plans of Ports in the East Indies* (1774–1775). One of his engravers, B. Henry, worked with a stone on the copperplate to scuff and shade the plate for depicting the views and hills. Effective though the technique was in that "nothing expresses distant land so well," Dalrymple had to admit that the shallowness of the engraving would not give more than two hundred good impressions. "It is more elegant than convenient."[56]

Whatever the technique employed, engravers were all imbued with certain qualities of character, described here by R. Campbell in his *London Tradesman:*

They ought to be acquainted with Painting, have a nice Judgment in the Works of the most famous Artists, and perfectly Masters of the Doctrines of Light and Shade, in which their Art consists: They ought to be early learned to draw, and keep in constant Practice; for there is nothing which the Hand is more liable to forget than the Performance of any thing relating to Pictures . . . They ought to have a fertile Invention, and a kind of poetic Fancy; They must have a delicate and steady Hand and a clear strong Sight, for their Work is very trying to the Eyes.[57]

Campbell warned that while engraving requires little strength, it does demand a sound constitution and that, like all businesses that require "poring and sitting," it was not recommended for "persons inclined to Consumptions."

Map publishers often used more than one engraving workshop. The inventory taken after the death of Nicolas de Fer in 1720, for example, described map plates left among four different engravers, and no fewer than nine other engravers' names appear on his maps.[58] By contrast, de Fer's competitor A. H. Jaillot employed a single engraver, Louis Cordier, with a contract for three years, guaranteeing a fixed rate (84 livres) for engraving each copperplate; the contract must have been renewed many times, for Jaillot housed Cordier and his family for many years.[59] Guillaume Delisle comments to a correspondent that he also employed his engraver year round and was looking for work to keep him busy.[60] Other geographers preferred to work with one or two engraving workshops. Gilles and Didier Robert de Vaugondy and J.-B. B. d'Anville consistently hired the workshop of Guillaume Delahaye to engrave their maps, while the geographer Rigobert Bonne frequently collaborated with the engraver and publisher Jean Lattré.

The complexity of the engraving workshop mirrors the wide variety of ways in which costs for engraving were ascertained. As shown in appendixes 1 and 2, they range from engraving plus the price of copper, engraving by the hour, by the day, by the size of plate, by square foot, by the type of engraving: word and lettering, line work, shading, topographical detail. All are represented by different values, which reflect the wide variety of sources that provide the figures. Many of the price estimates come from two cartographic institutions whose records still exist: the *Carte de France,* supervised by Cassini de Thury, and the navy's Dépôt des Cartes, Plans et Journaux de la Marine. In both cases, the price paid for engraving seems to have been determined by the institution, not the engraver. On the other hand, we find that engravers themselves usually determined the prices found in the inventories and contracts of the notarial archives of the Minutier central in Paris. While prices varied both between institutions and within institutions, the maritime folio chart engraved under the auspices of the Dépôt de la Marine in midcentury might serve as a universal average for engraving a French map: between 500–1000 livres, a price that included the price of the copper.

The predominant figure for map engraving seems to have been a rate per plate in its entirety. Though some of the costs are itemized into line work, lettering, water, mountains, it would appear that an engraver's estimate of the cost of engraving a plate was a total one, which included all the labor of the workshop. The costs of map engraving are not entirely comparable to the costs of print engraving of the same period. In the late seventeenth century, the price of print engraving was often given per square foot, a unit of measurement that was less useful for a map.[61] The plate for a print was almost entirely filled with engraving or etching; a map could have many areas left untouched by the burin or etching needle. In the first half of the eighteenth century, the cost of engraving prints ranged from 100 to 1,500 livres, with this figure almost always including the copper and the proofs, which were pulled at the engraver's expense.[62]

One area that sharply defined map engraving as different from print engraving was the lettering on the map. The customary charge was an amount per one hundred words; roman capitals generally cost four times as much as italic lettering. There was an inherent problem in this method of valuation. The Paris map engraver Guillaume Delahaye said it meant that engravers often hurried and became sloppy in their work because it was pay for quantity, not quality. He warned that such a system forced the letter engraver to go too fast. He then made mistakes in spelling, misplaced names, and made other "slips of the burin that degrade the purity of the contours and the shadings that figure on the plan, the roads, waters, mountains, woods, and terrain in their different states of cultivation."[63]

Delahaye himself advocated an unusual method for engraving maps. He proposed engraving the lettering of the map first, then the geographical outline, and adding the topographical detail last. His description of this method also explains in detail the process of transferring the manuscript design to the copperplate, and is worth considering in full.

> When I am given a topographic design to engrave, I take the transparent paper, well oiled, and place it on the design, holding it tight so it doesn't vary at all. Then I take the interior frame, drawing in China ink the plan of the mountains, rocks, woods, and rivers. Then I write the words, giving an agreeable arrangement to all of them, straightening up the badly placed words so that no letter is cut by the rivers or hidden by the woods.

> Having accomplished this procedure with great exactitude, I place blank tracing paper, reddened on the side facing the plate onto the varnished plate that is going to receive the design. Then I put my oiled design and the blank reddened paper on the plate, checking that it won't move, pulling the edges of the design down over the plate. Then, with a rounded steel stylus, I draw only the writing on the design, and I draw the interior frame exactly. Then I pull the oiled paper and the reddened paper off the plate and draw on the varnish with another steel stylus, the entire frame with a straightedge, very exactly. This is essential in order to replace the oiled paper and the reddened paper [onto the plate] after the writing is engraved, which I do with a burin.
>
> I have trained different students in [the engraving of] the plan and written words, who have become well known. There are four of them for the plan and two for the writing, but they do not combine all the aspects of this art. What happens is that very often those who engrave the plan do not know how to arrange the words, which are commonly very badly placed, running across the woods, mountains, etc. Others create blank places in the plan in order to put the words in, which are badly filled and in terrible taste, not having any good notions about their arrangement.[64]

Delahaye knew all too well that the quality of lettering could be important enough to cause a lawsuit. The Delahaye family brought the publisher of the *Atlas universel,* Antoine Boudet, to court over a dispute concerning the quality of their italics. The Delahayes maintained that their italic letters had been judged as perfectly adequate, but the Robert de Vaugondys, who had created the maps, disagreed, maintaining that the italics were too thin and squeezed (figure 8).[65] The plates had to be redone, the offensive lettering removed and reinscribed.

While Delahaye was concerned with the lettering on a plate, one of the engravers of the *Carte de France* outlined his views on the appearance of the engraved plan of a map. He proposed a visual hierarchy of features: woods should be first in prominence, followed by cities, rivers, and finally mountains "in proportion to their elevation, with all the rest such as the vineyards, the meadows, the marshes lightly treated." These natural and manmade features should be expressed, he thought, in the general way one sees them from a high mountain, looking down, with trees like brown bouquets and rivers cutting into the surface of the earth with an even, shaded tone.[66]

Correcting a plate incurred its own costs. The Delahaye workshop charged an extra 40 livres per plate to erase and correct the disputed italics on plates for the *Atlas universel.* The archives of the Dépôt de la Marine, where charts were regularly corrected with new information, show these

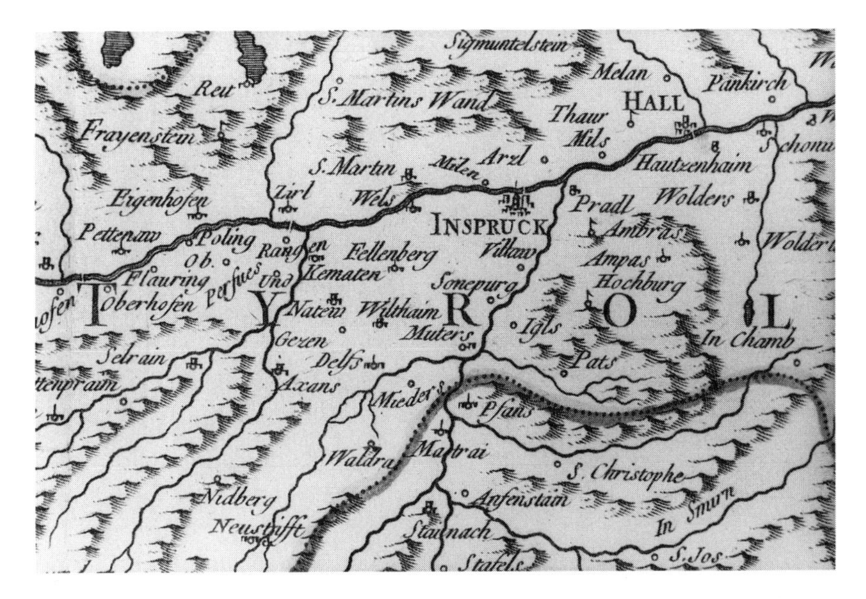

FIGURE 8. Detail of *Carte de Tyrol,* Didier and Gilles Robert de Vaugondy, *Atlas universel* (1757). The detail shows the squeezed italics, to which the geographer objected. Robert de Vaugondy's publisher, Antoine Boudet, refused payment to the engravers until the lettering was redone. WLCL, Atlas W-3-C. (William L. Clements Library, University of Michigan.)

costs could be as high as 250–700 livres per plate. Cassini de Thury tried to avoid such correction costs by investing in proofreading at a rate of 300 livres per plate.[67]

Budgeting for corrections tells us that checking proof states or early drafts of a printed map was an important part of printed map production. Guillaume Delisle's correspondence reveals that he regularly sent proofs of his maps to correspondents for comments.[68] Gilles Robert describes a similar process in his response to Philippe Buache's criticisms of the draft maps for the *Atlas universel.*[69] The minister of the French navy described the work of the dépôt as one of "acquiring, producing, correcting, and refurbishing maps daily"; the dépôt was in a state of "continual action" requiring designers, engravers, binders, and printers.[70]

While mapmakers sought corrections, their publishers saw expense. The map publisher A. H. Jaillot wanted to strictly limit the correction process. After purchasing manuscript map designs from Guillaume Sanson, Jaillot allowed Sanson to correct the maps only once. Sanson himself wanted to check them twice, in order to be assured that the corrections had been properly done and that other mistakes had not slipped in. Jaillot allowed his geographer to make the second set of corrections only if he kept the maps no more than two weeks. While Sanson was concerned for the maps' accuracy, his publisher knew that once the plates and the manuscripts were on the move between the compiler, the publisher, and the engraver, they were vulnerable to copying and plagiarism.[71]

A similar breakdown of costs for engraving in England remains much more elusive. Campbell's *London Tradesman* informs the prospective engraver that he or she, if "esteemed a tolerable Hand, may earn 30 shillings a Week, some that are very eminent are allowed Half a Guinea a Day. They are employed generally all the Year round."[72] The lack of public mapmaking institutions in England until the end of the century means no single archive maintains documentation for such costs, and it becomes difficult to verify Campbell's claim. Only the surviving records of private contracts reveal charges, and as a comparison of appendix 1 with appendix 2 shows, these are meager by comparison to the French listings, and they do not reflect the wide variance in methods of charging, especially for the detailed work on the plate. The closest we may come to a generalization of average costs in England for engraving are £10 – 50 per folio plate and £2 – 10 per quarto plate.

One source firmly accounts for engraving costs and payments: the receipt book of Arthur Pond in the British Library, which has been published by Louise Lippincott.[73] Pond was a London portrait painter, engraver, and print seller. He was a friend of and worked closely with J. Knapton, who published the account of the four-year (1740 – 44) global circumnavigation of Lord Anson: *Voyage around the world of George, Lord Anson* (London, 1748). The voyage had culminated in the swashbuckling capture of the Spanish galleon from Manila, loaded with over a million pounds' worth of spices, precious metals, and gems, "the richest haul ever made by an English commander."[74] The text was compiled by the ship's chaplain, R. Walter, under Anson's direction. Anson hired Pond to organize and supervise the engraving of the forty-two plates based on drawings and maps prepared by his men to accompany the work. Pond's daily receipt book shows the amounts paid to the engravers of different plates in the Anson work, allowing us to compare money paid with the plans themselves, their size, and their complexity immediately visible.

For example, the plan of Juan Fernandez Island (figure 9), with some topographical rendering and scales, cost £2 12s 6d. The plate, measuring 22.5 cm by 40 cm, was engraved by Richard Seale, a popular map engraver in mid-eighteenth-century London. We can compare the plan with the engraving of sea lions (figure 10), approximately the same size (24 cm by 40

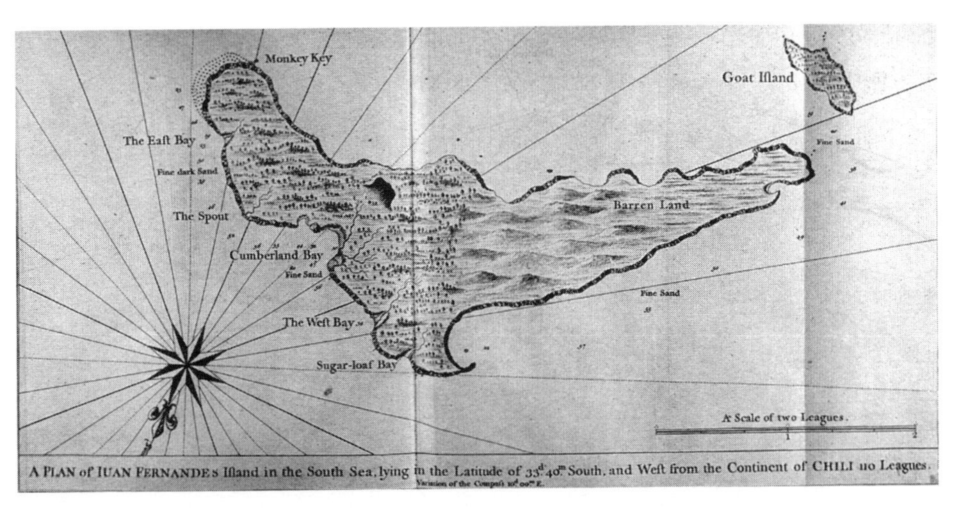

FIGURE 9. Juan Fernandez Island, engraved by Richard Seale for £2 12s 6d. Plate 15 in the atlas of forty-two plates accompanying Lord Anson's *Voyage round the World* (London, 1748). The engraving was supervised by Arthur Pond. WLCL, Atlas H-5. (William L. Clements Library, University of Michigan.)

cm) which cost Pond only half as much (£1 11s 6d), paid to Ignace Fougeron, one of the French emigré engravers living in London. While the sea lions may look like the more intricate engraving, they lack any lettering and are all line work and hachuring. The much higher price of the island map may be accounted for by the different skills required for lettering place names, topographical detail, and exacting line work for the compass. It also may be that Seale simply charged more than Fougeron or that Pond struck a different deal with each of them.[75]

[HOW LONG ENGRAVING TOOK]

The Pond receipt book also tells us that the forty-two quarto plates of Anson's *Voyage* took two and a half years to engrave. If another six months is added for printing, three years were required to produce the illustrations for Anson's book. For a folio map, much longer periods were required. In 1783, Cassini de Thury remarked that the most skilled engraver could engrave only two plates of the *Carte de France* in a year.[76] It took seven years for d'Anville to compile and supervise the engraving of the twenty-plus maps for Du Halde's *Histoire de la Chine,* during which time the terms were renego-

FIGURE 10. A sea lion and lioness, engraved by Fougeron for £1 11s 6d. Plate 19 in the atlas of forty-two plates accompanying Lord Anson's *Voyage round the World* (London, 1748). The engraving was supervised by Arthur Pond. WLCL, Atlas H-5. (William L. Clements Library, University of Michigan.)

tiated twice. D'Anville published the maps in a separate atlas in 1737. The engraver Guillaume Delahaye estimated that the topographic engraving on the *Carte des Chasses* (the map of the king's hunting grounds) of twelve plates would take four workers 2,897 workdays, amounting to about 241½ days per plate.[77] Even Napoleon could not speed up the process. In 1807, he asked for a printed copy of the map of the Battle and Siege of Danzig to be produced in three days. In fact, the engraving took sixteen days, even with the engravers working day and night and the lettering engraved on a separate plate (a time-saving technique that proved unsuccessful).[78]

Of course, the speed of engraving depended entirely on how much information actually had to be engraved onto the copperplate and what technique was to be used. Some engravers found that etching, with its use of acid to cut the plate's surface, was best done in the summer, when windows were open and ventilation better, reserving the work with the burin for the winter.[79] In the early nineteenth century, the Commission spéciale du Dépôt de la Guerre set up their budget for map compilation and engraving by assessing how quickly work could be done according to the scale of the map. A draftsman could prepare a sheet of the *Carte de France* at a scale of 1:50,000 in nine months, but it would require twenty-one months to prepare a sheet at

1:100,000. The commission heard that an engraver would require two and half years to complete the 1:50,000 plate. At a salary of 2,000 francs per year, the plate would cost 5,000 francs to engrave. The commission concluded that the best way to reimburse engravers was partly by salary, partly by price for the task at hand, paid as the work progressed.[80]

Further comparanda are provided by the records of the music and map publishing house of Artaria in Vienna. They equally demonstrate a range of engraving times, from a quick five weeks for a small plan of Vienna to ten months for very detailed maps. Johannes Dörflinger, who studied these records and the maps they cited, estimated that an engraver working a ten-hour day engraved between fifteen and fifty square centimeters a day.[81] This figure is similar to the thirteen square centimeters a day postulated by David Woodward for maps engraved in Italy during the sixteenth century and the estimates made by Günter Schilder for seventeenth-century Dutch engravers.[82] All these values are theoretical, for our incomplete picture of the organization of the map-engraving workshop muddies such comparisons. Speeds could obviously vary depending on the number of engravers available and on the quality of the engraving required for any given plate. Lettering with little topography would take considerably less time than an area requiring the array of textures and symbols for different types of terrain.[83]

In the Dépôt de la Marine, in 1737, Philippe Buache wrote to the minister of the navy, Maurepas, to tell him that three general maps of the Mediterranean would take three months to engrave, remarkably fast compared to other reports of the period. There were, however, some striking exceptions to the stately pace of engraving. William Faden, London engraver and map publisher, employed remarkable speed in engraving the plans of battles at the beginning of the American Revolutionary War. His plans of Fort Sullivan in South Carolina, from which the British commander Sir Henry Clinton was routed on the June 28, 1776, appeared in London on August 10, 1776, according to the copyright statement, barely six weeks after the event. Similarly, his plan of the victorious British engagements on Long Island, which took place between August 27 and September 15, 1776, was published in London on October 19, 1776, barely four weeks later.[84] The maps are not simple; they show considerable topographical detail, and the Long Island map has a great deal of lettering, which would normally require some time. It is not impossible that part of this map could have been prepared prior to the British landing on Long Island on August 27, but such an idea implies that Faden would have knowledge of British military intentions. Faden's ability to seize upon the opportunities of current events with a rapid response guaranteed his strong position in the market for battle plans, further fortified by reasonable prices: 1s for Fort Sullivan, 1s 6d for New York/Long Island. Similar speed of engraving and printing of a newsworthy event was evident in 1746 after the Battle of Culloden, fought on April 16. By May 1, *An Exact View of the Battle of Culloden* was published in London by Charles Mosley and distributed by John King and Mary Overton.[85]

[CARTOUCHES]

Before the engraving process was entirely complete, the final expense for the map publisher was the addition of a decorative cartouche. As other decorative nongeographic features slipped away from the surface of the eighteenth-century map, the title cartouche retained and even expanded its decorative and iconographic functions. Cartouche design, iconography, and taste await fuller study by map historians, but it is immediately apparent that many decorative cartouches expressed relationships of power and particular perceptions about geographical regions.[86] The archives occasionally preserve the considerations pondered by a map's designer. Some thoughts were even spelled out in the mapmaker's contract, as in the case of Philippe Buache's 1748 agreement with the États du Languedoc:

> One will display on these cartouches the arms of the Prelates and Barons of each diocese, those of the Province and those of the City. These coats of arms will be supported by their own devices or accompanied by geniuses or allegorical figures and attributes reflecting the products of the land, the province's natural history, and the commerce of the different cantons, and anything in particular concerning [its] ancient or modern monuments.[87]

In a recent study, Nelson-Martin Dawson analyzes the events surrounding the publication of Guillaume Delisle's 1718 map of Louisiana and its links to the ill-fated Company of the Mississippi (South Sea Bubble). Dawson reproduces the advice given to Delisle by the abbé Bobé, who was closely connected with some of the directors of the Mississippi Company and the court at Versailles. Bobé suggested that Delisle dedicate the map to the eight-year-old king, who "loves geography," and that he add the arms of the Louisiana Company to the map, placing around the cartouche "some Indians, especially the flat-heads, with their beautiful clothes."[88]

Bobé knew how evocative a cartouche could be on a map. Delisle's earlier *Carte du Canada* (1703) displays a cartouche (figure 11) that incorporates a miniature narrative of the efforts of missionaries to Christianize native North Americans, an image that Dawson carefully deconstructs. Though

FIGURE 11. G. Delisle, *Carte du Canada* (1703), cartouche. WLCL, Maps 4-B-2. (William L. Clements Library, University of Michigan.)

Delisle did not dedicate the map to the king directly, the royal crown and coat of arms surmount the cartouche. Along both sides of the title are representations of the clergy who led the vanguard of French colonization of North America. To the left, a Jesuit priest baptizes Indians; on the right, a Recollet missionary preaches to them. On the right, emerging from the thicket of prickly brambles, an unrepentant Iroquois holds in his right hand a freshly harvested scalp, no doubt from a French head. In a leafy glade on the left, a Huron woman, baby on her back, clasps her hands in prayer, in devout counterpoint to the bloody act of the Iroquois. Behind the Iroquois, a

FIGURE 12. Cartouche for Jean de Beaurain, *Carte du Port et Havre de Boston* (Paris, 1776), engraved by Charles-Emmanuel Patas (1744–1802). The horrified Amerindian grasps a bow from which emanates a phrase from Horace, "*Quo scelesti ruitis?*," the first line of Epode 7, a reflection on the civil war in Rome. WLCL, Atlas E-4-B. (William L. Clements Library, University of Michigan.)

hunter shouldering a rifle; behind the Huron, a noble savage. Underneath them, marked by rosary beads intertwined with pine cones, the fruits of North American harvest: the beaver, fish, and goose. The hierarchy here pushes religious orders to the top, not commerce.[89]

The French geographer Jean de Beaurain added a politically charged cartouche (figure 12) to his 1776 map of the port and harbor of Boston, copied from the chart of J. F. W. Des Barres. The vignette surrounding the title demonstrates French sympathy for the American revolutionary cause. A colonial militiaman, holding the banner of Massachusetts with its mighty white pine, grapples with British soldier, whose banner displays the British lion rampant. Watching them is a horrified Amerindian whose thoughts are mirrored in Latin on the shield above him: *Quo scelesti ruitis?*—Whither do you rush, o wicked ones? The eighteenth-century viewer acquainted with Horace would recognize the first line of his Epode VII, expressing the ancient poet's dismay at civil war in first-century Rome.[90] The map's dedication to the king of France in 1776 emphasizes the growing desire of some of Louis XVI's advisors to enter the North American conflict as a way to avenge the loss of Canada to the British in the Seven Years' War.

The importance of the appropriate cartouche did not escape le père Du Halde, author of *L'Histoire de la Chine.* After hiring the geographer d'Anville to prepare the maps to accompany the work, he requested that the cartouches be designed by a M. Humblot, "who is perfectly imbued with the taste of Chinese painting."[91] Humblot based his work on patterns sent to Du Halde by a M. du Velaer, many years resident in Canton as a director of the Compagnie des Indes.

As the military engineer Seguin finished his work on a reduction of the *Carte de France* for the provincial administrators of Burgundy, he sent the electors of Burgundy a questionnaire regarding their desires for the map's decorative features: its border and the form of the cartouche. The electors decided on two cartouches: one for the *avertissement* or information note with arms of the house of Condé and one for the title with the attributes and arms of the province, with a small plan of Dijon between them. At the top of the map appeared the arms of France, "dans une gloire de bel effet."[92]

Cartouches need not bear such heavy freight. Their attractiveness, however, required the skills of both master designers and etchers. The geographer d'Anville was fortunate in that his brother, Hubert Gravelot, was a well-established designer and engraver who produced many the cartouche designs for d'Anville maps.[93] The choice of a well-known designer for a cartouche could only add to the value and attractiveness of the map. Thus the Robert de Vaugondys (or their publisher Boudet) invited the engraver-

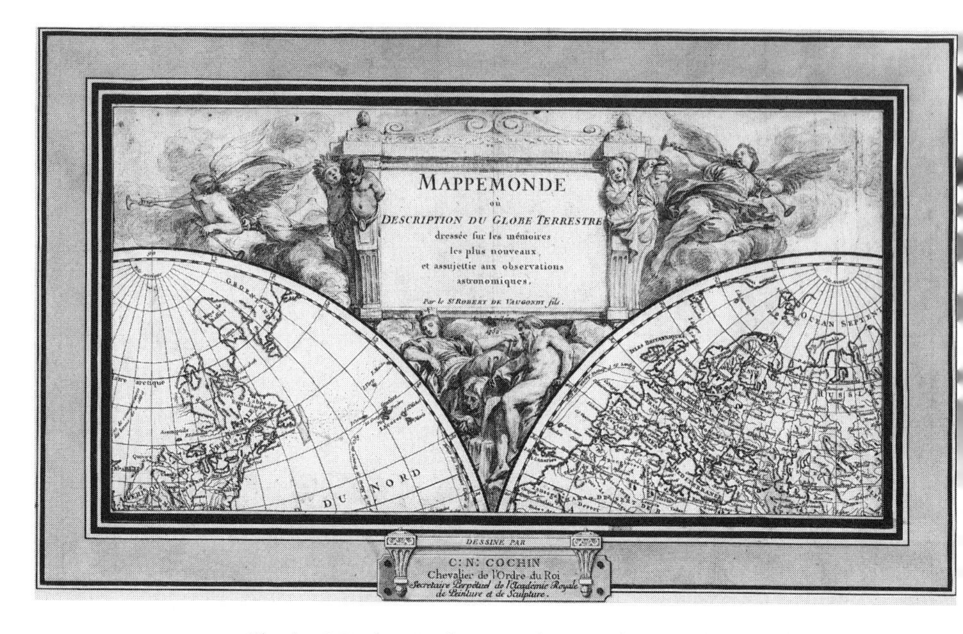

FIGURE 13. Charles Nicolas Cochin, graphite rendering of the title cartouche drawn on a proof impression of Robert de Vaugondy's *Mappemonde ou Description du globe terrestre* (Paris, 1752). (© Copyright The Trustees of the British Museum.)

designer Charles Nicolas Cochin to add the decorative scheme to the mappemonde for the *Atlas universel* (figure 13).[94] A design by Cochin could be expensive. He charged the Maison du Roi 72 livres for a small vignette (59 × 140 mm) that was used in the dedication of the *Catalogue raisonné des Tableaux du Roy* (Paris: Imprimerie Royale, 1752, 1754). Its engraving by Ingram cost 120 livres. For the less complicated royal arms for the same work, he charged 24 livres for the design and 80 livres for the engraving, done by his cousin, Jacques Tardieu. Both Cochin and Jacques Tardieu were cousins of Pierre-François Tardieu, engraver of the cartouche of the Robert de Vaugondy mappemonde.[95] This cartouche echoes those vignettes Cochin designed for the French edition of Anson's voyage.

To avoid this cost, a mapmaker might decide to try such decorative work himself. Georges-Louis Le Rouge attempted a Neptune emerging from the mouth of the St Lawrence river on his map of North America (figure 14); it is labeled "for practice" ("pour essai"). Though this kind of decorative motif of sea gods in the ocean was increasingly out of fashion at this date (1777), Le Rouge obviously felt it worth the effort.

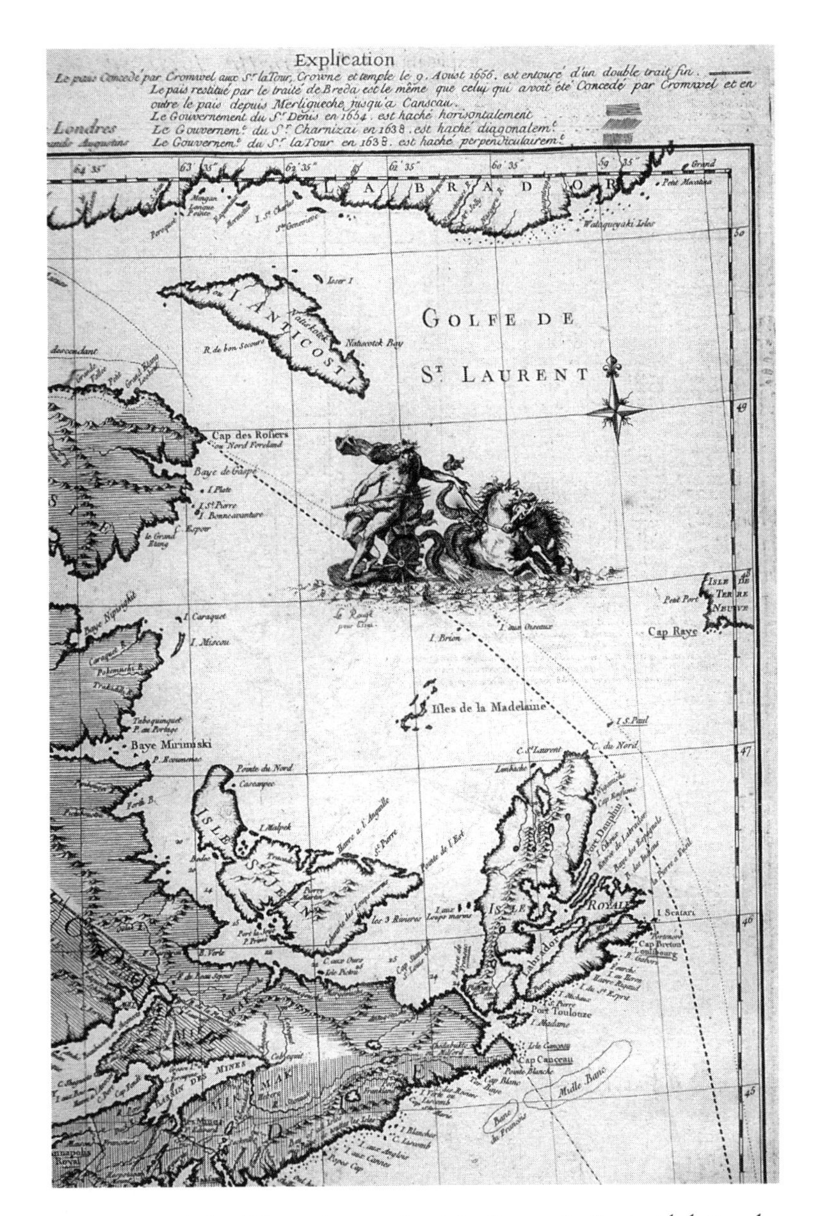

FIGURE 14. A detail from Georges-Louis Le Rouge, *Partie orientale du Canada,* in the *Atlas Amériquain septentrional* (Paris, 1778), plate 10. Note the Neptune signed "Le Rouge pour essai." WLCL, Atlas E-1-C. (William L. Clements Library, University of Michigan.)

FIGURE 15. Frontispiece from Bernard Romans, A *Concise Natural History of East and West Florida* (New York, 1775). Engraved by Romans. A Native American presents a map to a seated figure who holds a staff with a freedom cap and sporting "SPQA" (Senatus Populusque Americanus) on her shield. A winged figure at her feet measures distances on a chart. WLCL, C 1775 Ro. (William L. Clements Library, University of Michigan.)

The surveyor and mapmaker Bernard Romans similarly tried his hand at decorative engraving in the frontispiece (figure 15) to the *Concise Natural History of East and West Florida.* A Native American figure presents a map, symbol of land and possession, to the seated figure whom we might take for Britannia until we notice the SPQA on her shield: *Senatus Populusque Americanus,* a play on the motto of the Roman Republic, SPQR, and also perhaps, on the mapmaker's own name. In 1775 there was no doubting the political leanings of Romans, who was to join the American patriots after 1776 in their rebellion against the British.

We have very few figures for the cost of the cartouche as a separate element in map production. Perhaps the relevant documents do not survive, or perhaps the cartouche design and engraving were figured as part of the total cost without itemization. What we do have shows us that the cost could range from 25 to 65 percent of the cost of engraving the plate.

[PRINTING]

Once the copperplate was engraved, it was ready for the press (figure 16). First, proof copies were printed for corrections; after the corrections were

FIGURE 16. Printing from a copper plate. Diderot and d'Alembert, *Recueil de Planches sur les Sciences, les Arts Libéraux, et les Arts Méchaniques,* supplement to the *Encyclopédie* (1762), vol. 8, plate 1. (Harlan Hatcher Graduate Library, University of Michigan.)

made, the desired number of copies were printed for distribution. The location of the press itself varied. Many engravers seemed to maintain their own presses; for example, Alexis Hubert Jaillot in Paris kept his single press in his attic, where the copperplates were also stored.[96] Some geographers also owned a copperplate press: both Duval and Gilles Robert de Vaugondy record having a press in their workshops.[97]

Copperplate printers normally charged by the hundred. The figures in appendix 1 show that the cost of printing in midcentury Paris was around 15–25 livres per hundred; in London, the rate varied from slightly under a shilling (10 pence) to a high £2 per hundred (which may have included paper and binding). If paper were included in the price, printers in Paris might charge by the sheet, from 4 to 10 sous per sheet. Money alone did not guarantee results, as evidence of the occasional sweetener shows. For the seventy plates of Anson's *Voyage,* Arthur Pond recorded paying his printer £48 plus beer. However, demon drink, though always an important lubricant for the printer, could spell ruin for an engraving. The designer and engraver Charles Nicolas Cochin complained that his plates were ruined when the printer, "one of the best but who had to be watched," went drinking and allowed a "useless apprentice to do the work, who used up the plates before they had given even one hundred good proofs."[98] And if the printers were criticized for doing a bad job, as in the case of two printers of plates of the *Encyclopédie* who had dampened the paper too much, they simply responded with "haughtiness, insults, and even blows."[99]

These incidents of troubles with the journeymen in the printing workshop highlight several potential problems with printing maps. One was "using up the plate." A copperplate could sustain only a certain number of pulls before the engraved and, more particularly, the etched lines became faint and worn, as the soft copper was pulled hundreds of times between the rollers of the press. Several factors determined how many pulls the plate could tolerate. The quality of the paper and the type of finish it carried could wear on the copper in different ways. How the plate was prepared for printing also affected its life. After the plate was heated and inked, it was cleaned of excess ink either with a rag imbibed with urine (*au chiffon*) or by the palm of the hand (*à la main*). Cleaning with the rag was faster and allowed more maps to be printed per day, but the acid was hard on the copperplate, using it up more quickly and ultimately allowing fewer pulls. The slower method of cleaning the plate with the palm of the hand, especially good for highly detailed topographical maps that supported a good deal of etching, was gentler to the plate and allowed two or three times as many maps to be printed.[100]

The number of sheets that could be printed from a copperplate thus varied depending on the copper itself, the engraving method, and the manner of printing. Archival documents record varying amounts, from fifty to six thousand. Emanuel Bowen charged 6 shillings for a day's work of printing one hundred.[101] Occasionally an incredibly large number is encountered, such as the twenty thousand copies of a map of the West Indies cited by Edward Cave in the *Gentleman's Magazine.*[102] However, the average print run seems to have been two to three hundred. To print such a number would have taken five or six days from a single press operated by printers who averaged thirty to fifty sheets per workday.[103]

The printing process produced a slightly damp map that could take up to six weeks to dry.[104] To distribute them earlier left them vulnerable to smearing or offsetting the design onto another map, a common occurrence. Their humidity and subsequent shrinkage created another problem for the map designer or compiler, for the reduction in the map's surface altered the

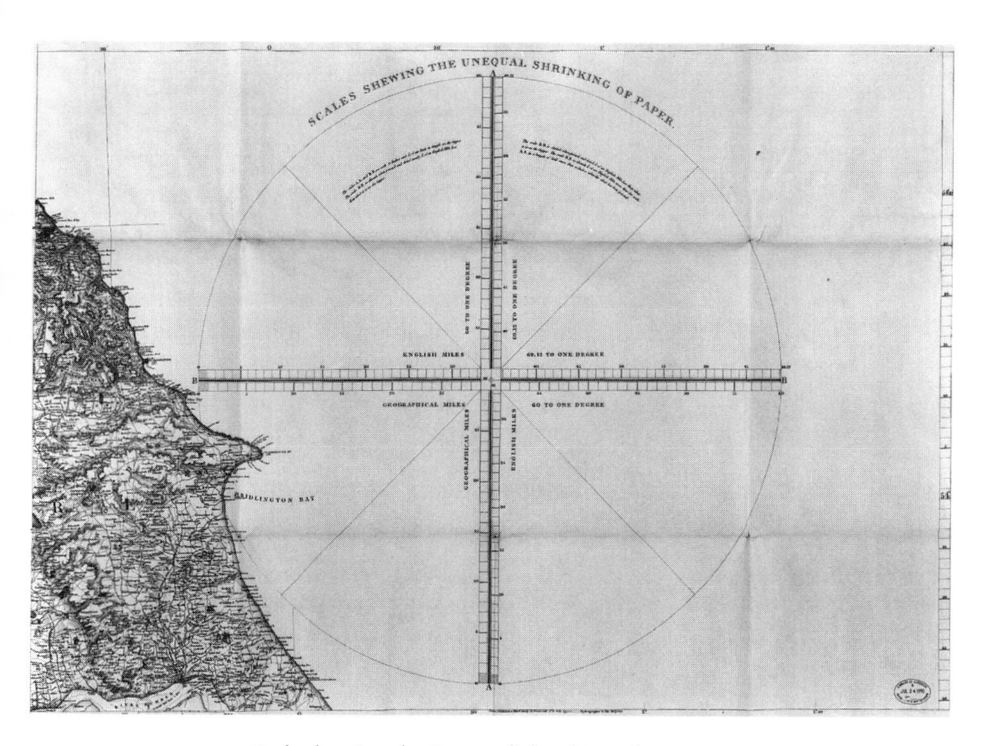

FIGURE 17. Scale showing the "unequal shrinking of paper" on Aaron Arrowsmith, *Map of England and Wales: the result of fifteen years labour* (London, 1815). G 5750 1818 .A7 Vault Oversize. (Library of Congress, Washington.)

scale of the map. Philippe Buache calculated that a designer must enlarge the manuscript plan on the copperplate by a sixtieth in order to keep the map to the scale to which it was drawn. In addition, the nature of the rag cotton from which paper was made caused the printed map to shrink unequally in two directions.[105] English geographers also tackled the problem of shrinkage. P. P. Burdett admitted to the Society of Arts, Manufactures, and Commerce that the contraction of the paper as it dried might have affected the accuracy of the scale of his map of Cheshire in 1777.[106] Aaron Arrowsmith included large scales on his 1815 map of England and Wales "shewing the unequal shrinking of paper." He demonstrates that after printing and drying, the width of the paper shrinks more than the length of the paper, thus making the east-west scale of miles more inaccurate than the north-south scale.[107]

[PAPER]

The criteria for choosing paper included its size, its weight, its thickness, its absorbency of the ink, and its propensity for taking the water color wash that might be applied after printing. The Auvergne was the center of paper manufacture, not only for France but for the whole of Europe.[108] Its high quality induced the Parisian map seller Roch-Joseph Julien to send *chapelet d'Auvergne* paper to the German map publishers Homann and Seutter for printing the maps they were to send to him. This caused their maps to be more expensive when purchased in Paris.[109]

Procuring a secure source of paper concerned the Dépôt de la Marine for their production of sea charts, which required the very large *grand aigle* (36 × 24 in., c. 105 × 70 cm.). A report to the minister of the marine, Gabriel-Joseph d'Oisy, in 1773, outlined the reluctance of some paper makers to manufacture this size unless they received secure, regular orders. To procure the special paper, the dépôt also had to budget for the cost of transport and duties along the way: 11 sous per quintal (100 kilograms) along the route and 12 livres per ream once the paper entered Paris.[110] The quality of paper occupied the geographer d'Anville throughout his life, and his maps are recognizable for the thick, white paper on which they are printed. In a letter to William Faden shortly after d'Anville's death, his clerk Demanne promises to imitate "the discernment that he uses as much for the choice of paper as for the printing and decoration."[111]

Paper was sold by the sheet, the quire, the ream, or the pound. Once again, appendixes 1 and 2 show the range of prices through the century. Roughly speaking, by the latter half of the century, the largest-size paper for maps (*grand aigle*) cost 120–30 livres per ream in France and £10–14 per ream in England. Produced in limited quantities yet indispensable for publishing, paper had its own economic world.[112] It is not surprising that the map publisher Desnos was indebted to various paper merchants to the amount of 39,729 livres in 1772.[113]

[INK]

As important as the quality of paper was the quality of the ink. In France, the most valued for map printing as for print engraving was the *noir d'Allemagne.* It was thought to provide the smoothest and most adherent base of any ink available. Even though it cost two times as much as other types of ink, some publishers bought it by the barrel.[114] The local Parisian black ink used for print and map engraving was regarded as inferior. It was a rough and gritty blend of mutton bones and wine lees boiled together in a malodorous process. The ink rapidly spoiled the plates on which it was used.[115]

[MAINTAINING THE PLATE]

Copperplates required care and maintenance while they were being engraved, while they were in the printing press, and during their storage after printing. An engraver for the *Carte de France,* M. Patte, recommended that *graveurs en lettres,* who exerted a greater pressure on the burin than did the etcher on his needle, keep a chamois skin under their elbows as they leaned on the plates; this would protect the plate as they worked.[116]

After printing, plates required maintenance that included cleaning the ink and the accumulated dirt from the press. The engraver Patte advised inspecting the plate after four hundred impressions, before the burin lines disappeared altogether. When the plate was stored, it accumulated verdigris built up by humidity on the copper. Cleaning the plate could then cost between 12 sous and 10 livres per plate. If a used plate was going to be reengraved, it had to be erased or replaned or some combination of both, adding further to the cost of maintenance.

[COLOR]

Color was so often an integral part of a map that it could not be dismissed as merely decorative. Color could convey pertinent information, for example, by distinguishing competing armies on maps of military campaigns or battles, or by outlining political possessions or administrative locations. In

Paris, *enlumineurs* or print colorers formed their own loose organization. Catherine Hofmann tells us that in the early years of printing, 1450–1600, there were forty-three *enlumineurs* in Paris. They carried on the traditions of manuscript illumination, adding color and gilding to printed texts, though their metier demanded that they use water colors, not oils. Their involvement in the map trade is evidenced by such Paris *enlumineurs* as Jean Boisseau and Nicolas Berey also functioning as map publishers.[117]

Watercolors were used for maps and prints. In the seventeenth century a distinction developed between the *lavis* and the *enluminure*. The *lavis*, or watercolor applied to a manuscript map or plan, added new elements to the map by shading, thus creating graphic information, while *enluminure* simply added color *over* information already engraved on the map. French map publishers generally employed color to highlight engraved elements only, defining the political or thematic limits of a map or highlighting water, towns, and certain topographical elements such as woods and hills. This style endured until the middle of the eighteenth century, after which map publishers might offer maps colored in four different styles. *Ordinary color* only emphasized the lines of political or judicial divisions and colored towns red and forests (if colored at all) green. *Color in the Dutch manner* produced darker divisional lines and a pale but transparent color throughout an area. *Color in the German manner* was a dense, opaque color throughout, as on maps published by the Homann firm in Nuremberg or by the Augsburg publisher Matthias Seutter. A fourth style of coloring was "new manner," which expressed, as the mapmaker Julien put it, "le nouveau goût des ponts et chaussées" (plate 1). Even though the newly founded (1747) École des Ponts et Chaussées apparently used no internal manual for guidance on coloring maps, the large-scale maps emanating from the school displayed their own style of coloring topographic and hydrographic features. The color attempted to imitate nature by highlighting in natural colors details displayed on the map itself.

Some maps, especially thematic ones, required coloring because of their map keys. Didier Robert de Vaugondy's world map from the *Nouvel atlas portatif* (1760) appeared four times in the atlas: once with its political divisions colored, again with distribution of world religions, a third time with the distribution of the world's inhabitants by skin color, and finally with the distribution of inhabitants by the shape of their faces.[118]

Map coloring has often been described as the gentle pastime of women. This view may be attributed to Hubert Gautier's remark in his essay of 1687 on the art of coloring. He expresses some disdain for illumination: "one can learn the complete essence of it in three or four lessons. Women and nuns can easily occupy themselves with this sort of activity".[119] But coloring was not always left to *les enlumineuses*.[120] The engraver Arthur Pond had one of his apprentices, Peter Maddox, devote much of his time to coloring, paying him on average 2s 6d per print. This rather high figure—more than the price of many maps—reflects the skill required for the detailed work in coloring a print, not the simpler effort of outline or wash for a map.[121] Thomas Jefferys's former apprentice, John Lodge, set himself up specifically as a colorer of maps and prints.[122] In London, coloring was also done by fan sellers.[123]

There was no shortage of printed advice on coloring maps. Buchotte's *Regles du dessein et du lavis pour les plans particuliers des ouvrages et des bâtimens* not only described the process of coloring but added a section that constituted a shopping list, including prices, for colors, brushes, and papers appropriate for water color.[124] John Smith, in *The art of Painting in Oyl, to which is added The whole art and mystery of coloring maps and other prints with Water colors,* advised that "The Art of colouring well may be attained to by Practice, and if a man does spoil Half a score Maps, in order to get the knack of coloring a map well at last, there is no man that is ingenious will grumble at it."[125]

While there is little evidence for the costs of coloring maps, coloring could increase the final price of the map by 50 to 200 percent. Delisle told his correspondent and erstwhile business partner Louis Renard that his maps normally sold in Paris for 12 sous uncolored and 15 sous with the political divisions colored.[126] Bellin's *Petit atlas maritime* in five volumes (1763) was priced at 96 livres, unbound and uncolored; 120 livres, half bound with the seas done in full wash.[127] Sayer and Bennett's catalogue of 1775 advertises "Large Maps," that is, wall maps, at two prices: the first for the uncolored map in sheets, the second for the map mounted on canvas, with rollers, and colored. The second price usually doubles the first.[128] Such price differences reflect the labor costs of coloring. Only two instances of wages for coloring have emerged for France, from either end of the eighteenth century: 6 deniers in 1706 and 3 sous in 1787. English records have produced only Arthur Pond's rate to his young apprentice, at 2s 6d per print, not a map.

What mattered most about coloring, as with engraving, was doing it well. Pierre Duval complained of colorists who mistook the lines of rivers for the division points on maps, and so, whether by caprice or mistake, enlarged or reduced sovereign territories.[129] Philippe Buache also noted the need for care in coloring. He warned that after printing, about one-tenth of printed maps were lost from either defects in the paper, or mishaps during printing, or accidents to the maps that occurred during coloring in the Dutch manner, "which we have to do in order to distinguish at a glance land from sea but without hiding the detail of the coasts, as the coloring of ordinary maps would do."[130]

Once the map has evolved from the manuscript, produced by survey and/or compilation, to the printed object, it is ready to be sold. We have seen that the names on the map of geographer and engraver do not include all the *petits gens* who have participated in the map's production. The titles on maps rarely include all the surveyors, travelers, sailors, engineers, astronomers, and savants whose reports, sketches, and manuscripts have been digested for inclusion, reduction, and amalgamation on the map. Only if the mapmaker has written a mémoire outlining his or her method of work and listing the sources can we be sure of knowing how the map was made. One thing is certain: it was not cheap. A map that required survey and analysis of materials and long compilation could cost up to 10,000 livres or £1,000 from start to print. César-François Cassini de Thury estimated each map of the 180-sheet *Carte de France* would cost 4,000 livres from survey to print.[131]

Once the map was ready for the market, it needed a price, a distribution network, and interested buyers with money in their pockets. What it did not need was competition. And its worst competition was itself: the copy.

PART TWO

SELLING MAPS

CHAPTER THREE

GETTING AND SPENDING

The way to spread a work is to sell it at a low price.[1] — Samuel Johnson

The decisions to print a map and then to sell it form two crucial joins in the financial mosaic of the map trade. Not all maps that were sold were printed, nor were all printed maps necessarily sold. Some maps were printed solely for private distribution. In France, the provincial government of Burgundy hired Seguin, an engineer from the Cassini survey, to produce a very large (eight feet by nine feet) map of the province, based on the measurements and topographical data assembled by the Cassini engineers. Four hundred copies of the map were printed, of which two hundred went to the king for his personal use and distribution and two hundred went to the provincial government for its own distribution, either as gifts or for sale at 48 livres each.[2] The geographer Rizzi-Zannoni described his participation in the survey and engraving of two maps, one of the kingdom of Naples and the other of Poland. Immediately after they were each printed, only a very few copies were available for sale; soon there was none, for the plates belonged to the monarchs of each of these nations.[3] Similarly in England, a map of the southern part of the city of York, surveyed by J. Dickinson in 1750, was "taken at the cost of the most Hon. Thomas, marquis of Rockingham, and reserved for the marquis' use and not to be sold."[4]

Such exceptions confirm that printed maps were usually destined for sale, if not for profit then to reimburse the mapmakers for the costs of production. Most geographers would have agreed with J.-N. Delisle, who described his sale of maps as the means to reimburse engraving and printing expenses and to produce sufficient profit for supporting future works.

The price of a map reflected several things besides profit (appendixes 4 and 5). The need to reimburse not only the mapmaker but also the printer, engraver, copperplate supplier, and papermaker was balanced against the price of maps offered by competitors. These two facets of the market are clear in an exchange of letters between the Delisles, Claude and Guillaume, and Louis Renard, a map seller and publisher in Amsterdam. The Delisles and Renard had been negotiating a wholesale price for maps that the Delisles proposed sending to Amsterdam for resale. Guillaume Delisle told Renard that "I profit by at least one sou on each map." He further asserted that he sold his maps in Paris for 10, 12, and 15 sous, depending on size, paper, and coloring, but that he would offer Renard his maps at the wholesale price of 8 sous, refusing to go lower than 6 sous.[5] In spite of Delisle's adamant stance on the price, Renard clinched the deal at 5 sous 6 deniers each, by throwing in an additional 6 deniers for coloring each map, bringing the final wholesale price up to 6 sous.[6] Renard had driven the bargain by agreeing to undertake the costs of packing, carriage, and customs. He also pointed out that Delisle maps would have to compete with other maps sold in Amsterdam at about a sixth the price of Delisle's. Renard told Delisle he was relying on the beauty and exactitude of Delisle's maps to give them the competitive edge in a Dutch market not just limited to the Low Countries, but, as Renard claimed, a market that sent maps by the thousands to Germany, England, and Scandinavia.[7]

Similar discussions of discounts and wholesale arrangements permeate the letters from Continental map sellers to William Faden in the third quarter of the eighteenth century. These correspondents gave each other a standard wholesale reduction of a sixth, though discounts as great as 40 percent were occasionally given.[8] When J.-B. B. d'Anville negotiated with le père Du Halde to prepare the maps for the latter's *L'Histoire de la Chine,* he stipulated that he be allowed to sell the collected China maps as a separate publication and to keep one-third of the market price as his *droit du marchand.*[9]

The amount of expected profit increased during the century. From Claude Delisle's expectation of a profit of one sou per map, his son Guillaume's widow claimed that she contented herself with a profit of only 2 to 3 sous per map, "instead of the usual 5 sous" during the twenty years following her husband's death in 1726. Believing she was protecting her husband's memory, she refused to sell many copies of his maps, not wanting to see them sold cheaply by *colporteurs* (street vendors) who hawked their wares along the sidewalks of the quais and along the walls of large buildings.[10]

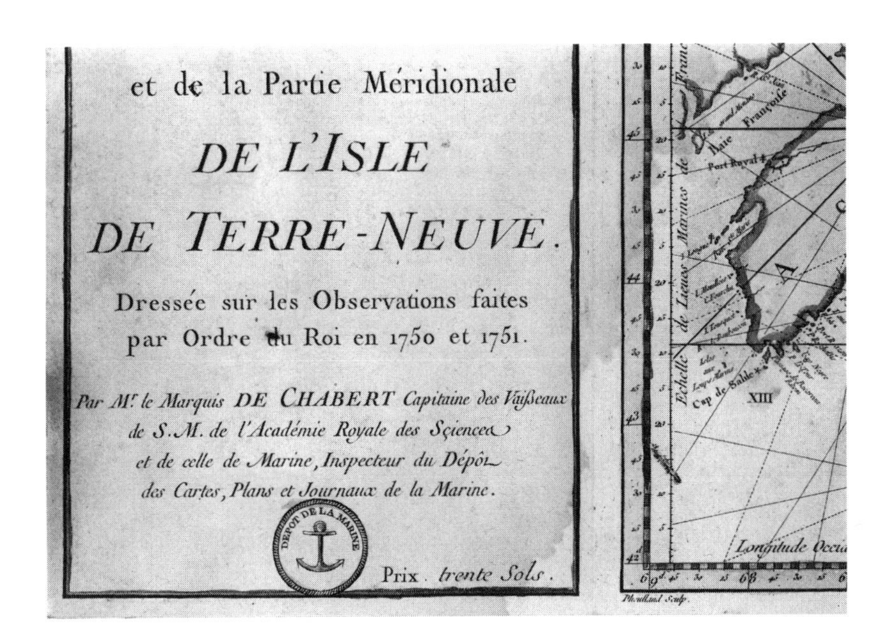

FIGURE 18. A detail from le Marquis de Chabert, *Carte réduite des Côtes de l'Acadie, de l'Isle Royale, et de la Partie Méridionale de l'Isle de Terre-Neuve* (Paris, 1753). Here in its fourth state (after 1775), distinguished by the additional engraving of the price ("Prix trente Sols") and the seal of the Dépôt de la Marine. WLCL, 3-D-10. (William L. Clements Library, University of Michigan.)

Pricing maps to vie with the competition concerned the Dépôt de la Marine as it prepared to sell its charts in the port cities of France. Philippe Buache, who worked in the dépôt briefly, recommended to the naval minister a government subsidy on charts, allowing distributors in the port cities to be able to sell charts printed on paper for 15 sous. This would undercut the ordinary price of 20 sous for charts printed on paper and 50 to 60 sous for those more durable examples printed on vellum.[11] The government would have to subsidize the distributors by selling them maps at roughly a third less than the retail price, i.e., about 10 sous. This would be low enough, according to Buache, to defeat cheap copies and foreign competition, especially Dutch sea charts. The prix fixe for charts gradually rose during the century, as the problem of copyists and counterfeiters continued. By 1775 the king had approved an ordinance setting a price of 30 sous for a large chart (on *grand aigle*) and 18 sous for a map on a half sheet. His intent was to undercut the copyist's price of 40 sous for counterfeit versions of the dépôt charts. Printing the inscription "*trente sous*" directly onto the chart and stamping the chart with the seal of the Dépôt de la Marine signaled the dépôt's authority and attempted to prevent counterfeiting (figure 18).[12]

State intervention in the price and sales of maps was not a new idea in France. In the late seventeenth century, the king's council was asked to protect the rights of customers, in this case, the officers of his majesty's troops. These men, who were required to supply their own maps, were obliged to buy an entire atlas when only one map was desired. The resulting royal ordinance (7 April 1688) declared that geographers and map sellers must sell maps separately, though allowing them to double the price in recompense.[13]

Catalogues, advertisements, correspondence, and the prices marked on the maps themselves all reveal an average range of prices for maps throughout the century. In England, early in the eighteenth century, a one-sheet map cost between 6d and 1s. By the latter half of the century, the same map could cost from 1s to 3s. Multisheet maps were more: two sheets for 4s; four to nine sheets for one-half guinea (10s 6d) to three guineas (£3 3s); an atlas, from £1 to £6. France saw a similar increase in prices. From the 5 or 6 sous for a one-sheet map in the late seventeenth century, the early eighteenth century saw similar maps priced from 10 to 20 sous; that is, up to 1 livre. By the latter half of the century, a map usually cost 1 livre per sheet, though the sheets of the Cassini *Carte de France* were consistently sold at 4 livres per sheet. Multisheet maps were priced according to the number of sheets. The price of atlases reflected not only the number of maps in them, but also the luxuriousness of the binding. An atlas could easily cost 100 livres, though 30 or 40 livres could buy a quarto-size atlas of sixty or seventy maps bound in paper or half-bound in leather.[14]

Imported maps cost significantly more than the homegrown variety. Paris map seller Roch-Joseph Julien commented in his 1763 catalogue that he would give no price for the English maps in his catalogue. They were much dearer than the others (mostly French, Dutch, and German) and "they would find few *amateurs* in France at the price they are sold in England." In order to be able to provide them at an affordable price, Julien explained that he acquired them by exchanging French maps with English mapmakers, in order to be able to supply them at a "plus juste prix."[15]

Giving value for money was clearly a consideration for Georges-Louis Le Rouge, a military geographer turned Paris map publisher, who specialized in copies of English maps. He published a collection of maps copied from English mapmakers as the *Atlas Amériquain* in 1778 to meet sharpening French interest in the War of American Independence. To maximize use of copperplate and to be able to offer his public value for money, Le Rouge combined two maps of disparate parts of North America into one space. The large, four-sheet map of South Carolina and part of Georgia taken from William De Brahm included a great deal of empty ocean to the east of the Carolina coastline. Into this space Le Rouge added a copy of Claude Sauthier's map of the Hudson River, a body of water nowhere near South Carolina (figure 19). Le Rouge explained his purpose: he was not only giving his public a two-for-the-price-of-one map but also providing them with something more cheaply than they could acquire the original. "Seeing that this sheet was nearly blank and recalling what I owe to the public who for forty years has so favorably welcomed my productions, I thought I ought to add this Course of the Hudson River, which is a masterpiece by Sauthier and sells for 3 livres 12 sous in London."[16]

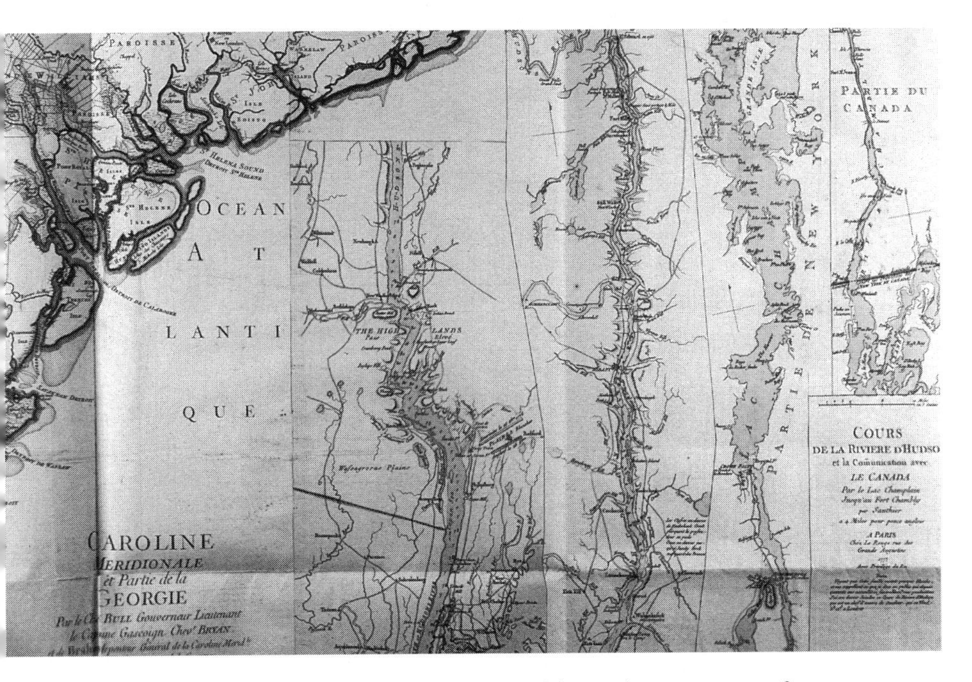

FIGURE 19. Georges-Louis Le Rouge, map of the Hudson River, inset from *Caroline meridionale et Partie de la Georgie par le Chevr. Bull . . . le capitaine Gascoigne, Chev. Bryan et De Brahm,* 1777. In Le Rouge, *Atlas Amériquain septentrional* (Paris, 1778), plate 22. WLCL, Atlas E-1-C. (William L. Clements Library, University of Michigan.)

[HOW MANY TO SELL?]

If these are the average prices for maps, how many maps did the producer need to sell to recoup the costs of production? If sold at a shilling or a livre each, between 500 and 1,000 maps would just recoup the expenses of engraving and printing. Such sales would not begin to cover the costs of survey and compilation.[17] Was it easy to sell 500 to 1,000 maps? No. Not to hear the adjudicators of the estate of the widow of Guillaume Delisle. After her husband died in 1726, she printed and stockpiled 18,000 of her husband's maps. This was enough "for twenty-five years of sales," said one of the experts analyzing the estate. In other words, a yearly turnover of about seven or eight hundred maps was what one might expect from the trade. Unfortunately, Mme Delisle was not an astute businesswoman. When she had the maps printed, she printed the same number of each title, leaving her estate "with 239 copies of 'The Retreat of the 10,000,' which will sell only a few a year, but 225 of 'Europe,' which will be sold out in a year."[18] Large numbers of maps left in stock can be found in the inventories of other map publishers, from the 5,000-plus left upon the death of A. H. Jaillot to the 20,000 maps comprising the *fonds* of Ambroise Verrier, successor to Roch-Joseph Julien.[19] These hefty inventories perhaps resulted from standard print runs being in hundreds rather than print runs calibrated to meet a specific market demand.

The business correspondence of Jefferys and Faden attests to relatively small numbers of maps sold or exchanged between the London map sellers and their Continental counterparts. Faden's French correspondents bought maps in single lots or in multiples of two, three, or six. Faden himself ordered slightly larger numbers from his French colleagues, three, four, six, or a dozen of each title being his standard order. While these small quantities may indicate that maps were bought to be copied, they also affirm that a map seller could not look to the international market to move large numbers of maps.[20]

[MAP DISTRIBUTION]

The sale of maps to places outside of London and Paris involved its own expenses and headaches. Even state agencies were not well equipped to deal with the complexities of distribution. Not until the end of the century did the Dépôt de la Marine undertake to incorporate the sale of its sea charts into the bureaucracy of production. The chief hydrographer, first Philippe Buache, then Jacques-Nicolas Bellin, had to arrange for distribution and sale using outside agents. In recompense, the hydrographer enjoyed the right to sell and profit from the charts produced by the dépôt. In the case of Bellin, the marine paid for the salaries of the hydrographer and his assistants and work space and for the copperplates and the engraving of the charts. Bellin paid for paper, printing, packaging, and transport of the charts. He was allowed, in fact encouraged, to sell the charts to map and print sellers or directly to customers from his home address. By the end of his career, Bellin would complain that this process had not benefited him at all. No one in the dépôt ever helped him; he had done all the compilation on his own. He had scarcely profited from this business, earning no more than 400 livres over thirty years, and had also suffered from the bad debts of retailers who had gone bankrupt.[21]

After Bellin's death, the marine briefly used the services of a bookseller, M. Merigot, on the Quai des Augustins in Paris. Merigot's complaints to the dépôt reveal many of the problems for the private wholesaler distributing maps for a government agency. Unlike Bellin, Merigot was obliged to pay in advance for the maps he was to sell. In only two years, he had accumulated a debt of over 17,000 livres, which he was unable to pay and some of which he felt he should not pay, for though he was billed for maps he ordered, he often had difficulties getting them. The maps were printed and warehoused in Versailles, twenty-three kilometers south of Paris. The clerk in the warehouse often sent the wrong maps, or they were badly packed, or were damaged in transit, torn, or wrinkled, or had holes punched in them. He was sent too many of one title and not enough of others, for the inventory numbers did not always match. Merigot paid for the costs of transporting maps to his customers in the ports, including the costs of customs, an expense that would be avoided if the dépôt would remember to put the seal of the ministry on the parcel; the dépôt routinely forgot. He reported that some customers complained that the dépôt's charts were too expensive and that others did not object to paying more for maps that were not produced by the dépôt.[22]

The marine soon abandoned the idea of using a bookseller to distribute their maps, and M. Merigot fades from the records. The dépôt's return to Paris from Versailles aided the distribution of charts. In 1776, an *entrepôt général des cartes* was established under the direction of Jean-Nicolas Buache de la Neuville, nephew of Philippe Buache. Buache de la Neuville became both a supplier and distributor of maps from the dépôt. He purchased maps from the dépôt for 22 sous and sold them to retailers in the ports for 24 sous, who then sold them, one presumes, for the stamped price of 30 sous.[23] This system became a franchise in 1780 when Buache de la Neuville sold his stock and business to the engraver Jean Claude Dezauche. Dezauche made a deposit of 20,000 livres to the dépôt, which gave him the right to purchase charts for the same low price of 1 livre 2 sous and sell them for a similar 10 percent profit.[24]

The costs of transport and the additional burden of import and export duties were also considerations in the distribution of maps. In France at the beginning of the eighteenth century, prints (including maps) were exempt from customs charges. In 1765, an export tax of 10 sous per quintal (hundredweight) and an import tax of 100 sous per quintal was levied, with an additional 25 sous on the hundredweight for entry to Paris. By 1771, a stamp act had been introduced, based on size of print or map, from 12 deniers for prints of six inches and under up to 6 sous on large ones of 24 by 30 inches (larger prints added 1 sou per additional six inches). These figures are very low compared to those in England, where imported prints were subject to a charge of sixpence each, about 50 percent of the value of the map.[25]

Such figures explain why William Faden and his correspondents were eager to avoid the customs officers when sending maps back and forth. The Dutch map sellers Covens & Mortier assured Faden that "the captain has asked that you would please have one of your servants pick up the roll [of maps] in order to avoid declaring it to customs." Jean Lattré of Paris also advised Faden to send maps "in a roll to avoid expenses."[26]

[FINANCING MAP PRODUCTIONS: THE GOVERNMENT AND THE CROWN]

Much research on eighteenth-century cartography has focused, naturally, on the state, analyzing the extent of government support and degree of interest in mapmaking endeavors. Certainly one finds centralized state support for reconnaissance survey and large-scale manuscript mapmaking of small areas, largely performed by the military. But it is rarer to find examples of the government or the state undertaking the entire financial risk of a printed map project. Even the great French achievement of the eighteenth century, the *Carte de France,* initiated by King Louis XV in 1747 and organized by César-François Cassini de Thury, lost its royal subsidy ten years later when the exigencies of war made its own demands on the state purse. Royal support continued in the guise of a thirty-year privilege for the map, with assistance in providing engineers for the surveying, and by allowing Cassini de Thury to keep the instruments purchased with moneys from the royal purse. But he was forced to create a private company, the Société de la Carte de France, to keep the project afloat; he relied on contracts with the provinces and on sales, both through subscription and through retail map stores like that of Roch-Joseph Julien, to supplement erratic government infusions.[27]

A similar situation was devised for Rizzi-Zannoni for his work on the map of the eastern part of the Turkish Empire prepared for the Ministry of Foreign Affairs to follow the Russian-Turkish conflict more closely. Though he was loaned money by the Crown to help defray expenses in executing the map, Rizzi-Zannoni was expected to sell 1,200 copies of the map to pay off the loan and reimburse his staff of designers, mathematicians, and engravers. However, as he could not sell 1,200 copies of the map, he was unable to pay his staff. When he offered the right of sale of map back to the ministry, it declined.[28]

Some support for the large-scale mapping of France came from the provincial governments, which contracted with Cassini de Thury's engineers to produce maps of their provinces. This contrasts sharply with the private enterprises of large-scale county surveys undertaken in England, largely in the latter half of the eighteenth century and underwritten by private subscription. No county governments existed with fiscal powers to authorize and sustain such surveys. And while the benefits of large-scale mapping might have been contemplated by individual government ministers, the constantly shifting locus of decision making and fiscal authority in eighteenth-century England meant that support from the Crown and/or Parliament was erratic and minimal.[29]

While several English monarchs were map users and collectors, their cartographic interests did not always translate into monetary support. Charles II was among the more generous. Along with Queen Catherine, he subscribed to John Ogilby's *Britannia* for £500 and exempted Ogilby from paying duty on the imported paper for the project (figure 20). He allowed the privy purse to pay £200 toward the *Map of London,* prepared by Ogilby's partner and heir, William Morgan. To his royal hydrographer, John Seller, the king granted a thirty-year privilege for the *English Pilot* and the promise of £200 for Seller's *Atlas Anglicanus.* He promised £300 to Moses Pitt for the proposed *English Atlas,* though nothing beyond the first volume appeared. He issued a royal order to John Adams "to go freely to all necessary places and view points" in prosecuting his proposed survey of England in 1681.[30]

But Charles's enthusiasms were not shared by his successor, James II, who, during his short reign, did not sponsor any national mapping efforts. Nor did Mary and William of Orange provide any significant central state interest in the mapping of the country or, with the exception of the East India Company, in establishing mapmaking agencies that would undertake

FIGURE 20. Detail from William Morgan, *London etc. Actually surveyed* (London, 1682). John Ogilby presenting subscription list to King Charles II. Map DA 675 .L84. (Map Library, Harlan Hatcher Graduate Library, University of Michigan.)

expensive cartographic or hydrographic projects.[31] Even George III, a collector of maps and scientific instruments on a grand scale and passionately interested in the scientific question of longitude, exerted little influence in promoting state mapping agencies. His personal generosity paid for a three-foot theodolite made by Jesse Ramsden, which was used for the triangulation between France and England under the supervision of William Roy in 1787.[32]

The inability and reluctance of monarchs to support large national surveys was by no means limited to France and England. Peter Barber points out that absolute monarchs had a better chance of succeeding in completing a national survey than those royalty who were limited by strong parliaments or landed interests.[33] In such nations, it was left to individuals and private institutions to encourage and reward mapping endeavors. British towns and corporations sponsored local mapping efforts, particularly along the coast, throughout the century.[34] In Scotland, increased attention was given to national and local mapping throughout the century, supported by landed gentry, town councils, or institutions such as the Society of Antiquaries.[35] From midcentury onwards, one sees a rapidly increasing movement towards projects of improvement, whether of a moral, educational, or commercial nature. It is exemplified in the Society for Encouragement of Arts, Manufactures and Commerce, which, from 1759, offered a premium of £100 or a gold medal for county surveys created with triangulation and accompanied by a "certificate of accuracy."[36] This prize stimulated interest in the large-scale mapping of the nation.

[PATRONAGE]

Into the vacuum left by Crown and state stepped the rich and the great. "And indeed, without the Patronage of the Rich and the Great, it is hardly possible that the Sciences should ever thrive much in any Place," wrote the geographer Bradock Mead.[37] The monied and titled classes in England and France had consistently supported surveys of their own landholdings, usually resulting in manuscript maps. Their exact role in printed map projects is more difficult to trace. It is clear that in France, the church actively supported mapping projects and its priests participated as surveyors and compilers, as the work of le père de Dainville has shown.[38] Very rarely did a single individual fund an entire project, however generous their contributions.[39]

An exceptional Mycenas lived in the house of Orléans. Philippe, duc d'Orléans (1674–1723), as regent of France during the minority of Louis

XV, reestablished the Dépôt des Cartes, Plans, et Journaux de la Marine. As its name suggests, the dépôt was initially an archival facility for ships' logs, pilots' journals, port plans, and sea charts acquired by sailors and officers; it soon became a significant producer of naval charts. An amateur scientist, the duc d'Orléans ensured that geography formed a significant part of the young king's education by hiring Guillaume Delisle as one of his tutors.[40] This began a tradition of Delisles or Buaches teaching young kings for the next two generations. The duc d'Orléans's own heir, Louis (1702–52), became the patron of the geographer d'Anville.[41] His map of Italy and accompanying *Analyse géographique de l'Italie* (1744) are dedicated to the duke, who is credited with initiating and facilitating the work by undertaking its expenses.[42] The duke's son, Louis Philippe (1725–85), whom d'Anville had tutored, also supported d'Anville's cartographic publications.[43]

Madame de Pompadour, mistress of Louis XV, was a prominent subscriber to the *Carte de France* and to the Robert de Vaugondys' *Atlas universel.* Her copy of the Robert de Vaugondy *Atlas portatif universel et militaire* bears her arms, and she commissioned a pair of globes from the young Didier Robert de Vaugondy in 1751, after his presentation of a globe to the king in 1750. In her portrait are the plans for buildings at Versailles, music, and, just visible, the edge of a globe. In the same portrait, on her desk are volumes bearing the titles of works of the *philosophes* whom she supported: the *Encyclopédie* of Diderot and d'Alembert and Montesquieu's *Esprit des Loix,* works to which the same geographer, Didier Robert de Vaugondy, had contributed.[44]

In England, the duke of Richmond stands out as a cartographic patron. He employed two surveyors on his Sussex estates, Thomas Yeakell and William Gardner, who undertook a "great survey" of Sussex to produce a printed multisheet county map at a scale of two inches to the mile by incorporating material from their estate surveys. The map remained incomplete, with only four sheets published, but the surveyors continued to enjoy the patronage of the duke when he became master general of Board of Ordnance in 1782. He appointed Yeakell and Gardner to posts in the ordnance, as draftsman and surveyor. The duke's abilities as an administrator and his reform of the salaries of the ordnance surveyors owed much to his earlier management of his estates and surveyors in regular service.[45]

[SUBSCRIPTIONS]

The patronage of the rich and the great is readily discerned in the subscription list. As a more stable way of acquiring capital before an enterprise began, the subscription method of financing cartographic projects was extremely popular in England. Some scholars construe the subscription list as an endorsement of the map itself.[46] The system was simple: offer the map or the atlas at a price below the postpublication sale price, requiring half the money upon subscribing, the other half to be paid on delivery of the map. Often publishers offered other incentives to encourage subscribers, such as the lure of a free copy with a multiple subscription or an even greater price reduction with a larger order. As a common and presumably effective method of acquiring advance capital, subscriptions financed most of the county surveys done in the latter half of the eighteenth century in England.

The subscription process offered to those who paid up or who paid a little extra the allure of seeing their armorial bearings printed either on the map or on separate sheets bound in the front. On the one hand, this sort of publication appealed to the vanity of those who liked to see not only their names but also their pretensions in plain view. On the other, the highlighting of nobility and gentry and their homes served another purpose, as Robert Plot explained in the preface of his *Natural History of Oxfordshire:* "Gentry will be influenced to keep their seats and arms lest their posterity see what their ancestors have parted with.... Vagabonds will be deterred from making counterfeit passes by putting false names and seals to them."[47] Thus the publication of the coat of arms secured the arms from pretenders and protected countryseats from sale (figure 21).

When Joel Gascoyne opened the subscription for his map of Devon, he offered subscribers the chance to have their coats of arms engraved at 5s each for the engraving, but only if there were enough to make up a border, thus placing design considerations ahead of social concerns. He did assure subscribers, however, that each family name would appear under the name of the countryseat.[48]

In France, few printed and published subscription lists have survived, suggesting that subscription was a less common means of funding cartographic projects. The subscription was more frequent in the book trade, which may account for its appearance with the Robert de Vaugondys' *Atlas universel,* financed by a bookseller (figure 22). The *Encyclopédie* of d'Alembert and Diderot ranks as one of the larger book subscription projects and many large graphic print projects were also funded by subscription.

The largest map project to be sold by subscription was the *Carte de France.* As mentioned above, when the king was forced by the exigencies of preparation for war to abandon regular monetary support for the national survey, Cassini de Thury almost immediately established a private company, the Société de la Carte de France. Founded in 1756, its core group of fifty members bought shares at an initial investment of 1,600 livres per person, in-

FIGURE 21. Portions of two pages of the coats of arms of some of the more than 700 subscribers to John Senex, *A new general atlas* (London, 1721). The top of each page (not shown) displays the arms of aristocracy and nobility. Two-thirds of the arms, however, belong to the untitled and unremarkable, such as Mr Smith, Mr William Burgis of New York, Mrs Sarah Fenwike, and Mr Thomas Arnold, Apothecary in Holbourne. John Warburton, Esq., Somerset Herald at Arms, F.R.S., was himself at this time soliciting subscriptions for his map of Yorkshire. WLCL, Atlas E-8-C. (William L. Clements Library, University of Michigan.)

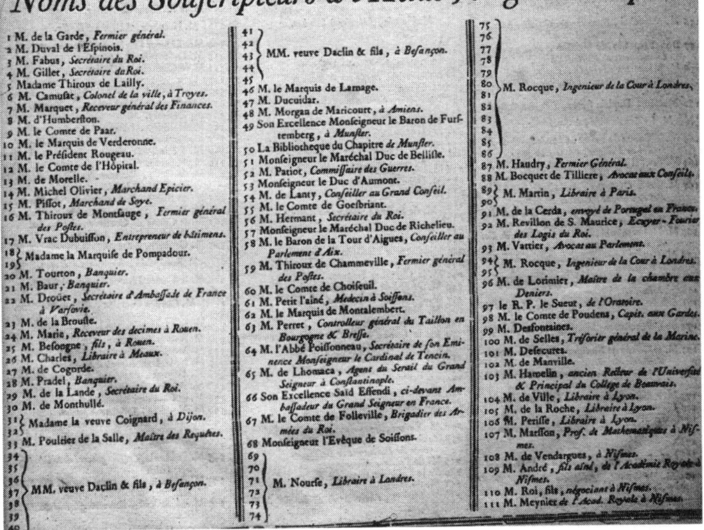

VOICI enfin la 5me & derniere Livraiſon d'un Ouvrage conçu & commencé il y a quinze années Nous ne nous étions engagés à nos Souſcripteurs que pour cent Cartes ; mais à meſure que nous avons exécuté, nous avons trouvé qu'il étoit utile, au plan que nous avions formé, d'y en ajoûter trois.

Ce plan étoit celui d'un Recueil de Géographie, qui ſans être cher & de plus d'un volume, pût ſuffire au plus grand nombre des perſonnes ſtudieuſes, de celles qui liſent l'Hiſtoire, comme de celles qui ne ſont attentives qu'aux évenemens modernes. Depuis nos différentes livraiſons, nous avons eu la ſatisfaction de reconnoître en effet que dans ce que les tems ont amené de digne de notre triſte curioſité, nos Cartes particulieres ont fourni un développement tel qu'on a toujours pû ſe tranſporter ſur la ſcene, & ſe trouver à portée de juger, pour ainſi dire, des coups, de raiſonner même ſur l'avenir.

Nous avons donné dans la précédente livraiſon (la quatriéme), avec le diſcours préliminaire imprimé en grand, la Liſte de ces cent trois Cartes : en ne demandant point de payement pour les trois non compriſes dans le projet de Souſcription, nous croyons avoir été plus que fidéles à nos engagemens, & moins par là encore que par l'exécution de tout l'ouvrage.

Des travaux d'habiles Ingénieurs pourront bien par la ſuite des tems fournir des productions, dont l'exactitude l'emporte ſur la nôtre dans les détails ; mais non certainement dans les points eſſentiels. Les Sciences univerſellement cultivées dans notre ſiécle, ont tellement aidé aux progrès de la Géographie & les correſpondans des Académies, les Ingénieurs, les Curieux de différens ordres ſont tellement multipliés, qu'actuellement le giſſement de tous les lieux d'un peu d'importance ſur la ſurface de la terre, ſe trouve déterminé ou vérifié par les obſervations aſtronomiques.

Reglé ſur ce point d'appui, infaillible comme on ſçait, de quelle correction importante notre Ouvrage ſeroit-il ſuſceptible & quel droit n'a-t-il pas, comme nous le diſions tout à l'heure, aux ſuffrages de nos Souſcripteurs?

Il ne nous reſte qu'à conſacrer, avec notre reconnoiſſance, leur goût & leur générosité ſans leſquels nous n'aurions pû faire éclorre cette grande entrepriſe, & nous le faiſons par la voie immortelle de l'impreſſion.

Noms des Souſcripteurs à l'Atlas, en grand Papier.

1 M. de la Garde, *Fermier général.*
2 M. Duval de l'Eſpinois.
3 M. Fabus, *Secrétaire du Roi.*
4 M. Gillet, *Secrétaire du Roi.*
5 Madame Thiroux de Lailly.
6 M. Camuſat, *Colonel de la ville, à Troyes.*
7 M. Marquet, *Receveur général des Finances.*
8 M. d'Humberſton.
9 M. le Comte de Paar.
10 M. le Marquis de Verderonne.
11 M. le Préſident Rougeau.
12 M. le Comte de l'Hôpital.
13 M. de Morelle.
14 M. Michel Olivier, *Marchand Epicier.*
15 M. Piſſot, *Marchand de Soye.*
16 M. Thiroux de Montſauge, *Fermier général des Poſtes.*
17 M. Vrac Dubuiſſon, *Entrepreneur de bâtimens.*
18–19 Madame la Marquiſe de Pompadour.
20 M. Tourton, *Banquier.*
21 M. Baur, *Banquier.*
22 M. Droüet, *Secrétaire d'Ambaſſade de France à Varſovie.*
23 M. de la Brouſte.
24 M. Marie, *Receveur des decimes à Rouen.*
25 M. Beſongne, *fils, à Rouen.*
26 M. Charles, *Libraire à Meaux.*
27 M. de Cogorde.
28 M. Pradel, *Banquier.*
29 M. de la Lande, *Secrétaire du Roi.*
30 M. de Monthullé.
31–32 Madame la veuve Coignard, *à Dijon.*
33 M. Poultier de la Salle, *Maître des Requêtes.*
34–40 MM. veuve Daclin & fils, *à Beſançon.*
41–45 MM. veuve Daclin & fils, *à Beſançon.*
46 M. le Marquis de Lamage.
47 M. Ducuidar.
48 M. Morgan de Maricourt, *à Amiens.*
49 Son Excellence Monſeigneur le Baron de Furſtemberg, *à Munſter.*
50 La Bibliotheque du Chapitre *de Munſter.*
51 Monſeigneur le Maréchal Duc de Belliſle.
52 M. Patiot, *Commiſſaire des Guerres.*
53 Monſeigneur le Duc d'Aumont.
54 M. de Lanty, *Conſeiller au Grand Conſeil.*
55 M. le Comte de Goeſbriant.
56 M. Hermant, *Secrétaire du Roi.*
57 Monſeigneur le Maréchal Duc de Richelieu.
58 M. le Baron de la Tour d'Aigues, *Conſeiller au Parlement d'Aix.*
59 M. Thiroux de Chammeville, *Fermier général des Poſtes.*
60 M. le Comte de Choiſeuil.
61 M. Petit l'aîné, *Medecin à Soiſſons.*
62 M. le Marquis de Montalembert.
63 M. Perret, *Controlleur général du Taillon en Bourgogne & Breſſe.*
64 M. l'Abbé Poiſſonneau, *Secrétaire de ſon Eminence Monſeigneur le Cardinal de Tencin.*
65 M. de Lhomaca, *Agent du Serail du Grand Seigneur à Conſtantinople.*
66 Son Excellence Said Effendi, *ci-devant Ambaſſadeur du Grand Seigneur en France.*
67 M. le Comte de Folleville, *Brigadier des Armées du Roi.*
68 Monſeigneur l'Evêque de Soiſſons.
69–74 M. Nourſe, *Libraire à Londres.*
75–86 M. Rocque, *Ingenieur de la Cour à Londres.*
87 M. Haudry, *Fermier Général.*
88 M. Bocquet de Tilliere, *Avocat aux Conſeils.*
89–90 M. Martin, *Libraire à Paris.*
91 M. de la Cerda, *envoyé de Portugal en France.*
92 M. Revillon de S. Maurice, *Ecuyer-Fourier des Logis du Roi.*
93 M. Vattier, *Avocat au Parlement.*
94–95 M. Rocque, *Ingenieur de la Cour à Londres.*
96 M. de Lorimier, *Maître de la chambre aux Deniers.*
97 le R. P. le Sueur, *de l'Oratoire.*
98 M. le Comte de Poudens, *Capit. aux Gardes.*
99 M. Desfontaines.
100 M. de Selles, *Tréſorier général de la Marine.*
101 M. Deſecures.
102 M. de Manville.
103 M. Hamelin, *ancien Recteur de l'Univerſité & Principal du College de Beauvais.*
104 M. de Ville, *Libraire à Lyon.*
105 M. de la Roche, *Libraire à Lyon.*
106 M. Periſſe, *Libraire à Lyon.*
107 M. Marſſon, *Prof. de Mathematiques à Niſmes.*
108 M. de Vendargues, *à Niſmes.*
109 M. André, *fils aîné, de l'Académie Royale de Niſmes.*
110 M. Roi, *fils, négociant à Niſmes.*
111 M. Meynier *de l'Acad. Royale à Niſmes.*

FIGURE 22. Part of the subscription list from Didier and Gilles Robert de Vaugondy, *Atlas universel* (Paris, 1757). WLCL, Atlas W-3-C. (William L. Clements Library, University of Michigan.)

creased later to 2,400 livres. Such a large sum meant that the société consisted of the wealthiest members of French aristocracy and politics. The marquise de Pompadour once again stepped up with her support, as did the prince de Soubise, the duc de Noailles, the duc de Luxembourg, and many *conseillers d'État* and members of the Académie Royale des Sciences, such as the comte de Buffon and Charles-Marie de La Condamine. The Société de la Carte de France provided the capital necessary for the surveying, engraving, and printing of the maps. Its members could recoup their investment only through sales. Subscriptions to the projected 173-sheet map were offered for 500 livres, if paid all at once, or 562 livres spread over five installments. For nonsubscribers, the purchase price was 720 livres.[49] By 1780, after twenty-two years, there were only 203 subscribers, among whom were foreigners, merchants, ecclesiastics, and provincial gentry, making up a more socially diverse and geographically scattered clientele than the société.[50]

When such a subscription exists, with names and occupations, it provides an exceptional picture of the map consumer. Few subscription lists to cartographic works in France were printed with the work. The *Atlas universel*, published by Antoine Boudet, a Paris bookseller, with maps by Gilles and Didier Robert de Vaugondy, does preserve its subscription list with occupations and addresses of the subscribers, allowing us to draw some conclusions about its distribution and marketing.[51] Booksellers, like Boudet himself, formed the largest group of subscribers, constituting nearly one half the list of nearly 700 names. The second largest group was the magistrate class of the *noblesse de robe*, followed by members of the church, the aristocracy, and the court, and a few military men, doctors, and women.

In England, subscription lists were printed with much greater frequency, and their thorough study would yield much consumer information.[52] County surveys in England often list the subscribers on the map itself, sometimes with the number of maps ordered, which could be in surprisingly large quantities. For example, in 1773 the duke of Queensbury and the earl of Shelburne each bought forty sets of the topographical map of the county of Wiltshire, priced at two guineas a set, a considerable outlay even for the aristocracy.[53]

English subscription lists are found most frequently in atlases. J. B. Harley highlighted their value in his brief study of two eighteenth-century English atlases, both of which provide a contrast to the bookseller-heavy subscription list of the French *Atlas universel*. John Senex published his *New general atlas* in 1721 with 1,047 subscribers. In Harley's analysis, 10 percent of the subscribers represented the nobility; 41 percent gentry; 21 percent clergy and professionals; and 28 percent craftsmen and yeomen. He compared these figures with those gleaned from the list in John Cary's *New and correct English atlas* of 1787 with 1,184 subscribers, of whom only 3.0 percent represented the nobility; 35.7 percent gentry; 17.5 percent clergy and professionals; and 43.8 percent "non-gentry" (i.e., craftsmen and yeomen). Lest these figures be misinterpreted, Harley pointed out that the two atlases were not representative of English atlases as a whole and that the Cary atlas was much smaller and cheaper than the earlier Senex atlas.[54] In addition, Harley's breakdown does not look at the number of orders for multiple copies or the geographic distribution of the subscribers. Nor does he disaggregate some of the more striking groups: women, foreigners, booksellers (3 percent), many of whom were outside London, clergy who had academic appointments and were ordering perhaps on behalf of their institutions (the mathematics master at Christ's Hospital ordered seven copies).[55] Indeed, one might look at John Senex's *Atlas maritimus et commercialis* of 1728, a large folio volume of 140 pages of text with fifty-three charts. It lists 431 subscribers, two-thirds of whom give a title or occupation. Borrowing Harley's categories, we find that, 13 percent were members of the nobility; 47 percent gentry, 19 percent clergy and "professionals"; and 17 percent "non-gentry" But many of the merchants whom Harley listed as "professionals" (along with the clergy, doctors, lawyers, and military officers) are those booksellers and print sellers who were deeply involved with the trades of printers, engravers, mathematical instrument makers, and watchmakers. The nongentry percentage would certainly grow even more were the categories to be drawn differently. Closer scrutiny of the professions and crafts represented on such lists will highlight the increased interest in cartography by the middling and working classes. Linen drapers, goldsmiths, instrument makers, painters, druggists, soap boilers, writing masters, and women are all represented on these subscription lists. An analysis of their presence would help to develop a broader view of consumption of atlases.

We should offer another cautionary note about the subscription list: a name on the list did not mean money in the publisher's pocket. Deadbeats nestled safe among the more stalwart. Bernard Romans alluded to this problem when he invited his subscribers to pick up their completed maps of the northern department of North America: "The Gentlemen who have subscribed for the map . . . and are really Lovers and Encouragers of the Geographical Science are requested to call for their maps. Those who have only wrote their names with a view to shine in a Subscription Paper may save themselves the trouble, as there are none printed for such."[56] Many of first subscribers to Cassini de Thury's illustrious *Carte de France,* a list headed by the Marquise de Pompadour, forgot to pay their installments.[57] Even monarchs could not be relied upon to make good on their commitments. Charles II and his queen each promised £500 to John Ogilby, ensuring that they would be placed at the head of the list of subscribers to his *Britannia,* but their money never materialized; they gave their support instead by relieving Ogilby from customs duty on imported paper.[58]

Nor did subscription discounts guarantee sales. Some folks thought even the subscription prices were too high. Potential customers for John Strachey's map of Somerset in 1736 jibbed at 7s 6d as "too much for two sheets of paper." They would wait until after it was published when they could get it cheaper; that is, when it had been copied by another publisher.[59] Strachey also offered his subscribers their coats of arms, but such an embellishment doubled the price from 7s 6d to 15s, a further disincentive.[60]

When subscriptions failed, lotteries were sometimes employed. John Ogilby's surveyor, Gregory King, organized two in 1673, one in London and one in Bristol, to raise cash for the survey work for Ogilby's *Britannia.*[61] One hundred years later, the same method was used to pay for the plates and engraving of Captain Abner Parker's map of Connecticut.[62]

[JOINT PARTNERSHIPS]

Another method of raising capital was joint partnership. Used chiefly in England, print and map sellers jointly owned the plates of large atlas enterprises, as for example John Senex's *New general atlas* (figure 23). While the sharing of costs had all the potential of being fractious and fiscally confusing, it seems to have been a workable system, at least for an atlas like Senex's. This perhaps is because all the partners were equally involved in the trade; they all had apprenticed as engravers and thoroughly understood the engraving and printing processes and all had diversified their stock. By the end of the eighteenth century, however, multiple plate ownership fades as stocks of plates and maps were consolidated into fewer and fewer hands both in London and in Paris.

A particularly complex example of this practice is found in Emanuel Bowen's *Royal English Atlas* (c. 1763), a volume noteworthy for being neither Bowen's nor royal, in the monarchal sense of the word. Though Bowen himself and Thomas Kitchin engraved the maps, the plates were owned by no fewer than fourteen different proprietors of whom Bowen was not one, though Kitchin was in for a quarter. The "royal" in the title referred to the size of the paper (18 in. × 12 in.), not to the crown. J. B. Harley and Donald

vations communicated to the *Engliſh* ROYAL SOCIETY, the *French* ROYAL ACADEMY of Sciences, and thoſe made by the lateſt TRAVELLERS: And the DESCRIPTIONS ſuited to the Courſe of each MAP, which has not been obſerv'd in any other ATLAS.

A. Johnston ſculp.

LONDON:

Printed for DANIEL BROWNE without *Temple-Bar*, THOMAS TAYLOR over-againſt *Serjeants-Inn* in *Fleet-Street*, JOHN DARBY in *Bartholomew-Cloſe*, JOHN SENEX in *Salisbury-Court*, WILLIAM TAYLOR in *Pater-Noſter-Row*, JOSEPH SMITH in *Exeter-Change*, ANDREW JOHNSTON Engraver in *Round-Court*, WILLIAM BRAY next the *Fountain-Tavern* in the *Strand*, EDWARD SYMON in *Cornhill*. M.DCC.XXI.

FIGURE 23. Detail of the title page of John Senex, *A new general atlas* (London, 1721). Under the trumpeting angel, engraved by Andrew Johnston, the joint owners of the atlas. WLCL, Atlas E-8-C. (William L. Clements Library, University of Michigan.)

Hodson's study of the changes in ownership of the plates of the *Royal English Atlas* shows how they were consolidated over thirty-two years into the hands of two firms from its initial ownership by seven firms.[63]

[PARTNERSHIPS IN FRANCE]

The contrasting situation in France, where partnerships between the print and book trade were rare, indicates that the regulations of the guilds prevented these economic liaisons. Not only did a working relationship exist between the print seller, the engraver, and the geographer but, as we have seen, their overlapping roles meant that it is sometimes hard to distinguish the engraver from the compiler. The late seventeenth-century example of the partnership of Hubert Jaillot and Guillaume Sanson further demonstrates that the owner of the plates was the one who grew wealthy from the sales of the maps. The geographer, Sanson, was in this case the hired hand and thus did not profit in the same way as the publisher, Jaillot. In the case where Jaillot and Sanson shared the plates, each owning half a map, the difficulties of agreeing on publication had to be surmounted before any profit could be realized.[64]

By the middle of the eighteenth century, the reluctance of geographers—that is the map compilers—to work for or with booksellers or print sellers is pronounced. When the Robert de Vaugondys first announced the publication of their *Atlas universel* in 1752, it became apparent that a bookseller, Antoine Boudet, was the financial force behind the project. The *premier géographe du roi,* Philippe Buache, accused the Robert de Vaugondys of betraying their profession and stripping their brother geographers of their business. Buache maintained that booksellers had no business selling maps, for they had no way of distinguishing good from bad, true from false, and their interest in making a profit prevented them from having a care for quality.[65]

But the Robert de Vaugondys responded that it was nearly impossible for geographers to produce an atlas such as the *Atlas universel,* 108 maps of the same folio size and of overlapping scales. They simply did not command the capital required for the copperplate and engravers by such a large project. They cited the association of the Mariettes with Sanson, and of Guillaume Sanson with Jaillot. Furthermore, the privilege acquired by the Robert de Vaugondys allowed them to sell their works as it suited them. Buache seemed to wish that the *Atlas universel* would not appear at all; if that were not possible, then he wished that the chancellor would at least introduce a law that prohibited booksellers from meddling in the map business.

Partnerships between mapmakers and booksellers could cause difficulty. Partnerships between engravers and geographers fared much better. Jean Lattré, *graveur du roi,* employed the geographers Rizzi-Zannoni, Bonne, and Jean Janvier to produce maps for his *Atlas moderne.* Engravers like Lattré were also print sellers and had access to greater capital and a wider market; they thus could afford the risk of a geographical enterprise. Partnerships among geographers were rarely formed in France. D'Anville and the Robert de Vaugondys never joined their talents, despite their proximity to each other across the Seine, nor did either of them work with Philippe Buache or Rigobert Bonne, although they were all known to each other. One significant exception was the globe and instrument maker Louis-Charles Desnos, who hired various geographers and engineers such as Rizzi-Zannoni and Buy de Mornas to execute his ideas for atlases. Desnos made a particular success in the sale of geographical materials by marrying well—his wife was the widow of the globe maker Nicolas Hardy—, by acquiring stock from the Jaillot family and Nicolas de Fer, and by diversifying the offerings in his shop to include not only maps, atlases, and globes, but prints, screens, writing materials, and books.[66] Roch-Joseph Julien seemed to enjoy similar success in running his map store in the Hôtel de Soubise, though part of his success may be accounted for by his title as "superintendant of the buildings of the prince de Soubise" which may have allowed him to keep his shop rent free.

Without support or subsidy of a patron, whether individual or institutional, or without the ability to diversify stock with a wider range of materials on offer, a mapmaker would find it difficult to be a financial success. The bankruptcies of geographers and engravers, like Didier Robert de Vaugondy and Thomas Jefferys, attest to the precarious balance of credit and debit accompanying the fiscal danger of devoting one's energy to mapmaking alone.[67] The successes in the map business were those who kept diverse stock, not specializing in maps but selling related materials and prints as well. Like Desnos and Julien in Paris, London map sellers Thomas Kitchin and William Faden left considerable estates upon their deaths.[68] The people on the selling end of the map trade had the better chance of success. For the folks who provided the raw material of maps and the labor of getting the map to print, life was more precarious.

[THE PROBLEMS OF GETTING PAID]

Appendixes 1 and 2 show the costs of engraving and printing; they also demonstrate the wide range of methods of paying workers. The feature typical of all methods of payment is the reliance on measurement of a piece of work, whether plate or worked surface. Long-term employment of an engraver or a printer involving a salary and job security is rare. It has been mentioned above that the publisher Jaillot maintained engravers on yearly or multiyear contracts, even providing them with lodging. Guillaume Delisle also mentions paying his engraver year round, as he looks for work to keep him busy.[69] In the early nineteenth century, when the Dépôt de la Guerre was in the process of taking over the publication of the *Carte de France,* the administration decided on a mixture of salary and predetermined rates to pay its engravers, depending on the progress of the work.[70]

By contrast, the Dépôt de la Marine employed varying numbers of draftsmen (only one in the early 1750s but as many as fifteen in the early 1770s) with salaries ranging from 460 livres per annum (*garçon du bureau*) to 1,200 livres (*ingénieur-géographe*). Many of these draftsmen were required to copy manuscript plans and to write précis of navigator journals. That the job was perceived as a sinecure may be detected in the notes of administrators in the daily running of the dépôt. After a draftsman and *sous-garde* of the dépôt named Claro died, the head of the dépôt remarked that solicitations to fill his spot were immediately in the air, fifty of them from the royal family alone.[71]

But being on the government payroll was no guarantee of being paid. In 1759, Bellin, hydrographer of the Dépôt de la Marine, complained that he was owed arrears from 1738 and 1739.[72] In a petition to the minister after his death, Bellin's widow maintained that he had never been reimbursed for the engraving costs for the *Petit atlas maritime,* a bill that totaled 20,000 livres, and for which she asserted that he received only a little over 8,000 livres.[73] The engraver Delahaye received only about one-third of his expenses for engraving the *Carte des Chasses,* the map of the royal hunting grounds around Versailles and Fountainebleau.[74] The designer and engraver Cochin told a similar story; he was owed 22,000 livres in arrears for his services to the Maison du Roi.[75] The head of the Imprimerie du Roi, M. d'Anisson-Duperron, joined the captious chorus after being accused of overly inflated prices for his printing work. He responded that these high costs were necessitated by the slowness of the Crown in paying its debts, and when payment did arrive, it was performed in "billets de Nouette, Contrats sur la Bretagne," and in shares of the tax, the *dixieme,* all of which were devalued in worth.[76]

Cash was ever in short supply and promissory notes regularly failed. Many geographers, when preparing maps for others, often asked for payment in kind, that is, copies of the printed map. Du Halde, the author of the *Histoire de la Chine,* paid d'Anville half his fee in cash and half in maps. When d'Anville settled on the price of 600 livres for designing the maps of China, he asked that the debt be paid in part with maps (viz., thirty copies of the complete collection, and one hundred of each of the general maps in the collection) and the right to sell the maps together in an atlas. He followed a similar pattern when negotiating terms with the publishers De Saint and Saillant for maps to accompany their *Notice de la Gaule.* His contract with the publishers stipulated that in exchange for a reduction in his honorarium, he would receive fifty copies of the maps free and others at the *prix de province.* He also wished to be allowed to sell his map for the work separately at 30 sous.[77] One of Cassini de Thury's engravers, Chalmandrier, was paid with seventy copies of the printed sheet for his work on the plate of the area around Tours for the *Carte de France.*[78]

Another means of ensuring payment was to refuse to continue working on projects until payment for the previous work was made. When the geographer Brion de la Tour stopped working on the maps for the abbé Expilly's *Topographie de l'Univers* in 1758, he was told that a month's delay of payment was nothing. The Paris publisher Bauche said that engravers were often not paid for three months after their plates had been delivered to the publishers, sometimes longer.[79]

Geographers in the Anglophone world endured similar plights. Peter Perez Burdett, the surveyor of Derbyshire and Cheshire, was dogged by debt. Despite a good education and connections with the business and intellectual elite in the north of England, he relied on his skills as a surveyor to make a living and as an engraver to avoid the expense of hiring one in order to produce his maps.[80] The surveyor general for the southern district in North America, William De Brahm, suffered for not having been paid for three years,[81] a situation echoed by J. F. W. Des Barres' long contretemps with the Admiralty over nonpayment for his surveying and engraving work on charts of the coast of North America.[82] J. H. Andrews cites the lament of the Irish surveyor Jacob Nevill, who inserted the following couplet in the *Dublin Journal* in 1761, after producing a topographical county map of Wicklow in 1754:

> As the county of Wicklow I did survey
> I must now quit my hold for want of pay.[83]

CHAPTER FOUR

PLAGIARISM AND PROTECTION

A modest price is the surest means of stopping the foreign counterfeiter.
—J.-N. Bellin, *Petit atlas maritime*[1]

[CUTTING COSTS]

Beyond the thicket of problems of acquiring up-to-date information for a map lay the minefield of sputtering cash flow, an unpaid and occasionally sullen, if not violent, workforce, and piles of unsold stock. Map publishers naturally looked for ways to cut costs. Survey and compilation were the two most labor intensive and skilled aspects of mapmaking, requiring the longest time, the hardest physical work, the most expensive equipment, and the greatest intellectual skill. As they constituted the largest expense in the mapmaking process, theirs were the most attractive costs to reduce. Mapmakers accomplished this by copying the work of others. Copying, reengraving, and selling someone else's labor were lifeblood to the map trade throughout the eighteenth century. By not spending money on compilation costs, one could sell a map at a price significantly lower than the original. Thus it was not the cost of engraving alone that influenced map production in this period. Expensive though engraving on copperplate was, it was the only means available for reproducing a wide tonal range of graphics and provided a surface that could be corrected and altered, replaned and reused over a long period without susceptibility to warping or rotting.

The savings to be gained by copying were well known in the print trade. Charles-Nicolas Cochin explained the economics of copying prints in simple terms. His hypothetical situation could easily be applied to the map trade:

> The print seller who has spent eight or nine thousand francs and is obliged to sell his print at ten or twelve francs, knows well that for twelve or fifteen hundred francs one could have the print copied by a young beginner of some

talent and that the counterfeiter can sell his print for fifty sous . . . and that this copy is sufficient for the general public who are not connoisseurs.[2]

The terms "copy," "plagiarism," and "piracy" were used indiscriminately in the rhetoric of eighteenth-century map criticism to describe the practice of copying and printing someone else's map. The problems of copying and plagiarism raised several important questions, all of them vexatious for mapmakers. Economically, they asked who should be rewarded for the hard work of compilation. On the proprietary level, they questioned whether mapmaking should be the domain of geographers and surveyors, people ostensibly qualified to do the work, or whether it should be open to anyone in any trade, such as engravers, print sellers, and booksellers. They also broached the moral question of whether maps could be trusted. Should maps be vetted or screened in some way to prevent error and mishap that could cost lives and to ensure a level of accurate representation? Though various answers to these questions emerged in contemporary discussion, none was translated into protective legislation or trade regulation.

The discussion was further complicated by the very nature of the map itself: can one copyright geographical information depicting what is already in nature?[3] The mapmaker relied on previous maps not only to save money in survey and compilation costs, but also because previously published maps might represent the best level of accurate information available. Part of a compiler's job was to assess the reliability of cartographic sources and use those that were the most appropriate. Thus it was, and still is, often difficult to determine what was a copy and what was a distillation of information that came to the same cartographic conclusion.

The problem of copying was exacerbated by the ease with which it could be done. Geographers and engravers both in London and Paris tended to cluster in locations chosen for ease of production and sale of maps. Manuscript drafts, proof prints, and corrected examples routinely passed from one shop to another.

In Paris, the map trade was found primarily in the area of the Left Bank of the Seine, the heart of the book trade, or on the Île de la Cité (figure 24). Geographers chose to live near the neighborhood's many schools and the Sorbonne, which provided a vibrant market for their wares and those of the scientific instrument makers who joined them on the quai de l'Horloge and along the quai des Augustins.[4] The quai de l'Horloge, on the north side of the Île de la Cité, was immediately adjacent to the law courts and within sight of the Louvre, providing a ready source of maps for lawyers and courtiers. The Galeries du Louvre offered state-owned apartments to royal

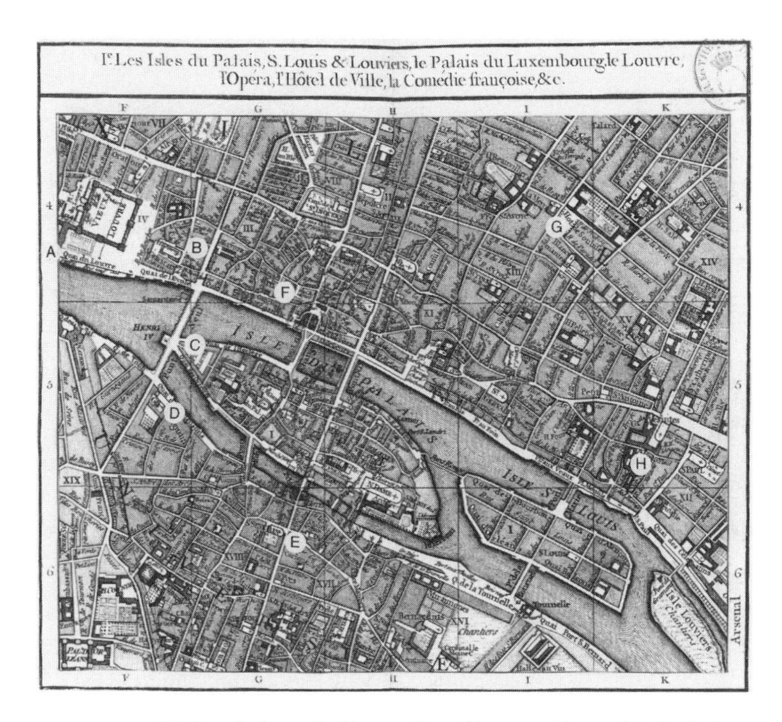

FIGURE 24. Didier Robert de Vaugondy, *Tablettes Parisiennes* (Paris: Robert de Vaugondy, 1760), plate 1. Plan of central Paris with locations of principal mapmakers, engravers, and publishers. They remained throughout the century primarily in the area of booksellers and printers on the Left Bank, south of the River Seine, around the rue St Jacques (E) where the shops of Jean Lattré and Louis-Charles Desnos could be found. The rue and quai des Augustins (D) hosted the shops of the Jaillot family and Georges-Louis Le Rouge. Guillaume Delisle, Jean-Baptiste Nolin, the Robert de Vaugondy family, and Philippe Buache (after 1745) could all be found on the quai de l'Horloge (C), close to the scientific instrument makers. D'Anville was lodged in the Galeries du Louvre (A) as had been the sons of Nicolas Sanson. Sanson himself, as well as his grandson, Pierre Moullart-Sanson, lived on the Right Bank near the church of St Germain l'Auxerrois (B), where the engraver Guillaume Delahaye later had his workshop. Delahaye later moved to a street nearer the Châtelet (F), behind the quai de la Megisserie, where Philippe Buache had an early address. Farther away from the trade, deep in the heart of the wealthy neighborhood of the Marais, Roch-Joseph Julien kept his shop at the Hôtel de Soubise (G). When the Dépôt de la Marine returned from Versailles in 1775, they stayed near engravers and printers, though on the north side of the river (H) in a building vacated by the Jesuits after their expulsion in 1762. BNF, Ge. FF. 4423. (Bibliothèque nationale de France, Paris.)

geographers like Guillaume Sanson and J.-B. B. d'Anville. Map engravers did not stray far from this neighborhood; many lived just across the river on the Right Bank. However, the entrepreneurial map seller Roch-Joseph Julien abandoned the crowded narrow streets of the Left Bank for the Hôtel de Soubise in the fashionable Marais neighborhood, which also became the home of the Dépôt de la Marine when it returned to Paris from Versailles in 1775.[5] The return of the dépôt to the capital caused some complaint among the copperplate printers and map sellers in Paris, for they stood to lose the chance of acquiring "advance copies" of the marine charts during the unsupervised transit between Versailles and Paris.[6]

The tight grouping of the cartographic community on or near the Left Bank in Paris had its parallel in London. Laurence Worms has noted the east-west axis of map sellers and producers north of the River Thames, stretching from the docks around Wapping to Westminster (figure 25). As in Paris, the zone shadowed the book and print trade as well as that of the instrument makers who were found on almost every street on which there was a map seller.[7] From Whitehall in the west, where regular map fairs attracted the trade, through Charing Cross, along the Strand to Fleet Street to the heart of the book trade around St Paul's Cathedral: map trade addresses changed little in one hundred years. The line continued from St Paul's to Cheapside and the Royal Exchange to Tower Hill and Wapping, following a book market supported by church and business interests in these neighborhoods. As in Paris, some variants stand out. Just north of the axis we find Emanuel Bowen in Clerkenwell, Thomas Kitchin on Holburn Hill, and Aaron Arrowsmith in the increasingly desirable Soho, an area packed with the Huguenot community of engravers.[8] Thus the physical world of map sellers, engravers, printers, and publishers was tightly contained within specific zones, as in Paris, or stretched along a crowded corridor, as in London. It is not difficult to see how easily partnerships were formed, engravers shared, plates carried from one shop to another, and maps copied.

Much copying was also practiced across national boundaries. For example, an English mapmaker like Thomas Jefferys, who copied the maps of the French geographer d'Anville, was out of any French jurisdiction if a case of plagiarism were to be brought to bear. Jefferys added authority to his map by giving credit to the French mapmaker in the title: "corrected and amended from," or "improved from the map of." Jefferys would translate the French place-names on the map into English and perhaps alter boundaries in order to suit English political mood.

The English esteemed d'Anville most highly of the French geographers. One copyist wrote: "the erudition of his maps, the abundance of objects, the

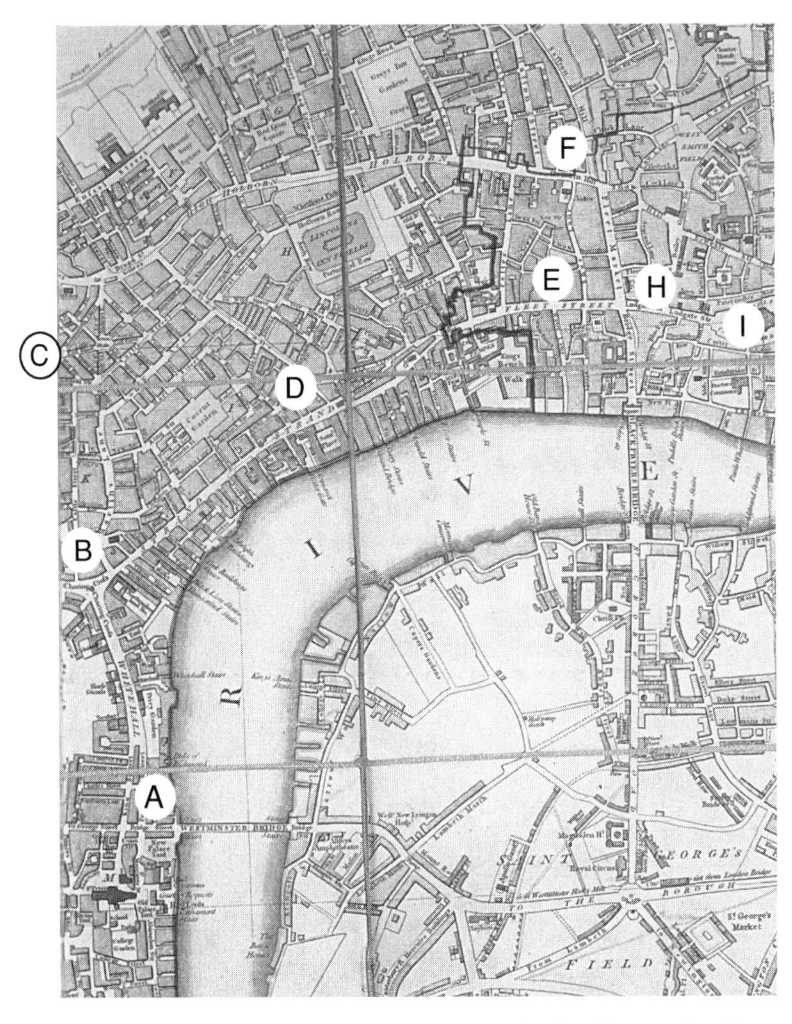

FIGURE 25. A detail from William Faden, *A new pocket plan of the cities of London and Westminster* (London, 1790). This plan is based on the work of Laurence Worms ("Location in the London map trade"), who points out the pattern of the map trade along the east-west artery linking Westminster to the old working heart of London north of the River Thames. Maps were sold at regular fairs in Westminster Hall (A). Just north of Westminster, the shops of Thomas Jefferys, William Faden, and Andrew Dury were located around Charing Cross (B). North and west of Charing Cross (and thus not really on the map) was Soho (C), attractive to many of Huguenot engravers and French immigrants, such as Hubert Gravelot. Jean Rocque had begun his career in Soho but moved south to Piccadilly, near Whitehall (A) until a fire destroyed his shop and stock, when he moved to the Strand (D). Herman

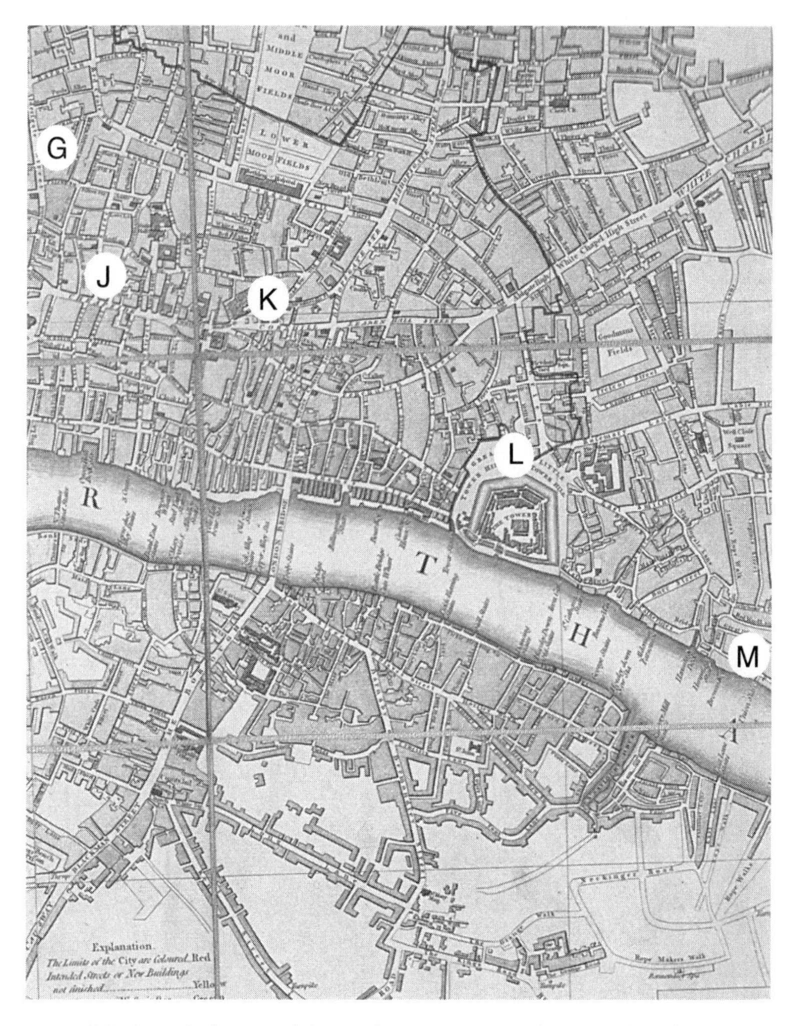

Moll had similarly moved from Charing Cross to the Strand. Robert Sayer could be found in Fleet Street (E), at an address inherited from Philip Overton. North of him was Thomas Kitchin in Holburn Hill (F). Emanuel Bowen kept his shop even further east in Aldersgate (G), then Clerkenwell, beyond the northern limit of this map. Joseph Moxon could be found on Ludgate Hill (H) with George Willdey nearby on Ludgate Street. The Bowles family moved from near the Royal Exchange (K), from Cheapside (J), and finally to St Paul's (I). John Senex moved from the Royal Exchange (K) to Fleet Street (E). Close to the river, Jonas Moore sold charts from Tower Hill (L), near John Thornton's shop of many years earlier. Even further east, to catch the chart trade, was John Seller in Wapping (M). G. 5754 .L7 1790. (Map Library, Harlan Hatcher Graduate Library, University of Michigan.)

scrupulous exactness of positions, the neatness and perspicuity of his designs and the beauty of their execution, give them a decided superiority over all that hitherto have been published. Hence the eagerness of the learned of Europe to possess them, of the most skilful geographers to choose them for models, of compilers of maps to copy them in preference to all others."[9]

D'Anville himself observed such imitations with *hauteur:* "More than once have I endured reproaches for not having a student who will carry on my work. I may not have a student but I have many copyists. . . . The esteem with which my maps are held in England has caused a universal copy of my work to appear in London, but it is a very crude execution, far removed from the elegance of the originals, an elegance that only adds to the value of their composition."[10] But d'Anville had no recourse against such copies; very little protected him from such practices. The print privilege he enjoyed in France applied only to his work in France, just as the copyright protection that covered works in England applied only in England.

[COPYRIGHT IN ENGLAND[11]]

In England prior to 1735, engravers or publishers who wished to protect their graphic work obtained privileges from monarchs or parliaments. Privileges offered the recipient the sole right to sell a particular map or atlas. It was a right of sale, not a protection of content.[12] Thus the Dutch mapmaker Blaeu received a privilege from the Scottish Parliament in 1647 for his maps of Scotland.[13] William Petty applied for and received a privilege from Charles II in 1661 for the publication of maps of Ireland and an atlas of Ireland, *Hiberniae delineatio.*[14] What the costs of these privileges were for map publishers has not been thoroughly investigated; Timothy Clayton suggests that they were costly, as they usually covered only expensive productions.[15]

The first copyright act, which protected only the book trade, was instituted in 1709, under Queen Anne. It gave control of the work, regardless of how it was printed, to the bookseller—that is to the printer or the person who paid the printer—not to the author. This situation was improved for designers, engravers, and mapmakers with the Engraving Copyright Act of 1734/35 (8 Geo. II, c. 13). It was also known as Hogarth's Act because the well-known engraver William Hogarth claimed a good deal of the credit for it, though he had been only one among seven artists and designers who exerted the necessary pressure on Parliament to pass the act. Hogarth's Act vested the property of designs, engraving, and etchings in the inventor and engraver (legally construed to be one and the same), not the owner of the plates, for fourteen years, a length of time that was also typical of privileges.

In the case of plagiarism, the act demanded the forfeiture of any pirated prints and plates to the copyright owner for destruction, and the imposition of a fine of 5s for every print (to be divided between the Crown and the plaintiff).

The 1734/35 act also required that the name of the print proprietor and the day of first publication be given on each print, an important feature since any case of copyright infringement had to be brought within three months of the discovery of the piracy. Though the publication date printed on the map provided a clear date after which a charge of plagiarism might be brought, the published date also revealed the act's crucial flaw. It offered no protection of a map or engraving prior to the date of publication. Thus viewings, proof states, and public exhibits of early versions of a print or a map all made it vulnerable to plagiarists. A further flaw was that the act protected only those engravers who also designed their own work.

Two subsequent acts of Parliament refined the definition of copyright. The Copyright Act of 1767 (7 Geo. III, c. 38) included the specific phrase "map, chart, or plan" in the wording of the act, which expanded protection from inventor-engravers to any engraver who produced a print. This act extended the duration of the copyright to twenty-eight years, and the limitation clause to six months after the publication date. In 1777, the Print Copyright Act (17 Geo. III, c. 57) further defined forgery or infringement of copyright as the engraving or causing to be engraved of "any Copy or Copies of any historical Print or Prints, or any Print or Prints of any Portrait, Conversation, Landscape, or Architecture, Map, Chart, or Plan . . . without the express Consent of the Proprietor in writing."[16] The Print Copyright Act of 1777 allowed the plaintiff to sue for damages and double costs and made it an offense to publish a copy even innocently. Further, it removed the time limits governing when a suit could be brought, though it restricted suits to cases where the copies were made for sale.

[PRIVILEGE IN FRANCE[17]]

The situation in France differed completely from that in England because of the absence of copyright. A central office in Paris, the Librairie, supervised and controlled the book and print trade. The bureau was headed by a court-appointed director whose agents looked out especially for antiestablishment, pornographic, and other unseemly prints. The protection for the work of an author or a graphic designer and/or engraver was found in the permission and the privilege.

The privilege for printed work had its roots in the sixteenth century. A declaration of King François I in 1536 required the deposit of a copy of any printed books in the royal library. The privilege was codified by royal decree in 1642; this granted rights given by print permission and obliged the author to deposit two copies of books or pictures in the royal library. Yet prints and the making of prints were regarded as a free art; engravers were not required to belong to any professional community or guild. Nonetheless, the community or guild of copperplate printers jealously protected their rights, and legislation required that engravers use the services of the Communauté des Maîtres-Imprimeurs en Taille-douce either as printers or as supervisors of presses maintained on the premises of engravers. After 1750, engravers were also required to pay 20 sous per week for every press they operated in their workshops.[18]

The permission was sought and paid for by either the engraver or publisher, who was required to deposit copies of the work in the royal library, which was increased to three copies by a royal decree in 1703. The king was usually the authority who granted privilege, but the Parliament, university, mayor, archbishop, and any of the royal academies also had privilege-granting powers. There were three types of privilege, which differed in their time limits, distribution rights, and protections against counterfeiters. Each of them cost the applicant different amounts. The most expensive was the *Privilège exclusif* or *Privilège général,* which granted exclusive rights to make, distribute, and sell a product throughout France for a period of from three to twenty years. It cost 18 livres early in the century and just over 101 livres in the last quarter of the century.[19] This privilege promised the confiscation of counterfeits and a fine imposed on the counterfeiter.

The least expensive was the *Privilège simple* or *Permission simple,* which allowed the owner to sell his work to more than one publisher within a three- to six-year period. It offered no protection against counterfeiters, though it did prohibit the importation into France of the same product. Its price was low: from 3 to 9 livres in first part of the century, but from 30 to almost 62 livres in the last quarter of the century.

Privileges did not protect the rights of authors; they only prevented the publication of unauthorized copies. In order to do this, the fines punishing infringement of privilege were hefty: from 1,500 to 10,000 livres. The fine was usually divided three ways, one-third going to the Crown, one-third to the plaintiff, and one-third to charity, the Hôtel-Dieu. The copies, copperplates, and tools of the counterfeiter were confiscated.

In 1714, Louis XIV granted a general privilege to the Académie Royale de la peinture et sculpture without time limit in his desire to support the beaux arts. With the consent of the academy, an engraver could publish his work "*avec Privilège du Roi*" and avoid paying the normal fees for the seal. He only had to deposit two copies with the academy; and if he were already a member of the academy he could use the title *graveur du roi.* The number of engravers in academy was limited to six.[20] Further refinement came as the royal decree of 1734 allowed engravers and print sellers to print "under their own eyes" plates they had engraved or already possessed. But as engravers were a free trade, with no community or guild to create a pressure group, enforcement of privilege for prints was weak. The copyright acts protecting engravers in England, such as Hogarth's Act, had come about as a result of concerted effort of engravers, who as members of London's community of guilds, had a powerful workforce behind them and could exert lobbying pressure on Parliament in ways that would be impossible in France.

Privilege, like copyright, had its loopholes. The procedure for obtaining a privilege involved depositing examples of a print or map in the office of the Librairie, two examples in the royal library, another for the Cabinet du Louvre, another for the chancellor or keeper of seals, another for the censor, and three for the booksellers' guild (*communauté de la Librairie*). These deposits created a potential source for copyists and plagiarists, for there was no way of keeping track of these preliminary examples. The king could do whatever he liked with his copies. The office of the Librairie was known to sell the prints deposited there at a lower price than the original author's. Thus it was a problem of who would guard the guards.

Both copyright and privilege protected authors and commerce only at home, not abroad. The international nature of the trade meant that foreign authors were an easy source for the copyists and were in many ways the staple of the map trade. It proved fruitless for a map seller to try to maintain an exclusive contract as the distributor of foreign maps when they were easily available to others. Guillaume Delisle's business partner in Amsterdam, Louis Renard, complained that his compatriot Pierre Mortier was copying and selling Delisle maps but using the name of Sanson on them, since he had a privilege for selling Sanson maps.[21] Renard reported that he had accused Mortier of acting fraudulently towards the public in giving them the same map under the names of different authors. Mortier however was not embarrassed,

> for he is rich and without honor. He told me and the whole company [present] that he would sell his maps to the taste of the buyers and that he would always provide them with the name of all the most famous authors, accommodating them thus with the same map those [names] whom they wanted—Delisle, Sanson, de Fer, Jaillot, Baillieu. He said he had more than fifty maps

under the name of Delisle. . . . He sells his maps under [Delisle's] name in Germany, under that of Sanson in England, under that of de Fer in Poland and Italy.[22]

Thus copying maps becomes not only a cost saver, but also a marketing strategy.

Against the foreign copyist a mapmaker like Delisle or d'Anville was powerless. Against the neighbor, he could bring suit. To enforce both copyright in England and privilege in France, the burden of proof always fell on the author of the original, as did the financial burden of taking the alleged copyist to court. A police commissioner could cost 3 livres a day plus an additional 15 sous for the clerk who recorded the affidavits.[23] For example, in a case discussed below, when Philippe Buache pursued the engravers Vallet and Dheulland for copying Delisle plates, the police commissioner charged 14 livres plus 40 sous for the paperwork and time spent by the court clerk in seizing the plates.

[PROVING THE CASE]

Prosecuting a plagiarism case was difficult because of the nature of the map itself: if a map purports to show accurately a given area, its correctness cannot be grounds for a charge of plagiarism. Fault can only be found in the copying of errors. A modern device for catching plagiarism employs the ruse of planting certain slight errors on maps that the unwary plagiarist will copy without realizing they are traps. The Ordnance Survey calls these tiny curves in roads or twists in rivers "fingerprints . . . not errors or faults, but subtle and secret ways of detecting plagiarism."[24] The American Map Corporation uses a similar technique of hiding "trap streets" on their maps. They place them "in areas that would be relatively harmless and would not mislead someone using the map—just a cul-de-sac at the end of some development. . . . This allows us to do a quick spot check of our competitors' maps to see if they have stepped on our toes."[25]

Many mapmakers simply suffered their work being copied, as in the cases of d'Anville and his English copyists or Delisle and Pierre Mortier's privilege for Sanson maps. However, when a mapmaker caught a colleague in flagrante delicto, the ensuing case revealed those aspects of cartography that set it against other forms of graphic print. The issues concerned the accurate representation of land and sea, the "ownership" of geographic information, and the reliability of maps for the general public.

[DELISLE VERSUS NOLIN: WHO OWNS "DISCOVERIES"?]

One of most public and long-running plagiarism lawsuits in France was that brought by the young geographer Guillaume Delisle and his father, the historian Claude Delisle, against the well-established map engraver Jean-Baptiste Nolin.[26] After training with the engraver François Poilly and a sojourn in Italy to polish his skills, Nolin had attracted the attention of the Italian geographer Vincenzo Coronelli during his years in Paris at the invitation of Louis XIV. Nolin engraved many of Coronelli's maps as well as the gores for his celestial globe. He was appointed geographer to the duc d'Orléans in 1694, and became *géographe du roi* in 1701. After an unsuccessful start on the rue St Jacques, he moved his business to the quai de l'Horloge, near other geographers and scientific instrument makers, where he specialized in large, decorative wall maps and atlases. A map collector and amateur geographer named Jean Nicolas de Tralage, who wrote under the name of Tillemon, supplied much of the material for Nolin's cartographic work. In 1700, Nolin published a very large and richly decorated world map, entitled *Le Globe Terrestre,* measuring 125 by 140 cm. The Delisles claimed that Nolin had copied the information for his world map from a manuscript globe that Delisle had prepared for the chancellor of France, Louis Boucherat. Four months after Nolin's map had been printed, Delisle published his own mappemonde, which naturally looked much like that of Nolin.

The controversy became public through the letters exchanged by Nolin and Delisle published in the *Journal des Sçavans,* the organ of the Académie Royale des Sciences. The letters outline the issues uppermost in Delisle's mind as he made his case against Nolin. Delisle describes Nolin's work as that of a "bad copyist" who "avoids appearing to be a plagiarist" by changing correct geography into incorrect geography. The telltale evidence of copying was to be found in North America: the location of the mouth of the Mississippi and the representation of California as a peninsula, not an island. Delisle claimed that these locations and landforms were *new* information, not to be found on other maps. They were Delisle's discoveries, the results of his long study of cartographic and literary sources. Nolin could only have copied them.

Nolin's response was simple: how could he have copied Delisle when his mappemonde had appeared four months before Delisle's? Delisle insisted that Nolin must have seen the manuscript globe in the chancellor's library. Nolin declared he had never set foot in the chancellor's library. Furthermore, he asserted, he was quite capable of preparing maps by himself from his own materials; he did not need recourse to Delisle's work. The bravado

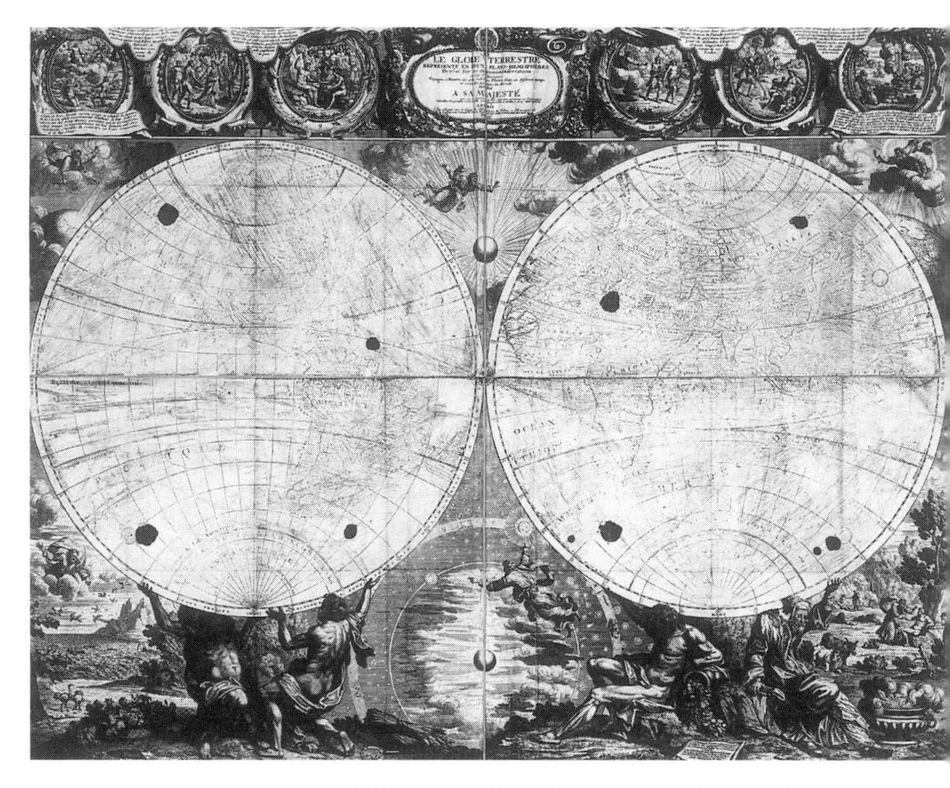

FIGURE 26. A print pulled from the effaced plate of J. B. Nolin, *Le Globe Terrestre* (Paris: engraved 1700, effaced 1707). The plate was ruined as punishment after Nolin was found guilty of plagiarism in the lawsuit brought by Guillaume Delisle. BNF, Rés. Ge. AA 1259. (Bibliothèque nationale de France, Paris.)

of Nolin's claim to geographic competence led to a cartographic duel played out in front of a commission specially formed to interrogate both men on how they compiled their maps.

Delisle's opening parry was a stinging account of the history of plagiarism in cartography:

> Since maps have begun to be multiplied and entered into pecuniary commerce, one has seen many ignoramuses who, by greed for profit, have mixed themselves up in doing what they don't understand, and some sellers of images who have wanted to make geography a subsidiary of print making. . . . The profession of geography has become a real den of thieves. . . . The pla-

giarists try not to copy line for line . . . making it a tricky business to discern who is the robber and who the robbed.[27]

According to Delisle, Nolin's plagiarism was twofold. First, he had copied Delisle's system of coordinates, the grid of longitude and latitude on which the map was based. He had copied it so exactly that the meridians and parallels passed through the same places on Nolin's map as on Delisle's globe. This was particularly important since Delisle had spent the prior fifteen years working to reform all parts of the globe by determining their exact shapes and precise locations.[28]

Second, Nolin had copied certain details from the Delisle globe and abandoned the cartographic ideas he normally used, acquired from the geographers Coronelli and de Tralage, whose maps he had previously engraved. Delisle claimed that Nolin had copied Arabia, Persia, the East Indies, China, and the interior of Africa "without even knowing that there are some dubious places there." But it was particularly in North America that Nolin revealed himself as a copyist by representing lower California as a peninsula, abandoning its representation as an island (which had increasingly become the convention). This information had only recently been verified by the travels of Father Kino and adopted by Delisle on his globe. In addition, Nolin published a Sea of the West on his map. Delisle claimed this "was one of my discoveries. But since it is not always appropriate to publish what one knows or what one thinks one knows, I have not had this sea engraved on the works that I made public, not wanting foreigners to profit from this discovery."[29]

Delisle here reveals the telling editorial moment, when the cartographer chooses not to reveal all information on his printed map. Though the Sea of the West was present on the globe prepared for the chancellor, Delisle pointed out that this sea was only conjecture. It was not supposed to be printed.

A council of experts was called in to determine the case. Several were well known to the Delisles: Jacques Cassini, the son of Jean-Dominique Cassini, the astronomer who had been Guillaume Delisle's tutor; the astronomer Giacomo Filippo Maraldi, who was Jacques Cassini's cousin; and Joseph Sauveur, a mathematician and member of the Académie Royale des Sciences, of which Delisle was already a member. They interrogated Delisle and Nolin closely on how they had constructed their maps. According to the report, Delisle acquitted himself well, explaining "in a neat and precise manner the use one ought to make of mémoires for the construction of maps."[30] M. Nolin, on the other hand, did not appear to be very well informed on how he had used the same mémoires; indeed, he appeared not to know even the first principles of geography.

But most important was the panel's decision that certain geographic knowledge is the property of everyone. This public knowledge could be the observations made by the members of the Académie Royale des Sciences or the relations of missionaries, such as the Jesuits, as well as the maps already published. Nolin could not be reproached for using these materials. As for the details, such as the Sea of the West and California, the experts thought it would be unlikely that two authors working from the same mémoires, vague as they were and often contradictory, could arrive at the same conclusions. The panel found against Nolin, who protested, not without reason, that the panel was partial to the Delisles. The panel ordered that Nolin's mappemonde be seized, the plates effaced, and a fine of 50 livres levied. Perhaps not wishing to inflict such financial loss on his neighbor on the quai de l'Horloge, Delisle suggested that the punishment be reduced to simply effacing the plates, which they were. (Figure 26)

One result of this suit was to establish the young Delisle's authority as a mapmaker through his command of a map's sources. The charge of plagiarism did not hinge on *when* Delisle's own map was published, for, as the judiciary council noted, "one can have access to the works of an author as soon as they are in the hands of the engravers and printers or when the originals have been passed to someone."[31] Rather, the true author of a map was the one who knew *how* to make the map.

[THE DANGER OF COUNTERFEIT MAPS]

Delisle's considerable authority as a mapmaker continued well after his death. As we have seen, his widow was reluctant to distribute his maps very widely, restricting their circulation in a well-intentioned attempt to prevent their being copied. After her death in 1745, Delisle's son-in-law Philippe Buache sued the family for possession of the geographic stock, claiming it through his marriage to Delisle's daughter. He won his suit, and on 30 April 1745 obtained an exclusive privilege for fifteen years to design, print, and sell the Delisle maps. Eighteen months later Buache learned that two engravers, Dheulland and Vallet, were engraving maps they had copied from those of Delisle, and adding to the titles "sur les observations les plus récentes."[32] Buache visited the workshops of both engravers and recognized the maps, comparing them with Delisle maps in his possession. He found Vallet engraving places, cities, and titles on these new plates and Dheulland in possession of plates still awaiting titles and authors. All the plates in question had been corrected by J.-N. Bellin, ostensibly moonlighting from his work as hydrographer in the Dépôt de la Marine. The plates proved to be exact copies of Delisle maps drawn in the early years of the century, with only minor adjustments, such as an altered scale or reworded title. As the counterfeited maps contravened Buache's privilege, the maps, plates, and proofs were confiscated and turned over to the police.

In his petition to the King and his council for redress in respect of the plagiarism, Buache commented on what he considered the particularly reprehensible aspects of the affair. The maps in question were simply tracings of maps by Delisle on which nothing was changed except the title, the name Delisle, and the address where the maps were sold. The only exception to this was on the map of Spain, where the islands of Minorca and Majorca were erased and copied from a more recent plan of 1740. Buache had found that the plate for the Denmark map was worked from an engraved map of Delisle on which the title was effaced, part of Pomerania hidden, and the name of Delisle removed, but the words "sur les observations les plus récentes" added. That copies of maps originally published in 1710 by Delisle, using the same coordinates for cities, rivers, and coasts, should be sold as "based on the most recent observations" when they were thirty-five years out of date was nothing but a deceit of the public.

Though he wished the plagiarism stopped, Buache was as accommodating as his father-in-law had been with Nolin in regard to the fine. A penalty of 6,000 livres could have been imposed on the offending engravers, but Buache agreed to drop the charges if the engravers would pay half the expenses of the case and Buache kept all the plates and maps.[33] This in fact caused further problems, for the bookseller Antoine Boudet appeared at this point, claiming that the maps and plates were his, in preparation for a projected atlas. Boudet undertook to pay half the costs and reimburse Dheulland and Vallet for their work.

[COPYRIGHT LAWSUITS IN ENGLAND]

The adjudication of lawsuits in England regarding maps that infringed on copyright also required careful study of the contents of the maps. The print sellers and map sellers Robert Sayer and John Bennett brought suit against John Hamilton Moore, chart seller and teacher of navigation, in 1785 before the lord chief justice. Sayer had been a business partner of Moore, owning one fourth of the copyright in Moore's *Seaman's New Daily Assistant.* Sayer and Bennett, suing Moore for £10,000 in damages, complained that Moore had taken four charts published by Sayer and made them into a single map.

Moore responded that he had also made alterations and improvements to Sayer's maps, thus making his a different map from Sayer's four charts. The judge was left to decide whether Moore's alterations were significant enough to make the alleged copy a different map. In an opinion similar to that voiced by the panel of experts in Delisle's case against Nolin, Lord Chief Justice James Mansfield ruled that "whoever has it in his intention to publish a chart may take advantage of all prior publications. There is no monopoly of the subject." The judge further pointed out that the "plaintiff's charts were drawn upon a wrong principle, inapplicable to navigation." Thomas Jefferys had compiled the Sayer charts from maps of Cook, Holland, and Lane of North America. They were drawn on a common cylindrical projection. Moore had adapted the material onto a Mercator projection, particularly useful for navigation, and, as he pointed out, he had added information from marine journals. Moore's witnesses, Admiral John Campbell and naval hydrographer John Stephenson, attested to the usefulness of his charts and emphasized the defects of the Sayer map projections. The jury found for Moore; in the words of Lord Mansfield, "The defendant therefore has been correcting errors and not servilely copying."[34]

Not all—probably very few—cases of copying or plagiarism went to court. Cries of fraud had the advantage of increasing the possibility of sales. The engraver and publisher Jean Lattré accused the globe maker Louis-Charles Desnos of copying a map printed by Lattré in two colors that showed the visibility of the solar eclipse, due to occur on 1 April 1764, from different places in Europe (plates 4 and 5). On both maps, Europe is printed in red or sepia with the gradually thickened black hachures covering those areas where the eclipse would be the most visible. These maps are unusual in their use of two-color printing, a technique not widespread in eighteenth-century copperplate printing. Lattré's map is further notable for being an entirely feminine production. It was published with the imprint of his wife, the engraver Mme Lattré, prepared by Mlle Le Paute, a well-known astronomer, and engraved by Mlle Croisey. This singular alignment of female talent was remarked by an anonymous reviewer in the *Journal de Trévoux:* "[this map is] new proof that the knowledge of science and the beaux arts is not an exclusive privilege reserved for only one gender; women also will distinguish themselves in this noble career, when they will desire to busy themselves usefully, or rather, when we see the abuse of giving them only a frivolous education."[35]

Lattré insisted that his privilege protected this map from copying; the police had even seized Mlle Le Paute's map on Desnos' premises, and the engraver was very suspicious that his former employee, the Italian geographer Rizzi-Zannoni, might be responsible for the theft. Rizzi-Zannoni had been paid for map designs by Lattré but had never completed them and was, at the time of Lattré's complaint, working for Desnos.[36] For his part, Desnos merely made Lattré's complaint a source of advertising for his map in the *Journal de Trévoux.* M. Desnos declared that he found Lattré's charges of plagiarism to be sheer calumny; the public should be assured that his map continued to be available at 1 livre 4 sous, less than half the price of Lattré's map, which cost 3 livres.[37] Thus the Cochin formula worked once again—copy cheap, sell low.

Copying was worth the risk of being accused of plagiarism. The costs of going to court weighed against the length of time a case could be tied up (six years in the case of Delisle versus Nolin), during which plates could not be printed nor maps sold. Sometimes the speed with which maps had to be

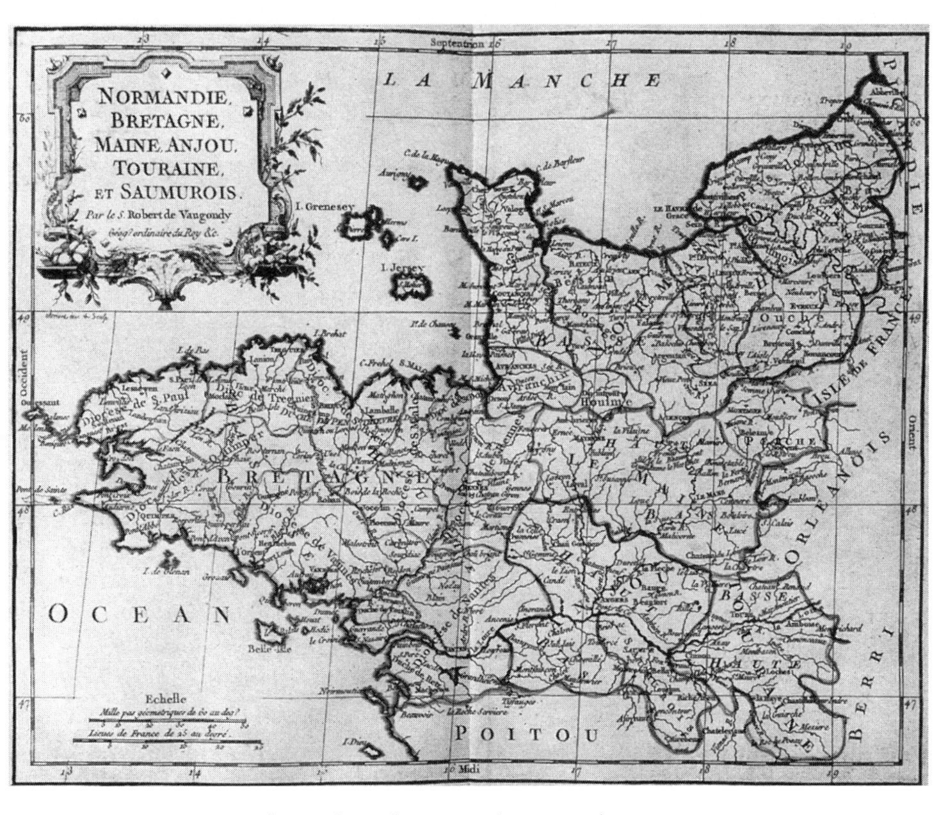

FIGURE 27. Didier Robert de Vaugondy, *Normandie, Bretagne, Maine, Anjou, Touraine et Saumurois,* plate 16, *Nouvel atlas portatif* (Paris, 1762). Greenlee 4891 R 64, no. 16. (Newberry Library, Chicago.)

FIGURE 28. J.-N. Bellin, a cut-up copy of Didier Robert de Vaugondy's *Normandie, Bretagne . . .*, used in preparation for republication as Bellin's *Carte de la Normandie et de la Bretagne* in the *Petit atlas maritime.* AN, CP MAR 6/JJ/76/A pièce 34. (Archives nationales, Paris.)

produced to meet a particular demand necessitated a certain cutting of corners. J.-N. Bellin at the Dépôt de la Marine had already helped the publisher Boudet prepare the copies of Delisle maps for the engravers Dheulland and Vallet. Nor did he have scruples about using a map by Didier Robert de Vaugondy in his preparations for the *Petit atlas maritime,* made by order of and at speed for the duc de Choiseul in 1764 (figures 27–29). Bellin was required to produce more than 600 maps to make up this elegant five-volume set in a very short time. According to his widow, he incurred many of the costs himself, for which he was not reimbursed.[38] Evidence of the speed at which he worked may be found among the drafts of maps he prepared for the engravers. A map of Brittany and Normandy taken from Robert de Vaugondy's *Nouvel atlas portatif,* published in 1760, was, happily, just the size needed by Bellin; it required no reduction or change. Bellin merely moved the scale and altered the title cartouche, in a manner reminiscent of the cut-and-paste job done by the engravers Vallet and Dheulland on the Delisle maps. The resulting map was engraved and printed as number 21 in volume 5 of the *Petit atlas maritime.* There is no record of Robert de Vaugondy's complaint.

In his preface to the *Petit atlas,* Bellin stated that the ministry undertook part of the considerable expense of this large, costly work—the paper, printing, and coloring—but kept the price low for the public benefit and to prevent foreign counterfeits.[39]

The practice of copying maps would explain the fact that so few of the foreign correspondents of the London geographers Jefferys and Faden ordered more than a few maps at a time. Rarely do they order more than a dozen of any single title, and more often request just one or two examples of

FIGURE 29. J.-N. Bellin, *Carte de la Normandie et de la Bretagne,* from *Petit atlas maritime,* vol. 5, plate 21. WLCL, Atlas S-1-B. (William L. Clements Library, University of Michigan.)

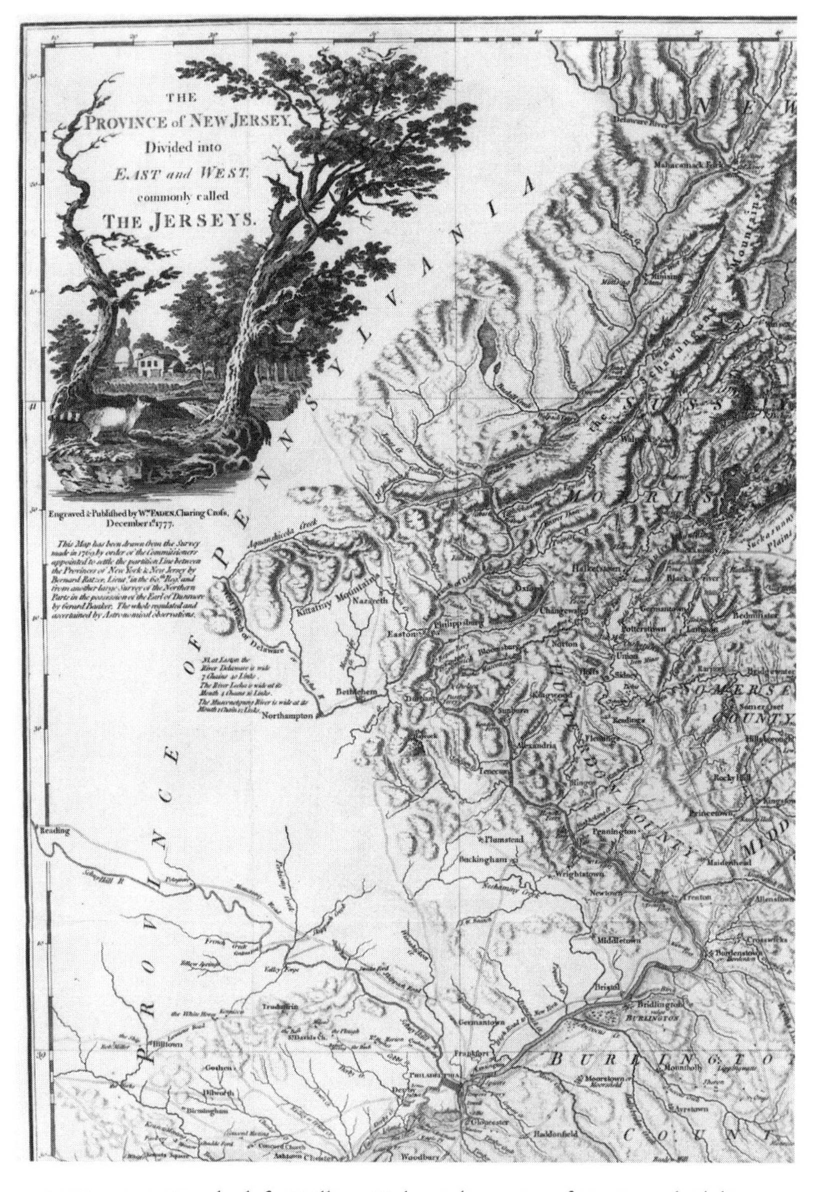

FIGURE 30. On the left, William Faden, *The province of New Jersey divided into East and West* (London, 1777). The detail shows the cartouche and the latitude scale on the left edge of the map, with the mistaken 39° where 40° should have been. On the right, Perrier and Verrier, *The province of New Jersey divided*

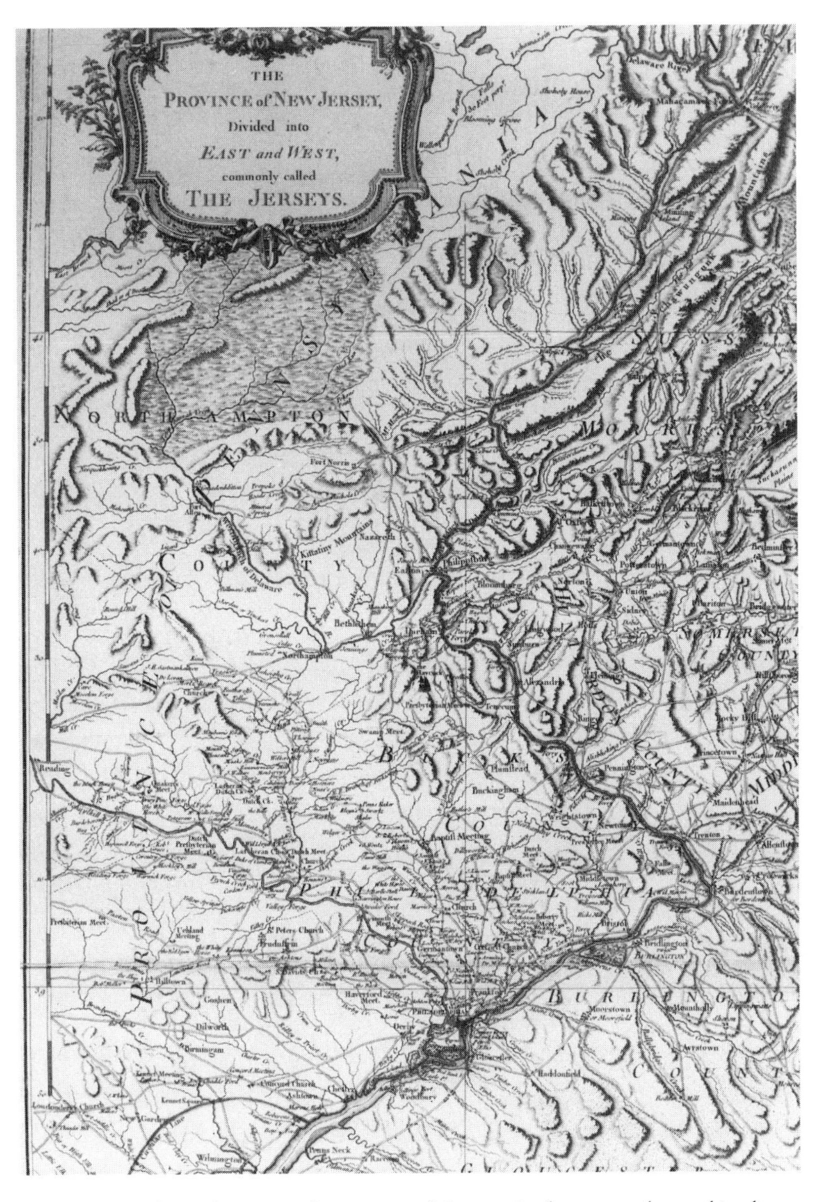

into East and West (Paris, 1777). In spite of changes in the cartouche, and in the geography and topography of regions just below the cartouche, the French engravers have nonetheless copied the mistaken 39°. WLCL, Map 5-A-9, 10. (William L. Clements Library, University of Michigan.)

a title. Studying the maps published by the French correspondents reveals occasions when English maps were copied. For example, the map firm of Perrier and Verrier, two engravers who succeeded to the Julien business, published their version of William Faden's map of the Jerseys in North America (figure 30).[40] The French engravers modified the design of the cartouche and, using other maps they had ordered previously from Faden, they altered the topography and geography of eastern Pennsylvania, an area left blank on Faden's Jersey map. But their use of Faden's map as their base is tellingly revealed by the error on the left margin of the map: the latitude number 39 appears twice on each map, rather than 39 and 40.[41]

In spite of Bellin's rosy view that international copying could be prevented by lowering the price of a printed map, the practice was inevitable given the national differences in access to geographic information and the uneven distribution of skilled geographers. French geographers were held in great esteem throughout the first three quarters of the eighteenth century; their works dominated the European map market. By the last quarter of the century, however, English map publishers began to rise to prominence as they benefited from the favorable results of the Seven Years' War, which brought enlarged territorial holdings in North America, the West Indies, and India. Though the surveyors who measured and mapped these territories were often trained on the Continent, their works were published in London. The geographers and hydrographers of France now turned to England for material to copy.

CHAPTER FIVE

MULTIPLYING MAPS: THE SURVEY AND PRINTED CHARTS OF NARRAGANSETT BAY, RHODE ISLAND

The international copying of maps had been a staple of European cartographic production since map printing began in the Renaissance. The movement of people, plates, and skills meant that there was little to protect a mapmaker or publisher from the foreign copy of his or her own work. Copying was cheaper than original work, and the copy might be reproduced in a style more appealing to a particular public.

But in a more particular way, the foreign copy also reflects the manner in which geographic information about remote areas entered the marketplace, making such places accessible to a reading public. This process is most clearly seen in the way in which map publishers of various nations acquired or assimilated through printed maps the results of surveys of colonial territories, information otherwise unavailable to them. Printed maps of these areas assume especial importance in time of war, when strategic information about territory could benefit an enemy. Printed maps also served (and still serve) propagandistic purposes by their exaggeration or diminution of features that may highlight or inflame or justify action.

War is one of history's prolific generators of maps. Because wars are often about possession and control of space on earth, the depiction of space and the distribution of phenomena on it are necessary requirements for planning the strategy for the war and for the analysis that informs the later historicizing of war. During the long eighteenth century, the rhythm of wars beat the march time of passing years; it was also a period in which the graphic explanation of battles was a large part of the visual print culture. The last quarter of the century in Europe was dominated by two great conflicts that demanded the Continent's full attention. The war of independence of Britain's colonies in North America not only established the United States of America as a sovereign nation; it also drew France once again into war with Britain, a state of conflict that had ebbed and waned between the two nations throughout the century. America's conflict was soon followed by a revolution in France that disestablished the monarchy and paved the way to Napoleon's empire and yet another war between France and England. As these wars engaged public attention and spent economic capital, a series of printed maps, based on the same sources, show how disputed territory was depicted by the warring nations, whose mutual commerce, at least in the trade of geographic information, often remained intact.

The War of American Independence, unlike previous European conflicts, brought European tactics and strategic skills to a land that was still terra incognita to Europeans. Three groups were involved in the struggle, each with access to different sets of information required for purposes of strategic planning and public persuasion: the American colonists, their French allies, and the governing British. The British dominated the public presentation of the war in mapping terms. While American colonials had participated in the survey of their land, their access to printing facilities, particularly from copperplate, was limited; more skilled engravers and the means to wider distribution of a printed map were to be found in London.[1] After their loss of Canada at the end of the Seven Years' War, the French were also reliant on the British map trade for public information about the coasts, interior, and infrastructure of eastern North America and the Atlantic coast.

Three maps of Narragansett Bay in Rhode Island demonstrate how these three groups gained access to and presented information about a crucial area in the War of American Independence. These printed maps demonstrate national and local attitudes towards cartographic information and the capacity of the printed map to alter, refine, and reinvent spatial information. The mapping of this area emerged from the conclusion of one war between France and England and the anticipation of another. The various copies of the printed map demonstrate how the marketplace filled the administrative and military gaps on both sides of the conflict. Briefly, in order to make up for their lack of knowledge and access to the North American shoreline, the French relied on the availability of British printed maps sold through the map trade.

[NARRAGANSETT BAY]

Narragansett Bay is endowed with one of the outstanding natural harbors in North America. Oriented south and thus protected from north and northeasterly winds, the bay extends in two long arms northwards, embracing a long central island at the southern end of which is the town of Newport. The colony took full economic advantage of its natural features. Its mild climate and lowland meadows with good soil allowed the breeding of sheep, cattle, horses, and hogs, products which were traded in Britain and the West Indies as well as to the other mid-Atlantic colonies. Trade connections with South Carolina, the West Indies, and Africa brought wealth to Newport and environs from the assets of the triangular trade of slaves, sugar, and rum. Miles of irregular shoreline, bitten by bays, harbors, and estuaries, offered sailing and shipping advantages to such a trade, along with ideal conditions for avoiding the customs inspector. By 1774, Newport boasted over 9,000 year-round inhabitants,[2] while its salubrious climate offered summer retreat for wealthy residents of these southern establishments, where the summer heat could be unbearable.[3]

However, in the early history of Rhode Island, Narragansett Bay had not commanded consistent attention from either of the supervisory institutions in London: the Board of Admiralty or the Board of Trade and Plantations. More appealing in their eyes was Halifax in Nova Scotia. With a large, well-protected natural harbor, Halifax lay on a latitude convenient to England and also lay near the vital entrance to the interior of North America, the mouth of the St Lawrence River. It was well situated for the cod-fishing industry, one of the initial lures to North America for Europeans. Thus, it is not surprising that despite its attractive natural features, Narragansett Bay had not been charted or mapped in any great detail until the third quarter of the eighteenth century, when the British undertook a careful survey of their North American holdings.[4] The results of their work were made available to the public in four printed maps, two English and two French, all drawing upon the same manuscript source, which was the work of a European surveyor in the employ of the British: Charles Blaskowitz. We will see how and under what circumstances Blaskowitz made his map and how its information was modified in the subsequent published versions.

[PETER HARRISON, SURVEYOR]

The midcentury impetus for mapping Narragansett Bay came not from the colonizers in London but from the colonists themselves in Newport. North Americans recognized the harbingers of the Seven Years' War in the skirmishes between British and French throughout the Ohio River valley during the early 1750s. The General Assembly of Rhode Island voted funds for the restoration of Fort George, on Goat Island, lying opposite the town of

Newport, protecting the entrance of Newport's harbor from possible French incursions. A committee of concerned citizens was formed, among whom were the ship captain–merchant–architect brothers, Joseph and Peter Harrison. Peter Harrison (1716–75) not only designed new fortifications for Fort George but also prepared "A Plan of the Town and Harbour of Newport on Rhode Island," the manuscript plan for which he sent to the Rhode Island Assembly before the end of 1756.[5] Though drawn to scale, Harrison's map shows little topography; it does show detailed soundings, reflecting his life as a ship's captain. The features that particularly stand out on the map are the fortifications on Goat Island and the proposed lighthouse at the southern end of Cononicut Island, known as the Beaver Tail. Harrison had designed both monuments for the community of Newport. The map was sent to the Board of Trade in London, where Joseph Harrison and the agent for Rhode Island used it in their petition to the board to provide guns and equipment "to render the Fortification compleat." The survey map showed that Newport harbor was so deep the "the largest Ship may anchor within 200 yards of the Shoar."[6] The map did not win any hearts on the board, and monies were not forthcoming for constructing the whole of the putative fort.

The successes enjoyed by the British in the Seven Years' War engaged administrative attention in a way that colonial pressure had failed to do. Serving the British in the war was the Royal American Regiment, or 62nd (later 60th) Regiment of Foot, raised in 1756. Made up largely of non-British soldiers from various Continental European countries, the Royal Americans included military engineers who had been trained in surveying and reconnaissance techniques more sophisticated than their British counterparts. One of their number was a Dutch military engineer named Samuel Holland (1728–1801), who had served in the Dutch artillery during the War of the Austrian Succession. His work during the Seven Years' War in French Canada and in the northern tier of the British colonies made Holland realize the "want of exact surveys of those countries, many parts of which have never been surveyed at all, and others so imperfectly that the charts and maps thereof are not to be depended upon." Holland proposed to the Board of Trade and Plantations, who in turned proposed to the king, that "no time should be lost in obtaining accurate surveys of all your Majesty's North American dominions" in order "that your subjects may avail themselves of the advantages . . . of settlement."[7] Royal assent resulted in the board dividing North American holdings into Northern and Southern Districts for the purposes of survey. The Northern District was under the charge of Samuel Holland, and the Southern District under William De

Brahm (1718–99). Like Holland, the German De Brahm was a foreigner trained as a military surveyor on the Continent, where he had similarly served during the War of the Austrian Succession in the 1740s. De Brahm had come to North America in 1751, accompanied by 156 German immigrants to settle in Ebenezer, under the auspices of Samuel Urlsperger, senior of the Evangelical Ministry of Augsberg and one of the non-British trustees for the colony of Georgia.[8] De Brahm's skills quickly became apparent and by 1754 he was appointed surveyor general of lands in Georgia's colonial administration. The appointments of Holland and De Brahm as surveyors for the Northern and Southern Districts reflect both their previous training in surveying techniques and their familiarity with the terrain of North America in the areas that most interested the London government: Canada, the Ohio Valley, and the southern region of Florida.

Pari passu with the Board of Trade's efforts to codify and visualize the newly acquired territories, the Board of Admiralty sought to chart in greater detail the North American coastline. They entrusted this task to another foreigner, Joseph Frederick Wallet Des Barres (1721–1824). While Des Barres' birthplace has not been creditably established, his early education in Basle, mentored by the mathematical Bernoulli family, is attested.[9] In contrast to Holland and De Brahm, his military training was British, for he attended the Royal Military Academy at Woolwich. But like Samuel Holland, he joined the Royal American Regiment in 1756, arriving in North America by the spring of that year. Also like Holland, his energies during the French and Indian War were concentrated on the fortifications at Halifax and on surveys of the St Lawrence Bay and River. After the war, Des Barres coordinated coastal survey work with Holland's surveys along both banks of the St Lawrence, the islands in the St Lawrence Bay, and the lands lying southwards along the east coast of North America, from Nova Scotia to Long Island.

The survey of the Narragansett Bay area evolved from the efforts of foreign surveyors, trained in careful triangulation and the use of the plane table and the theodolite, operating on behalf of a victorious power in territory far away from their native lands. But neither Holland nor Des Barres was responsible for the running of lines, taking of angles, or measuring of depths of Narragansett Bay. That task fell to Charles Blaskowitz.

[THE SURVEYOR: CHARLES BLASKOWITZ (C. 1743–1823)]

Charles Blaskowitz served Samuel Holland as a deputy surveyor; unlike his commander, his origins remain unclear. Even his birth date is unknown, though a British War Office note of his death in 1823 gives his age as seventy-one, making his year of birth 1751 or 1752. However, the Army Index of 1783 gives his birthplace as Prussia and his age as forty, making his year of birth 1742 or 1743.[10] This latter date accords more closely with the assertion that he entered the Tower Drawing Room in 1753, at age twelve.[11] He was officially salaried as a member of Samuel Holland's North America survey team as of 24 March 1764.[12] But he may well have arrived in North America earlier, for he signed as a draftsman the map of the St Lawrence surveyed under the supervision of General James Murray during 1761.[13] He also signed as draftsman a manuscript copy of a survey of Lake Champlain in May 1765.[14] He rose in salary from 1s per day to 1s 6d per day by December 1767. Over the next eight years, he would continue to climb the ranks from volunteer surveyor to assistant surveyor to deputy surveyor by 24 December 1775. In 1777, during the War of American Independence, Blaskowitz joined the Guides and Pioneers, a provincial Loyalist regiment of which Samuel Holland was one of the founding officers.[15] Blaskowitz served this unit in the rank of captain as a draftsman and surveyor throughout the rest of the war.[16]

In the late summer or autumn of 1764, Blaskowitz seems to have left his work along the St Lawrence and in upstate New York to survey Rhode Island (now known as Aquidneck Island) and the environs of Narragansett Bay. According to an article published in *The Rhode Island Historical Magazine* in 1885, this survey was executed at the request of the Admiralty. The article reproduces an unsigned letter in the form of a report to an unnamed lord; it reported that the Board of Admiralty had ordered a survey "to illustrate Newport's potential as a naval base."[17] The author of the report described two months of surveying work that had produced a "large map of this Island and Bay with accurate soundings as far as it is navigable for Ships of War of the second class. [It] will designate to your Lordship the locality of the different positions for the contemplated works." The author promised that surveys and drafts accompanying the map would display "the positions for docks, ship yards, hospitals, &c and also the points of defense by forts and batteries, against the attack of an enemy—in conformity to your Lordship's directions . . . very explicitly noted in my instructions by desire of the Board of Admiralty." The author noted that "Mr Blaskerwich, though young, is an able surveyor"; he had completed the drafts and map, "drawn from actual surveys; all the roads are laid down, and seats of the principal farmers designated, a list of whose names are annexed, and also a correct plan of the town of Newport."[18]

While I have not yet found a map that precisely matches the description of this date, a map signed by Charles Blaskowitz now in the Hydrographic

Department of the Admiralty archives in Taunton could well have been copied from the putative 1764 map. *A plan of Rhode Island with the Country and Islands adjacent* is a large (1,205 mm × 970 mm), carefully colored manuscript map (plate 6). It is signed under the title as "surveyed by Charles Blaskowitz" and includes an inset with a plan of Newport, accompanied by a long key to buildings, streets, and houses. The map does not encompass the entire Narragansett Bay; only Rhode Island, Cononicut, and Prudence Islands to the west, and the north-south channels on either side of Rhode Island are shown. Though undated, internal evidence suggests it was made or copied after 1770.[19] This map lacks soundings in the channels, but otherwise matches the 1764 report's description: "on the map . . . all the roads are laid down, and seats of the principal farmers designated, a list of whose names are annexed, and also a correct plan of the town of Newport. The roads on the island are bordered with a variety of ornamental trees; nearly every farm has its orchard of engrafted fruit of every description."[20] The Taunton manuscript contains precisely these three features: the roads, many with bordering trees, the dwellings of the principal farmers and a keyed list of the farmers, and the plan of Newport, also with a key to major buildings.

The 1764 report is positive about the potential of Newport and Narragansett Bay as a naval station. The suitability of the harbor for the man-of-war may be found in its good anchorage, sheltered in every direction and devoid of ice in winter, offering room for naval maneuvers. Its natural configurations make it well suited for defense and its central position in North America was ideally located en route to the West Indies: "a whole fleet may go out under sway and sail from three to five leagues on a tack; get the trim of the ships, and exercise the men within the bay, secure from attack."

The report continued with a rapturous description. Rhode Island's climate was "the most salubrious of any part of His Majesty's possessions . . . the Garden of America." The colony boasted a "refined and polished society . . . among them many men of science and erudition [and] a very extensive and well selected public Library, given by an opulent individual." The 9,000-plus inhabitants were "celebrated for their hospitality to strangers and extremely genteel and courtly in their manners—engaged extensively in navigation and commerce" who enjoy "perfect religious freedom. These different societies . . . live together in perfect harmony. The Jews are a highly respectable class, some of the most opulent and respectable merchants are the followers and exact observers of the laws of Moses. I attribute their high standing to their perfect freedom, religious and political."[21]

This is not the Newport that was feeling the effects of tightening economic constraints imposed by the British government in the guise of the

Revenue Act of 1764 (also known as the Sugar Act, much despised by colonists) or the Stamp Act of 1765. Mob actions in Newport in the summer of 1764 encouraged the Rhode Island General Assembly to establish a committee to demand the British Parliament's repeal of the Sugar Act. Several of the principal farmers listed on the Blaskowitz manuscript were among the outraged merchants who protested the swingeing taxes on molasses as "schemes to effect our Ruin."[22] Groups of angry Newporters who "engaged extensively in navigation and commerce" broke into British customs warehouses and retrieved cargo of smuggled molasses that had been seized by the customs officials for lack of duty paid. The threat of impressment by the British navy further inflamed the "refined and polished society" to fire upon HMS *St John* in the summer of 1764 and burn a boat from HMS *Maidstone* the following summer.[23] The "genteel and courtly" Newporters did not always show respect to the wealthy summer visitors from the southern colonies and West Indies. "To be a gentleman was sufficient to expose the bearer of that name to mockery and rudeness."[24] The local gentry celebrated the repeal of the Stamp Act in 1766 with an eight-by-fourteen-foot painting showing "the Advantages which Liberty gives to Commerce" including a "Prospect of the Harbour of Newport."[25]

Such acts of violence and mob rule continued in Newport and surrounding territory in the early 1770s in defiance of the Tea Act; British ships were captured, some of which were set on fire in the bay. In 1774, these acts of rebellion culminated in the dismantling of forty-four cannons from Fort George, located on the island opposite Newport, and their transport to Providence on the mainland to prevent British forces from using them.[26]

[THE FURTHER SURVEY: 1774]

The 1764 report and Blaskowitz's initial survey of Newport and Rhode Island did not result in a British naval station for Newport. Blaskowitz would return to the area in 1774 as Samuel Holland's survey of the Atlantic coast progressed from Nova Scotia towards New York. In May 1774, the *Massachusetts Gazette and the Boston Weekly Newsletter* announced that the "exact Survey of our Sea Coast from the Bay of Fundy to this Port . . . is at length happily accomplished under the direction of Capt. Holland, and . . . a further survey, under the same Direction from Boston to Plymouth by Mr Wheeler and from Plymouth round the Cape to Rhode Island Government by Mr Blaskowitz, is intended which when done will finish the whole coast of this Province. Mr Wheeler and Mr Blaskowitz came to Boston from Portsmouth last week and are immediately to proceed on said survey."[27]

Blaskowitz was in Newport by October.[28] "This day the Kings Surveyors began to take the plan of the Town of Newport," noted the Rev. Ezra Stiles in his diary on 25 October 1774.[29] Stiles makes no further mention of the surveyors' work, but from parallel resources concerning Holland's surveys in the northern tier of the colonies and Canada, we can reconstruct the makeup of the survey team and the costs accrued by the survey.

[COSTS OF SURVEY]

A team of eleven men supported Blaskowitz and Wheeler. These would include two chainmen, a porter, two "color men" to hold flags (or colors) for triangulation sightings, a naval midshipman to help with soundings, four oarsman, and a guide.[30] By using the estimated expenses submitted by Samuel Holland in 1764 for budgeting allocations for the survey of the Northern District, it is possible to calculate, albeit roughly, the expenses of the survey of Narragansett Bay.[31] The 1764 survey had taken two months. While the 1774 survey may not have taken as long, it is reasonable to surmise that a further two months may well have been required to check measurements, redo soundings, and add information to earlier maps.[32] Estimated expenses for the two months follow, with the daily expenses (one-sixtieth of the total expense for every item but the last one) in parentheses:

£21 for Blaskowitz (7s)

£21 for Wheeler (7s)

£7 10s for the surveyors' team: two chain men, a porter, and two color men (6d per man, 2s 6d total)

£6 for four oarsmen (6d each, 2s total)

£63 for a midshipman (£1 1s)

£21 for a guide (7s)

c. £2 10s to £5 for a draftsman (5s per day for 10–20 days)[33]

These putative costs add up to about £145 or £150 for sixty days' work. This amount does not include instruments or the cost of drafting work required after the survey or any other miscellaneous but important expenses, such as lodging and fuel. Such additional outlay could significantly increase survey costs.[34] We can compare this to the 100 guineas granted to Holland by the New Hampshire Assembly under the leadership of Governor Wentworth for a survey of that province. The New Hampshire moneys would pay "the sole charge of transporting his partys their provision Cloathing & necessary Instruments without any charge or payment for his or their time Provisions & Labour."[35] Thus our estimate of £150 for the survey could be conservative.

Two versions of this survey may be seen in manuscript maps based on the survey of 1764 and the follow-up survey of 1774. That of Thomas Wheeler, deputy surveyor on the Holland survey team, parallels the Blaskowitz map in the depiction of roads and farms, the outline of the islands, and the topography of the land, but adds soundings to all three north-south channels that form Narragansett Bay (plate 7). A manuscript map attributed to Blaskowitz in the Faden collection of the Library of Congress[36] depicts the entire area of Narragansett Bay, following the same geographic outline and topographical features as the Taunton survey but extending coverage north to Providence in the west and the Taunton River in the east (plate 8). This manuscript adds the placement of batteries erected by the American rebels in the period from 1774 to 1775, information found on the later printed version published by Faden.

From Rhode Island, Blaskowitz moved on to New York. His commander, Samuel Holland, left for England in November 1775, but had already sent copies of the manuscript survey of Narragansett Bay to London.[37] He would share topographic information with Des Barres, who was already in London compiling maps and charts of the Atlantic coast for use by the British military. For war between the colonists and the British was imminent. By the summer of 1775, Boston Harbor was closed and the town, though still controlled by the British, was under siege by colonial militia until March 1776, when the British were forced to evacuate to Nova Scotia. In July 1776, the thirteen colonies declared their independence from the British crown, making much of North America hostile territory.

In December 1776, a contingent of 7,000 British and Hessian soldiers under the command of Maj. Gen. Sir Henry Clinton arrived in Newport from Nova Scotia, seeking a winter supply base for the army and an ice-free port for the twelve warships and fifty-one transport ships that would patrol the harbor.[38] Once his troops were securely installed on the island, Clinton returned to England in January 1777, leaving Rhode Island under the command of Hugh, Earl Percy. The engineers associated with Clinton's forces in

Newport, such as Edward Fage, clearly had access to the Blaskowitz map, as copies of sections of it formed the basis of their own cartographic and engineering work.[39]

[FROM MANUSCRIPT TO PLATE]

The works that Holland and his surveyors carried out for the Board of Trade and Plantations were not intended for immediate publication and distribution, though it is through the published versions that their names and work are known to us. Some of Holland's surveys performed during the Seven Years' War and from the early 1760s were published by a small group of London engraver-publishers, chief among them Thomas Jefferys, engraver and geographer to the king, though some were also engraved and published by the map seller Andrew Dury. Jefferys's plates were acquired after his death in 1771 by Robert Sayer, print and map seller, and William Faden, engraver.

The survey work done for the Board of Trade and Plantations from 1764 to 1774 was undertaken for internal departmental use in planning and laying out tracts of newly acquired land for future settlement. The board had not expressed any interest in a publication program. By contrast, the Atlantic coastal surveys of J. F. W. Des Barres were designed to be printed and distributed first to the navy itself and then to a larger market. It was presumed that sales from the charts would reimburse the expenses of engraving and printing. Des Barres returned to London in 1774 to supervise the engraving and publication of his charts. The large expenses he incurred in financing this process become the subject of a protracted disagreement between him and the Admiralty.[40] The Admiralty was slow to reimburse Des Barres for the survey work and the publication costs of what ultimately became known as the *Atlantic Neptune*. The study of Des Barres' expense reports reveals that the costs of engraving and publishing the *Neptune* outweighed the surveying costs by a ratio of about eleven to one.[41] Yet since his publication of coastal surveys coincided with the growing tensions and ultimate hostilities between British and colonists in North America, a pressing demand emerged from the British navy for his charts or, indeed, any charts. To meet this need, Des Barres printed the charts seriatim, in various states of completion. The printed version of Narragansett Bay found in the Clinton's papers, for example, lacks a title and the soundings, rendering it of little use at sea, though the topography and bearings on land reflect the survey of Blaskowitz and Wheeler. The Admiralty did little to hasten the process of publication, allowing Des Barres to make all the arrangements by himself for printing the charts, which meant he had to rely entirely on the availability of labor and materials in the print trade.

[FROM NORTH AMERICA TO LONDON]

It was common practice in eighteenth-century London for military engineers to have their maps engraved and printed privately. None of the government agencies that supported the survey work necessary for maps maintained engraving and printing facilities of their own.[42] Not the Board of Admiralty, under whose aegis the charts of the North American coasts were surveyed, nor the Board of Trade and Plantations, who ordered the surveys of the interior, nor the army, whose engineers regularly drew plans and maps of strategic areas and forts. When officers or ministers required maps for tactical, strategic, or intelligence purposes, maps were copied by hand, a practice much cheaper than printing, as we have seen. However, the engineer or surveyor might enter into a private arrangement with an engraver like Jefferys in London who would be willing to take the financial risk of engraving and printing a map for sale. In this way, the surveyor had an opportunity to display his survey skills, glorify his unit's prowess, and dedicate his work publicly to those who could help to advance his career. During the preparation of the manuscript of the survey of the St Lawrence River in 1762, Holland and the army engineer John Montresor competed for authorial credit, knowing how important the signature on a map could be for their careers.[43]

Both Holland and Montresor saw their work published in London. In a letter to the lieutenant governor of Canada, Samuel Holland described how charts of the Gulf of St Lawrence and its islands reached London:

> Another chart of the [St Lawrence] River, including Chaleur and Gaspé Bays, mostly taken from plans in Admiral Durell's possession, was compiled and drawn under your father's inspection [Capt. John Simcoe] and sent by him for immediate publication to Mr Thos. Jeffreys [*sic*], predecessor to Mr. Faden. These charts were of much use, as some copies came out prior to our sailing from Halifax for Quebec in 1759.[44]

Similarly, Montresor brought his own work from North America back to London in 1766 and had it engraved by Mary Ann Rocque, widow of Jean Rocque. Montresor's journal records his visit to the engraver's shop and "assisting them in the Executions of the several Draughts I have given them to

Engrave for me vizt. one of Nova Scotia, one of the province of New York, one of Canada from the first Island to Montreal and one of the City of New York and Environs with the Boston Harbour and Channel from the Hook."[45]

[THE PRINTED MAPS]

The arrival of British troops in Newport in December 1775 no doubt encouraged the publication of the two printed versions of the Blaskowitz survey of Narragansett Harbor a little more than a year after its completion. J. F. W. Des Barres, already hard at work completing the engraving of the coastal surveys for the Admiralty, published *A chart of the harbor of Rhode Island and Narragansett Bay* in July 1776. One year later, in July 1777, William Faden, successor to Thomas Jefferys, published *A Topographical Chart of the Bay of Narraganset.* Both maps, large and strikingly similar—not surprisingly, given their common source—were ostensibly aimed at a similar market. Their soundings and titles as "chart" appealed not only to the nautical market of warships and commercial merchant vessels, but also to a general public following the current events in North America. At the same time, neither map purported to describe actual events in Rhode Island nor to lay out the strategy of British warfare. Instead, they presented two quite different pictures of the region by including or highlighting markedly different features. The Faden chart further ingrained the printed image of Rhode Island by being copied by the French navy's chart-making department, the Dépôt de la Marine. The dépôt included the chart in its wartime publication, the *Neptune Americo-septentrional,* which was prepared and issued for the French fleet as it entered the American war and for sale to a larger public. While copying Faden chart, the dépôt work further modified the Blaskowitz survey to present yet a third image of revolutionary Rhode Island.

J. F. W. DES BARRES AND THE CHART OF THE HARBOUR OF RHODE ISLAND

Physical Description

The chart is printed from an engraved copperplate on two large sheets of paper (each c. 572 mm × 820 mm) joined to form a rectangle. The plate marks joined measure 1,065 × 742 mm. The scale in the lower right corner measures one statute mile equal to 30 mm, denoting a scale of c. 1:54,000. The state of the chart illustrated here is uncolored, typical of many charts in the *Atlantic Neptune.* Longitude (71° to 71°26′ west of Greenwich) and latitude (41°23′ to 41°51′ north) are printed between the neat line and the border of the chart.

Title: Claim to Authority

A Chart of the Harbour of Rhode Island and Narraganset Bay Surveyed in pursuance of Directions from the Lords of Trade to His Majesty's Surveyor General for the Northern District of North America. Published at the request of the Right Honorable Lord Viscount Howe, by J. F. W. Des Barres Esq. 20th July 1776.

The title of the chart establishes several things beyond the simple declaration that it is a chart of a particular area. Des Barres names himself as the author of the chart. He alludes to the work of the surveyor Blaskowitz only by asserting that the chart has been "surveyed in pursuance of Directions from the Lords of Trade to His Majesty's Surveyor-General for the Northern District," a direct reference to Samuel Holland, under whose direction Blaskowitz performed his survey. The "Lords of Trade" were the members of the Board of Trade and Plantations, the administrative unit responsible for the Holland survey. "Published at the request of the Right Honorable Lord Viscount Howe" refers to Howe's position within the navy. Admiral Richard Lord Howe was the commander-in-chief of the British navy's forces in North America; as will be shown below, he was thinking about the importance of Narragansett Bay even while contemplating his departure with the fleet from Boston. Thus, Des Barres draws attention to two authorities whose imprimatur reinforces the authority of his map: it is made from a survey (ipso facto, made from direct observation) under the aegis of not one but two civil authorities.

The date in the title has some historical significance, for in July 1776 the British colonies declared their independence. For copyright purposes, Des Barres had added an earlier imprint notice in the lower right-hand corner of the map: "Published according to the act May 3, 1776 by JFW Des Barres, Esq." This suggests that the material for this map had been in his possession for some time, perhaps from the early days of 1775, after Blaskowitz had completed his survey of Newport and Rhode Island. Upon his return to London in 1774, Des Barres was authorized by the Admiralty to publish the charts of North America. He was also allowed to use surveys and other materials found in the offices of the Board of Trade. Holland himself reported providing survey materials to Des Barres in the course of 1773 and his intention to share further surveys with him.[46] Such access may have permitted Des Barres to see the papers and reports of 1764 with its earlier survey per-

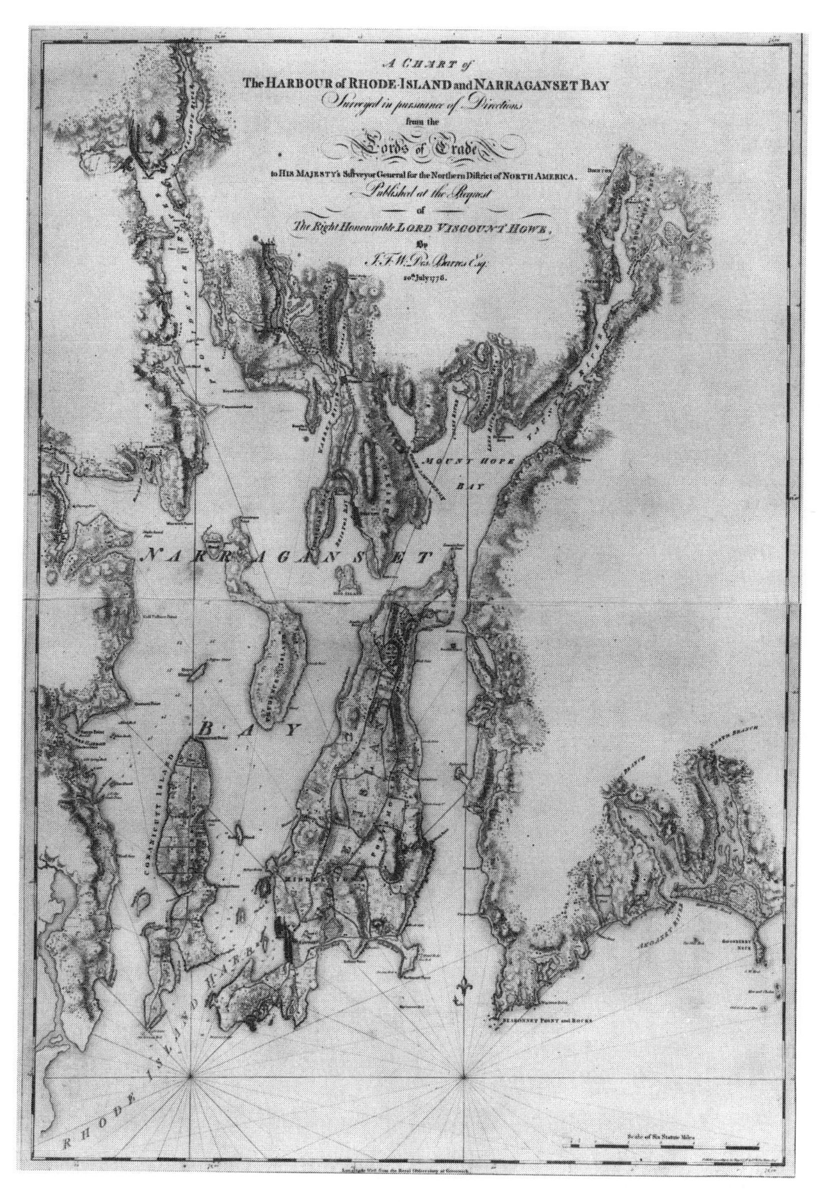

FIGURE 31. J. F. W. Des Barres, *A Chart of the Harbour of Rhode Island and Narraganset Bay* (London, 1776). Printed on two large sheets of paper, the large map measures 1,065 mm × 742 mm when joined together. Scale c. 1:54,000. WLCL, Map 3-J-15. (William L. Clements Library, University of Michigan.)

formed by Blaskowitz. In addition, Holland's return to England in late 1775 would perhaps have brought another copy of Blaskowitz and Wheeler's work to London.

Des Barres and the Blaskowitz Survey

The size of Des Barres' chart is similar both to the manuscript attributed to Blaskowitz in the Faden collection of the Library of Congress and to the Wheeler manuscript in the Clements Library (drawn at scales of c. 1:63,000 and c. 1:50,000 respectively, i.e., about one inch to one mile). Des Barres adds bearings of longitude west of Greenwich and latitude along the border of the chart, providing a reference grid for the chart; these don't appear in the Blaskowitz manuscript. The compass rose shows both true north and magnetic north, with a magnetic variation of 6° west, matching the compass rose on the Blaskowitz manuscript from Taunton and the written notation on the Library of Congress manuscript.[47] The outline of the islands in the bay and the coastline, the pattern of the roads (including ornamental trees), the topography of the hills and the grid of the field systems match the Blaskowitz and Wheeler manuscripts. Des Barres highlights the physical description of the islands and mainland shore in the distinctive engraving style associated with the *Atlantic Neptune,* using a recognizable pattern of etching and engraving, as well as the more innovative techniques of the roulette and drypoint or needle engraving.[48] By emphasizing the relief of the shoreline, he reiterates the chart's usefulness to sailors: the shoreline is what he sees; he has no need of interior information.

The soundings marked in the Des Barres map, though they parallel many of the figures on the Library of Congress Blaskowitz manuscript, are much more numerous and more closely resemble those found on Peter Harrison's chart of 1755. As early as March 1776, when the British left Boston, Admiral Lord Howe was already thinking about Newport and Narragansett Bay. He specifically requested "an accurate chart of Rhode Island with all soundings . . . by Mr Harrison" be furnished to Major Holland, who was preparing to return to North America and serve with the British forces in New York.[49] Des Barres, too, could have used the Harrison map in the offices of the Board of Trade, where it remains in the Colonial Office holdings in the Public Record Office.[50]

The Des Barres chart differs markedly from the Blaskowitz manuscripts in that no use is made of the details of property ownership on Aquidneck Island, though the network of roads and the symbols for dwellings are left intact. Nor does the Des Barres chart contain any information about the batteries erected by the Americans, as noted on the Blaskowitz manuscript in the Faden collection. Thus, there is no sign of anyone living on Rhode Island or that it had been the scene of rebellion.

Aesthetic Details and Costs of Publication

As mentioned above, Des Barres supervised the engraving of the charts of the *Atlantic Neptune* with a close eye on the rendering of landforms. His engravers employed a variety of techniques to approximate topographic relief. This emphasis is unusual for a chart, which traditionally emphasises water. Des Barres adds no decorative motif to his chart but incorporates elaborate calligraphy in the title with considerable flourish in all its letters, a style much in vogue at this period.

The Admiralty allowed Des Barres 35 guineas per plate for engraving.[51] As the chart of Narragansett required two plates, the engraving would have cost 70 guineas (£73 10s), a sum which may or may not have included the price of the copperplate itself.[52] It is likely that his costs went well beyond mere engraving; the cost of paper and printing may be estimated from other costs of English map publishing (see appendix 2). Des Barres planned to sell the maps at 1s per plate; the Narragansett Bay chart would be priced at 2s.[53] To recoup the engraving costs alone, he would have had to sell 735 maps. Part of this quota would be filled by sales to officers serving in the navy. For example, Admiral Howe himself received 36 sets of charts of New England (41 plates), which included Rhode Island, and 36 charts of Nova Scotia (72 plates) for which Des Barres received £264 5s. This sum works out to nearly 47s per plate, demonstrating the high cost of binding the sets of charts into volumes.[54]

Distribution

The maps of the *Atlantic Neptune* were often printed and distributed before they were finished in order to meet the pressing demands of warfare. Consequently, there are numerous states for each of the charts in the volumes of the *Neptune,* which themselves vary in composition and content. For the chart of Narragansett Bay, four different states may be discerned. Two early states in the collection of the Clements Library belonged to Thomas Gage and Sir Henry Clinton, both British generals in North America. Both maps are incomplete. The Clinton copy lacks the title and the imprint. Perhaps it reached Clinton before his December 1776 landing in Newport.[55] The Gage copy lacks soundings in the channel to the east of Rhode Island (unnamed on Des Barres chart but called Seakonnet Passage or Saconot River on later maps). A third state adds soundings as well as shoals along the east coast of Aquidneck Island and along the shores of the Warren River along the channel northwards to Providence. A fourth state of the chart shows the movement of the French fleet in its arrival and engagements off Cononicut Island during its brief stay in Newport in August 1778.[56]

Des Barres faced little competition for his chart of Narragansett Bay, as did his other maps in the *Atlantic Neptune.* The print publishers and sellers Robert Sayer and John Bennett had assembled the *North American Pilot for New England* from maps engraved by Thomas Jefferys and augmented by other maps drawn from surveys done on the spot, but it contained no Rhode Island map. Though the Surveyor General of the Southern District, William De Brahm, also published a book of coast charts and directions, his *Atlantic Pilot* only contained three small scale maps, of little use for coastal sailing in the North American waters. Des Barres' sharpest competitor was William Faden.

WILLIAM FADEN'S *TOPOGRAPHICAL CHART*

Almost twelve months to the day after the appearance of the Des Barres chart, the young map engraver and publisher William Faden (1749 – 1836) issued his own printed version of Charles Blaskowitz's work:

> *A Topographical Chart of the Bay of Narraganset in the Province of New England. with all the Isles contained therein, among which Rhode Island and Connonicut have been particularly Surveyed. Shewing the true position & bearings of the Banks, Shoals, Rocks &c. as likewise the Soundings: To which have been added the several Works & Batteries raised by the Americans. Taken by Order of the Principal Farmers on Rhode Island. By Charles Blaskowitz. Engraved & Printed for Wm. Faden, Charing Cross, as the Act directs July 22d. 1777* (figure 32).

Physical Description and Orientation

Faden's map is printed on one sheet of paper (965 mm × 655 mm) from a plate measuring c. 955 mm × 650 mm; its border measures 930 mm × 633 mm. It is therefore smaller than Des Barres' chart. The scale measures one statute mile to 32 mm, or c. 1:50,000. It has no longitude or latitude markings, though the bearings of the colony (between 41° and 42° N and 71° and 72°) are given in the text printed on the map. The compass does not show magnetic north. The chart illustrated here is uncolored, though many versions of the map are often colored with a green wash for the landforms, blue for the water.

Title: Claim to Authority

Faden calls his printed work "a topographical chart," a nomenclature that moves this map into hybrid territory, half topographical map, half nautical

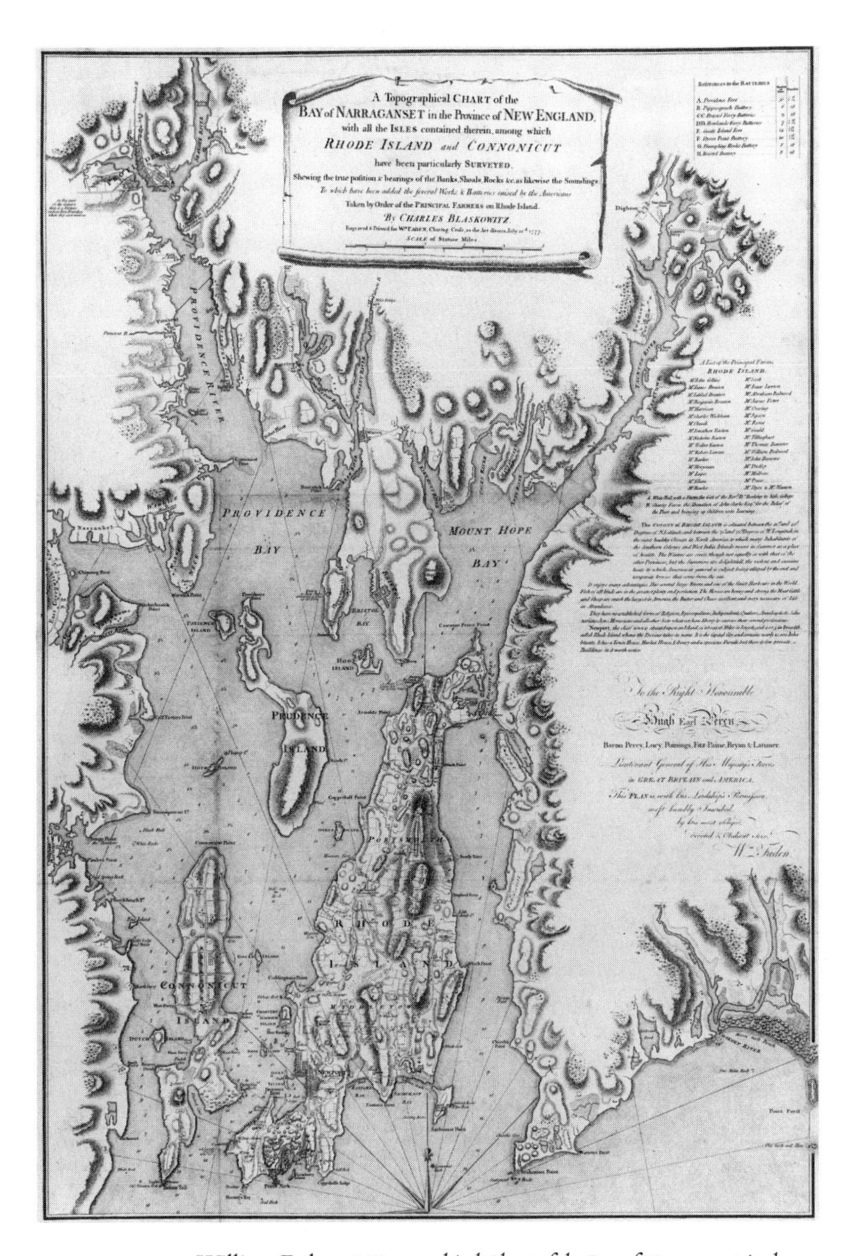

FIGURE 32. William Faden, *A Topographical Chart of the Bay of Narraganset in the Province of New England* (London, 1777). Faden's map was printed on one large sheet, measuring 930 × 633 mm. Like Des Barres' chart, it was based on the survey of Charles Blaskowitz, at a scale of c.1:50,000. WLCL, Map 3-J-11. (William L. Clements Library, University of Michigan.)

chart. Landforms and water have been given equal attention, the maritime areas having been engraved throughout with undulating lines instead of dotted with soundings only.

In his title, Faden stresses the accuracy of the chart by claiming that the bay and the islands therein "have been *particularly* surveyed, shewing the *true* positions and bearings of the Banks," etc., "as *likewise* the Soundings" (my emphasis). This established the chart's origins in observed fact. Faden's authorities, however, were not driven by the governmental departments noted in Des Barres' title; rather, he declares that the survey was "taken by order of the Principal Farmers on Rhode Island." Such an authoritative group—landowners living on Rhode Island—would have closer access to information than a government agency in faraway London. Authority is reinforced by the inclusion of Charles Blaskowitz's name: we are told who made the survey and at whose behest. Faden draws attention to the conflict between the British and the American colonists by adding "several Works & Batteries raised by the Americans."

Dedication

The dedication of the map also points to the colonial troubles:

> *To the Right Honourable Hugh Earl Percy Baron Percy, Lucy, Poinings, Fitz-Paine, Bryan & Latimer, Lieutenant General of His majesty's Forces in Great Britain and America. This Plan is, with his Lordship's Permission, / most humbly Inscribed, by his most obliged devoted & obedient Servt. Wm. Faden.*

Faden gives Percy all his personal titles as well as his military title as "Lieutenant General of His majesty's Forces in Great Britain and America," emphasizing the British presence in the colonies. Yet except for the distribution of 127 eighteen- and twenty-four-pounders around and about the bay, little on the map itself suggests war. Even more curious, at the time Faden published his chart in July 1777, Percy had resigned from his post in North America and returned home to England.

Faden and the Blaskowitz Survey

There is no doubt that Faden used the survey material by Blaskowitz and Wheeler for his topographical chart. The size and scale of his chart match theirs, as do all the other elements: the rendition of landforms and placenames, the location of the batteries raised by the Americans, the "Principal Farms" and their locations on Aquidneck Island, and the soundings. All are derived from the Blaskowitz manuscript in Taunton, the manuscript in the Faden collection of the Library of Congress, and the Thomas Wheeler manuscript in the Clements Library.

Faden places the letter key for the "Batteries raised by the Americans" in the upper right-hand corner of the chart. Immediately below the key, filling space along the right-hand side of the chart, are other noncartographic features. First, a list of the principal farms of Rhode Island, the names written in the same order as on the Blaskowitz manuscript in Taunton. Beneath this lies another letter key to two charitable farms endowed by Rhode Islanders, one a gift to Yale College and the other for the relief of the poor. Finally, there follows a text describing the colony of Rhode Island in terms similar to those of the 1764 report to the Admiralty, discussed above. Faden writes on the map:

> The Colony of Rhode Island is situated between 41st and 42d Degrees of N. Latitude, and between the 71st and 72d Degrees of W. Longitude, in the most healthy Climate in north America, to which many Inhabitants of the Southern Colonies and West India Islands resort in Summer as a place of health. The Winters are severe, though not equally so with that of the other provinces, but the Summers are delightfull, the violent and excessive heats to which America in general is subject, being allayed by the cool and temperate breezes that come from the sea.[57]
>
> It enjoys many advantages, Has several large Rivers, and one of the finest Harbours in the World. Fish of all kinds are in the greatest plenty and perfection. The Horses are boney and strong, the Meat Cattle and Sheep are much the largest in America, the Butter and cheese excellent, and every necessary of Life in Abundance.
>
> They have no established form of Religion, Episcopalians, Independents, Quakers, Annabaptists, Sabatarians, Jews, Moravians and all other Sects whatever, have liberty to exercise their several professions.
>
> Newport, the chief town is situated upon an Island, of about 16 Miles in length, and 4 or 5 in breadth, called Rhode Island, whence the Province takes its name. It is the Capital City, and contains nearly 10,000 Inhabitants. It has a Town House, market House, Library and a spacious Parade, but there is few private Buildings in it worth notice.

Faden is not describing here a colony in revolt against Mother England but rather an idyllic corner of the New World: a place of religious freedom and toleration, enriched by a balmy climate, a civilized population, and a fecund environment. There is no hint of trouble in paradise suggested by the "Batteries raised by the Americans," a group of people who do not seem to live in Narragansett Bay.

Another disjunctive note is struck by the contrast of Faden's title, which claims that the survey of Narragansett Bay was "Taken by order of the Principal Farmers," and the list of "Farms." Faden's phrase "taken by order of" suggests a uniformity of purpose in the construction of the map. The thirty-three farms listed on the map are the same thirty-three included on the Blaskowitz manuscript in Taunton and mentioned as a group in the 1764 report.[58] But there is no evidence that these farmers ordered the survey of Narragansett Bay or had anything to do with it.[59] In fact, if this particular group of thirty-three had anything in common, it was only their wealth, derived from the staples of Newport's trade in slaves, molasses, rum, and whales.[60] Their prosperity is evidenced by the fact that twenty-one of the thirty-three were assessed £2 or more in the 1772 tax list. The wealthiest, Aaron Lopez, who paid £37 11s in tax, represented Newport's rich trade. Joining him among the top taxpayers were the Wanton brothers, George Rome, Samuel Dyre, John Tillinghast, Simon Pease, and John Collins, all of whom paid more than £10 in ratable tax. Twenty of the principal farmers held slaves in their households, some in large numbers, like John Collins, who owned thirteen.[61]

Although all of these farmers were wealthy, their politics divided them. Five of the thirty-three were signers of the Rhode Island proclamation of 4 May 1776, which declared support for the incipient revolution: Barker, Collins, Nicolas Easton, John Jepson, and Tillinghast. Ten regarded themselves as sufficiently loyal to the Crown that after the war they claimed for damages from the British government.[62] Their political allegiance was denoted by the Rev. Stiles in his diary with a system of stars, from one to four, designating their degree of Toryism. The four-star Tories among the principal farmers were Samuel Dyre, Isaac Lawton, and James Honyman.[63] Other farmers, like Joseph Wanton, governor of Rhode Island in 1774, and Metcalf Bowler, chief justice in 1776 and 1777, vacillated. Bowler espoused the American cause, yet acted as agent and spy for General Henry Clinton.[64]

Would Faden have known the diversity of this group? His map does not distinguish their loyalties, nor does it even show which farms the British were using as military headquarters. The landowners on Faden's map appear as worthy denizens of Newport. The map records their history of charitable works: John Clarke, "father of Rhode Island," established the Charity Farms for "Relief of the Poor and Bringing up Children unto Learning." The philosopher Bishop George Berkeley who had resided in Rhode Island for a brief period in the 1720s, turned over his Whitehall Farm to Yale College. Abraham Redwood, one of the principal farmers, was the founder of the oldest private library in America. Yet some of these same persons were among that "set of lawless piratical people . . . whose whole business is that of smuggling and defrauding the king of his duties."[65] Many of the merchants listed on the map were importers of sugar and molasses, both legally from the British West Indies and illegally from the French islands, and exporters of rum, running it past the custom house officers as often as stopping for them. Among the smugglers were the Malbone brothers, Aaron Lopez, and Wanton, whose family boasted a governor of Rhode Island who was known to aid and abet the rum runners on occasion.[66] "This noble smuggling colony" is undetectable in the neat array of roads and homesteads on the Faden map.[67]

Aesthetic Details

Whether or not Faden had access to Des Barres' chart for comparison or copying, many aesthetic details of Faden's chart parallel elements on the Des Barres chart. The only decorative element on the Des Barres chart is its title, engraved in elaborate copperplate script. The title of the Faden map is enclosed in a trompe l'oeil frame that replicates a piece of paper hanging from tacks on the blank north end of the map, conveniently located between the east and west arms of the bay; the imaginary paper curls up gently at the bottom. The lettering is executed in mix of roman capitals and italics, as if it were a notice printed from typeface tacked to the wall.

Just as Des Barres encouraged his engravers to expend extra energy and skill on various techniques to portray landforms, so too did the Faden workshop. While Faden uses fewer techniques for depicting landforms, the effect is no less dramatic. Wetlands, fields, trees, and elevations are all given distinct symbols, creating a varied landscape. Faden has taken particular care and expense in rendering the water, with multiple parallel lines that effectively contrast water with the land. The undulating lines create a rhythm of whorls around islands and along coasts, giving the water its own shape and energy. Though the line work does not indicate depth, the soundings provide the variability of the sea. Unlike Des Barres' chart, which devotes all its engraving energy to the depiction of landforms, leaving the water only randomly speckled by soundings, Faden's chart allows the sea to dominate the space.

Finally, Faden has matched Des Barres in fine calligraphy in the dedication of the chart, which is performed with an elaborate flourish directly below the descriptive text regarding Rhode Island. The dedication to Hugh,

Earl Percy, is engraved in a particularly elegant fashion, unusual for Faden, who usually eschewed such florid script on his maps. The layout of the dedication is strikingly similar to the arrangement of the title of the Des Barres map: the centering of the script, the contrast of roman capitals and fine italic, the elaborate rendering of "Board of Trade" matching a similar rendering of "Hugh, earl Percy." The use of this ornate style at parallel points within the texts of these two charts prompts the question whether Faden, or his dedicatee, Lord Percy, intended Faden's chart to contrast sharply with Des Barres' chart, made at the order of Admiral Lord Howe.

Costs of Publication and Distribution

The Faden map differs from Des Barres' in price. At 5s, as noted in his catalogue of 1778,[68] Faden's map cost 2½ times as much as the Des Barres' map, which sold for 2s. This higher price might be accounted for in part by the extra large paper (965 mm × 655 mm) on which it is printed, more expensive and difficult to obtain than Des Barres' two sheets (790 mm × 570 mm).[69] Yet the higher price is also attributable to the detailed engraving employed by Faden, especially of the water. Unlike Des Barres, Faden was not limited to a budget determined by a government agency; he could decide for himself how much to invest and risk in the production of the map. Nevertheless, we do not how much of the engraving Faden did himself or if he was helped by apprentices (for he had finished his own apprenticeship only in 1771).[70] We know nothing of his costs of printing or paper. We do know that the chart was listed in his catalogue issued for 1778 as one of fifteen maps of America (not included among twenty military plans). Of these fifteen, seven were prepared/engraved by Faden and published in 1776 or 1777 (Faden lists himself as "author" in his catalogue, which can mean compiler or publisher). Of the fifteen, the eight-sheet Mitchell map of North America at one guinea is the most expensive; Narragansett Bay is next at 5s. Six maps are listed at 4s, two at 3s, one at 2s 6d, one at 2s, and one at 1s 6d. These high prices suggest that these maps are not meant for use on ship's deck or in the battlefield, but are aimed at armchair warriors and news-hungry bystanders. With large maps painting an attractive picture of North America, Faden may also have had in mind future emigrés to the New World should the British be victorious in the conflict.

Access to Information

We know Des Barres had access to materials in the Board of Trade and Plantations, under whose auspices Blaskowitz had performed his survey, but we do not know how Faden acquired the Blaskowitz material. He was some years away from being appointed geographer to the king (1783). His business connections with Thomas Jefferys may have introduced him to Samuel Holland and thus to Charles Blaskowitz. Certainly he derived information from the published Des Barres chart, itself clearly taken from the Blaskowitz survey, but a comparison of details suggests that Faden and Des Barres used different manuscript versions. For example, the north point of Rhode Island is labeled "Common Sense Point" on Des Barres, "Common Fence Point" on Faden. It is "Common Sense Point" on both the Blaskowitz and Wheeler manuscripts. The southwest tip of the island is "Brenton's Ledge" on the Blaskowitz Taunton manuscript and the Des Barres chart, but "Brenton's Key" on the Blaskowitz-Faden collection manuscript and on Faden's chart. Des Barres' chart lacks any mention of the American batteries. Faden's information, taken from the Blaskowitz manuscript in the Faden collection, also includes a British battery at the north end of Rhode Island just south of Bristol Ferry, information not on the Blaskowitz manuscript that would have come to Faden after the arrival of British forces in December 1776.

The other maps he engraved of battles and places in North America provide a clue to his encounter with Blaskowitz's work. At the time the chart was published, in July 1777, the map's dedicatee, Hugh, Earl Percy, had resigned his position in North America and returned to London. He was accompanied by his private secretary, the engineer Claude Joseph Sauthier (1736–1802), an architect and surveyor from Strasbourg, who had worked for Governor William Tryon of North Carolina and later of New York. Sauthier had worked with Percy mapping military operations in New York prior to the British seizure of Rhode Island; Faden also published several of his New York maps.[71] Sauthier was with Percy on Rhode Island in the winter of 1777.[72] He may have brought some of the surveys of Narragansett Bay with him, including those copied from Charles Blaskowitz's original work, such as the plan of the city of Newport. Could the manuscript map of the bay and the islands therein[73] that is attributed only to Blaskowitz actually be in the hand of Sauthier?

Percy's return to England was surrounded by much gossip. He had been left in command of the troops on Rhode Island by General Sir Henry Clinton, who had himself returned to England on leave in January 1776. Percy's tenure during the winter months of 1777 were marred by continuing disputes with General Sir William Howe, who was in charge of British land forces in New York. Howe was the brother of Admiral Lord Howe, admiral of the British fleet in North America, who had ordered the Des Barres chart. General Howe had sharply criticized Percy for not supplying wood and fodder with due speed from Rhode Island to the army in New York. He broadened his complaints against the junior officer to include failures for which Percy had little or no responsibility.[74] By the spring of 1777, Percy had had enough and resigned, leaving Rhode Island in high temper early in May 1777. By the time he reached England on 2 June 1777, the London press was already "castigating General Howe for driving away an able subordinate."[75] When Faden's map appeared in July 1777, even though he no longer had anything to do with the army on Rhode Island, Percy's name in the dedication, in the most elaborately engraved lettering on the entire map, could only have drawn attention to his quondam presence on Rhode Island. His name rendered in elaborate script parallels the equally florid presentation of Admiral Lord Howe's name on Des Barres' map of Rhode Island. By 1778, responding to mounting disagreement about the conduct of the war, the Howes had resigned their commands: Sir William Howe in February and Admiral Lord Howe in September. In choosing to dedicate his map to Percy, Faden was riding the wave of Percy's popularity, helping him sell a large, expensive map of a colony hitherto unknown to the London market. The connection between Percy and Faden may have been more than cartographic. Faden's address on "Charing Cross at the corner of St Martin's Lane" was directly across the Strand from Northumberland House, the London home of the Percys, suggesting a further possibility for Percy as the source of manuscript material for Faden.[76]

The image of Rhode Island displayed in Faden's *Topographical Chart of the Bay of Narraganset* evokes a mode of cartography that was enjoying increased interest, investment, and popularity in England: the large-scale county map based on survey. Later in the century, Faden himself would be publishing such maps, even winning a gold medal for his effort.[77] His predecessor in business, Thomas Jefferys, had spearheaded the movement towards large-scale surveys using trigonometric techniques to map the counties of England.[78] In keeping with a tradition that extended two hundred years earlier to the county maps of Christopher Saxton in the sixteenth century and John Speed in the seventeenth, such maps located and described not only the geographical and topographical features of a county but also private property and landowners, their large estates and more modest homesteads. County maps often included information about the county's history or unique features. Faden's list of "principal farms" and the repetition of their names on the landscape itself, next to their homes, exudes a familiar aroma of well-being, stability, and the noblesse oblige of charity and good works. Missing from view is the fact that these farmers were a mixture of patriots and loyalists and that by the time the map was published many of their homesteads had been destroyed by the British occupation.

THE COPY BY THE DÉPÔT DES CARTES PLANS ET JOURNAUX DE LA MARINE

As we have seen in numerous instances, and in the case of Narragansett Bay, even when the government was willing to fund a survey, the cash flow stopped with the manuscript. The engraving, printing, and distribution were left to the private sector, and it was confidently assumed that sales would pay for the engraving and printing costs. As we have further seen, it was very difficult to generate sales sufficient to cover the costs of publication—this required selling hundreds of maps. And even though wartime increased demand, the requisite demand could only be found through the export market. One of the chief beneficiaries of the English export of maps was the French Navy.

French Interest in Narragansett Bay

Narragansett Bay not only was a significant port for the British in North America, it was also the site of the France's entry into the American Revolution. The Treaty of Amity was signed in Paris in February 1778, and formally announced in March, marking France's official support for the American rebels. Sixteen French warships commanded by Charles-Henri comte d'Estaing arrived in Newport Harbor in late July 1778. Having established contact with the American army, d'Estaing agreed to the American plan of a two-pronged attack on the British defenses on Rhode Island. The Americans would descend from the northeast while the French would attack from the west, bombarding the fortifications at Newport from the water. The initial foray of French vessels up the Sakonnet Passage sent sufficient fear into the British that they ran their ships aground and set them on fire in order to block further French entry into the channel. The attack on Rhode Island was further stymied by the arrival of more British ships under Admiral Lord Howe from New York. D'Estaing's attempt to engage Howe's fleet in battle off the coast of Rhode Island was interrupted by a late summer hurricane that blew for forty-eight hours, scattering both fleets, completely demasting the French flagship, preventing a true engagement between the two fleets. After the standoff, Howe withdrew to New York and d'Estaing removed to Boston. Meanwhile, miscommunication between French and Americans had caused the northern attack to falter. The British remained in Rhode Island until late in 1779, while the primary military action of the War of Independence moved south, from New York to Virginia and the Carolinas.

A Proposal for the Neptune Americo-septentrional

The French navy had no reconnaissance of American coastal waters to rely upon for their first forays into North American regions. Instead, they benefited from the work of the British surveyors by buying and copying maps published in England. Records in the Dépôt de la Marine show that standard issue for the French ships in this period were the *Neptune François,* the *Hydrographie Française,* and the *Coasting Pilot* of Greenville Collins, despite the last of these being nearly sixty years out of date. Ship captains heading to American waters were also given the *North American Pilot,* published in London by Robert Sayer and John Bennett in 1775. By June 1778, the chief entrepreneurial officer in the Dépôt de la Marine, Buache de la Neuville, proposed putting together an atlas of charts based on English works that would be entitled the *Neptune Americo-septentrional.*[79]

In his proposal, he pointed out that the dépôt could take advantage of an ongoing English survey of the coastline of North America by procuring all the small- and large-scale maps published in England in the past fifteen years. From these, the dépôt would prepare new maps, profiting from recent English surveys and adding information from the dépôt's own archives. The atlas would consist of fourteen maps: two general maps of the whole coast and twelve larger-scale maps of ports and harbors, "whose plans have been surveyed at different times by order and at the expense of the English government."[80]

He may have been responding to a market demand encouraged by the mapmaker and publisher Georges-Louis Le Rouge. Le Rouge had recently published the *Atlas Américain septentrional* (Paris, 1778) and the *Pilot Américain septentrional* (Paris, 1778), both of which consisted of copies of English maps. Not only was Le Rouge copying maps published in London, such as the map of North America by Dr. John Mitchell, he also had contacts with Americans resident in Paris, such as Benjamin Franklin, for whom his command of the English language must have been an asset.[81] Le Rouge had published his own version of Narragansett Bay in 1778: *Port de Rhode Island et Narraganset Baye publié à la Requête du Vicomte Howe. Par le Chevalier des Barres Londres 1776. Traduit de l'Anglais et augmenté d'après celui de Blaskowitz Publié à Londres en 1777. A Paris. Chez Le Rouge Ingr. Géographe, Rue des grands Augustins 1778* (figure 33). On this large two-sheet map Le Rouge incorporated information from both the Des Barres and Faden charts, with an inset of the plan of Newport, taken from that

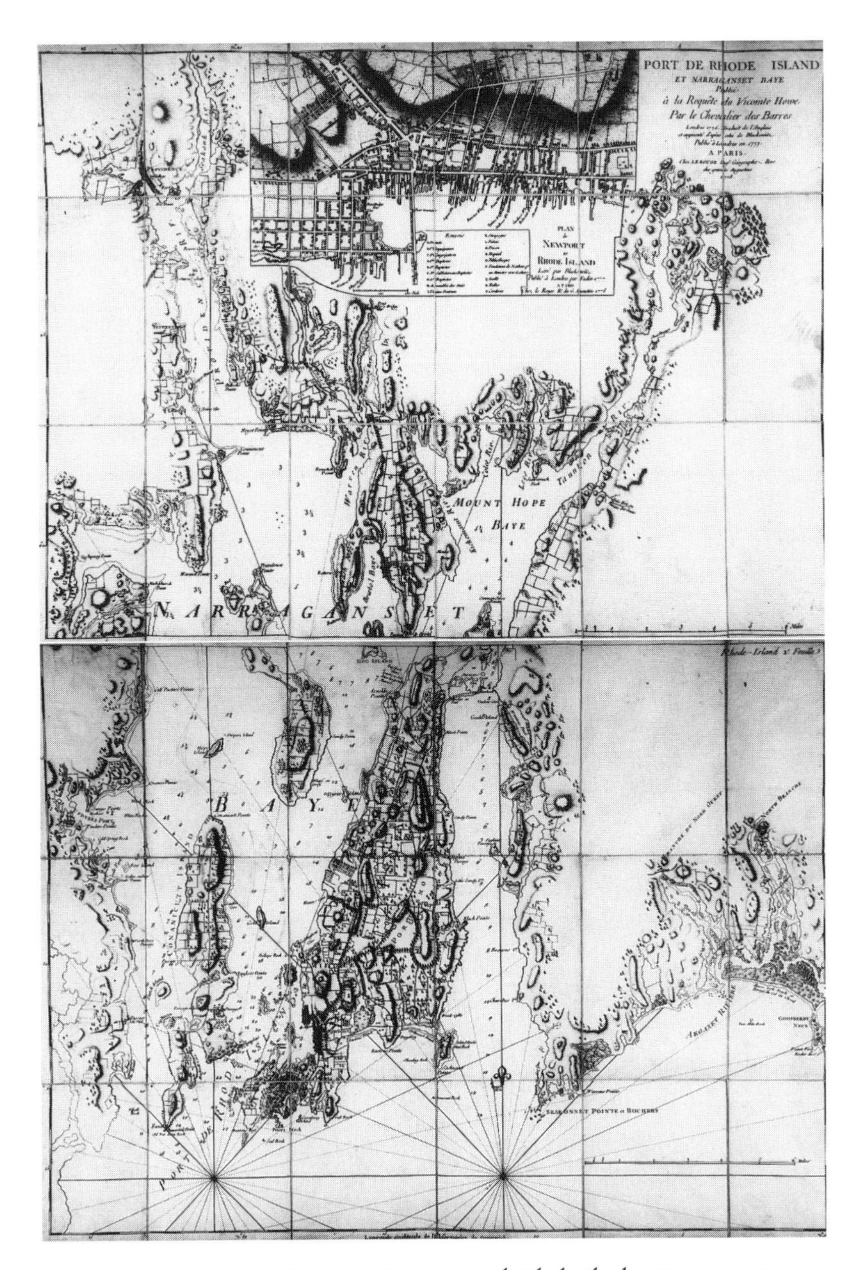

FIGURE 33. Georges-Louis Le Rouge, *Port de Rhode Island et Narraganset Baye* (Paris, 1778). The map copies both the Des Barres chart and the Faden map, as is clear from the title. Each sheet measures 530 mm by 765 mm. Joined together, they form a map roughly the same size as the Des Barres chart. Ayer 133 D44 1778 Box. (Newberry Library, Chicago.)

published separately by Faden in 1777, based on Blaskowitz's survey. Le Rouge included the map in the *Pilote Américain Septentrional.* We can guess at the price of the map by comparing it to similar two-sheet maps advertised in his catalogue of 1773: *3 livres en blanc, 6 livres lavés* (colored).[82]

Whatever the stimulus, Buache de la Neuville further noted that the dépôt could not fund such a project out of its annual budget because the new edition of the *Neptune François* was absorbing all its cash. Slow sales caused by the international agitation had reduced proceeds from all the maps sold by the dépôt. Therefore, he requested extraordinary funds: 1,000 livres each for engraving the two general maps and 600 livres for each of the twelve large-scale maps, a total of 9,200 livres. Buache de la Neuville's estimated engraving cost is about the same as Des Barres' budgeted amount, 35 guineas per plate, approximately 735 livres. The king approved his budget on 4 June 1778.

The list of English maps procured by Buache de la Neuville for the dépôt includes titles taken directly from the 1774 sales catalogue of William Faden and Thomas Jefferys. Buache's purchases continued throughout 1778 and 1779. Among the maps Buache bought from London was a "Carte Angloise Baye de Narragansett," for which he paid 6 livres, roughly equivalent to Faden's price of 5s.[83]

THE DÉPÔT DE LA MARINE MAP

Plan de la Baie de Narraganset dans la Nouvelle Angleterre avec toutes les îles qu'elle renferme parmi lesquelles se trouvent Rhode-Island et l'Ile de Connonicut. Levé par Charles Blaskowits et publié à Londres en 1777. Dressé au Dépôt Général des Cartes, Plans et Journaux de la Marine. Pour le Service des Vaisseaux du Roi. Par Ordre de M. de Sartine Conseiller d'Etat Ministre et Secrétaire d'Etat ayant le Département de la Marine 1780 (figure 34).

Physical Description and Orientation

Printed on one sheet of paper (625 mm × 470 mm), the map measures 582 mm × 435 mm at its border (the plate mark measures 606 mm × 410 mm). The scale is 23 mm to a mile or c. 1:70,000. Its scale is thus about two-thirds that of the Faden and Des Barres maps. Like the Des Barres map, the dépôt chart has longitude and latitude bearings printed between the border and the neat line, from 73°12′ to 73°33′ west of Paris, from 41°29′ to 41°52′ north. The engraver has signed the map in the lower left corner, outside the border: "Petit Sculp."

Title: Claim to Authority

The title of the map gives credit to "Charles Blaskowits" as author of the survey on which the map is based and further mentions its publication in

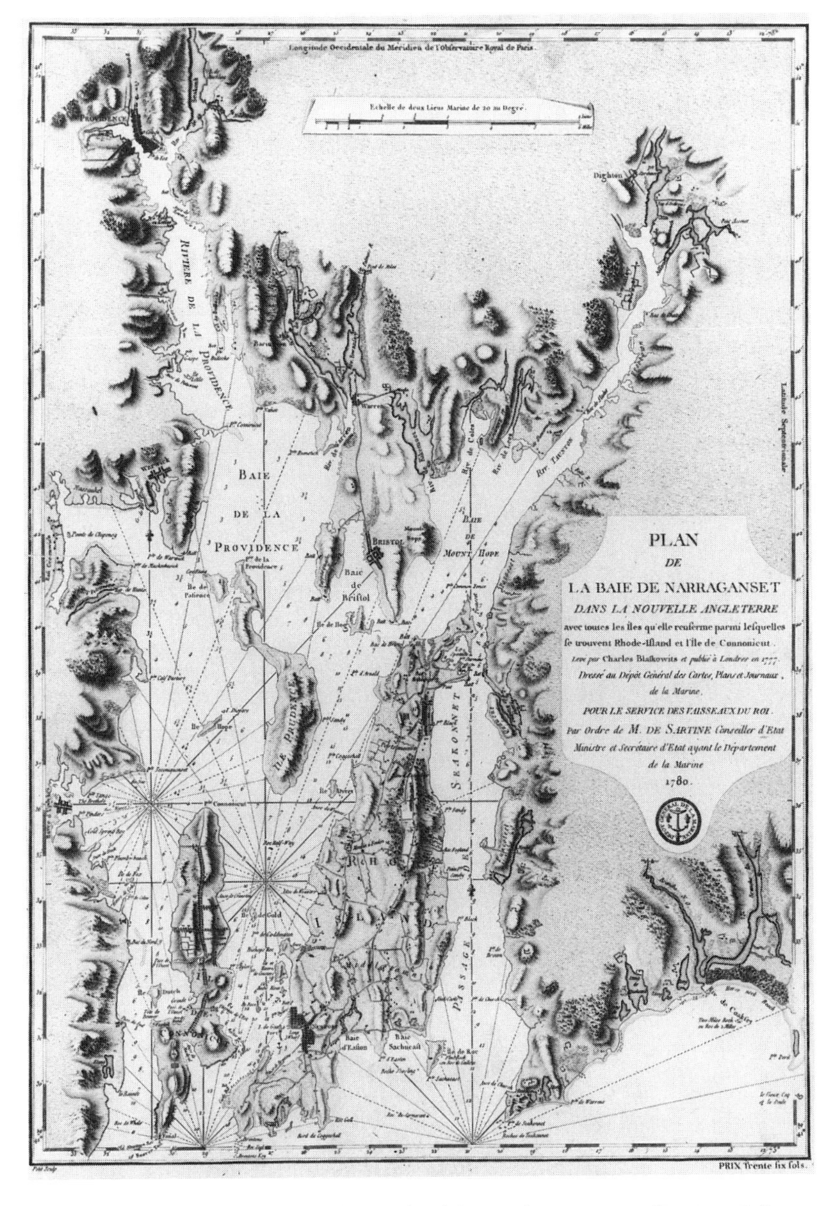

FIGURE 34. Dépôt de la Marine, *Plan de la Baie de Narraganset* (Paris, 1780). Measuring 582 by 435 mm, the French navy's chart is taken from the Faden map, but reduces it by one-third. Its scale is thus smaller at c. 1:70,000. WLCL, Map 4-J-29. (William L. Clements Library, University of Michigan.)

London, though William Faden is not included in the citation of sources. That the map is prepared "by order of M. de Sartine, counselor of state, minister and secretary of the naval department," assures the user that the preparation of the map has been supervised in the same way as maps prepared for all His Majesty's ships. This imprimatur recalls the Des Barres map, which invoked the Admiralty and the order of Admiral Lord Howe.

Dépôt de la Marine and the Blaskowitz Survey

The dépôt's map is French both in its choice of prime meridian (the longitudes having been determined from the Observatoire de Paris) and in its language. However, only the geographical terms have been translated into French; the place-names themselves have been left in English. Thus Hope Island becomes Île Hope (not Île d'Espoir); likewise, Cold Spring Rock becomes Cold Spring Roc, and Black Point, Pointe Black. In this way the map preserves its original qualities of English survey and makes no effort to dominate the landscape with colonial pretensions.

The map preserves soundings and landforms of Faden's map (figure 35). It includes the batteries raised by the Americans, abbreviated "batt.," though they are here reduced to labels on the land itself; no key or legend explains their location or elucidates their various sizes. Though the houses of the "principal farmers" are marked as on the Faden map, the owners' names have been omitted. The roads are taken from the Faden map, including the grand allée of trees on Cononicut Island. The note regarding the iron works near Providence, "where cannons may be cast," has been eliminated; surprisingly, for one might have expected such information to have been of interest to the arriving French force.

Aesthetic Detail

Along with reducing the scale of the Faden publication by about one-third, the dépôt's staff also reversed the representation of topography and water. While Faden had engraved the sea with parallel wavy lines, the French version leaves the sea blank except for the sounding numbers and the radiating compass lines. The land, by contrast, is carefully rendered by a combination of stipple, etching, and burin engraving to show the variety of landforms that make up the Narragansett landscape: hills, fields, marsh, and bog.

This reversal may be attributed to the talents, or lack of talents, of the dépôt's engraver, M. Petit. An assessment of Petit's work written in the 1790s by the Marquis de Chabert, as part of a proposal for the reorganization of the dépôt, reveals that Petit often subcontracted his work from the marine to "unknown workers," which proved costly and inconvenient to the

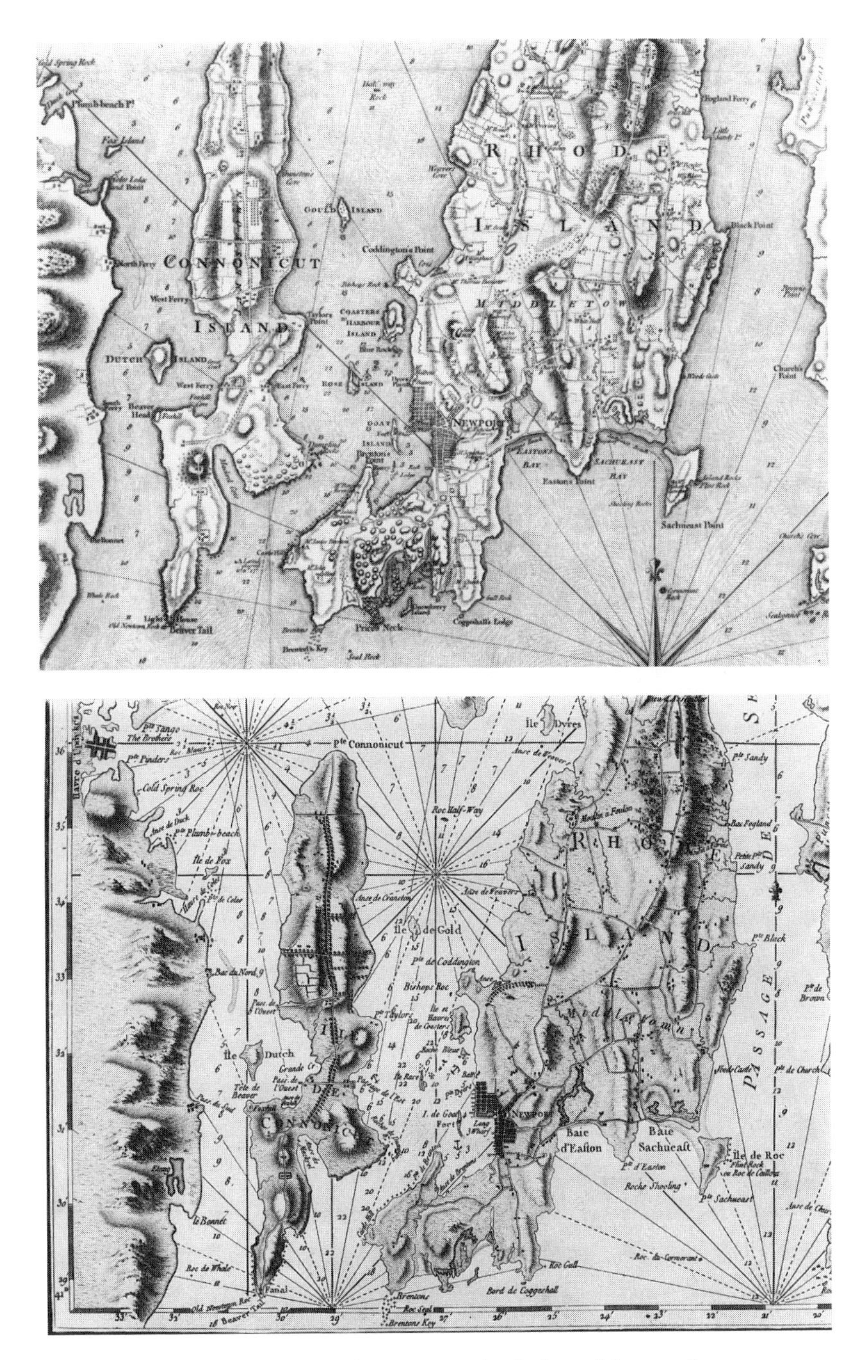

FIGURE 35. Detail from Faden's *Topographical Chart of the Bay of Narraganset* (top) and the Dépôt de la Marine's *Plan de la Baie de Narraganset* (bottom).

dépôt. Chabert thought Petit's work was only really good for engraving scales; that he was "hardly capable" for line work and "nothing" at relief, so much so that from about 1786 his work had been replaced by that of workers who were more "exact . . . docile . . . cheaper."[84] In the case of the Narragansett Bay map, either Petit has hired on some very skilled engravers or was at that moment particularly en gambe.

Publication Costs and Distribution

The *Neptune Americo-septentrional* was not published for the sole use of the French navy; it was also designed for the general public.[85] The individual maps of the *Neptune* were priced at either 3 livres or 36 sous. The Narragansett Bay map sold for 36 sous, which made it less than half the price of Faden's map (c. 2s against Faden's 5s) and roughly the same price as Des Barres' map. Like Des Barres and the Admiralty, the Dépôt de la Marine hoped to recoup its publishing costs through sales both internally and externally. The dépôt stood the best chance of making a profit, as it had neither the expenses of the survey to undertake nor the initial expenses of drafting a copy for the engravers.

If Faden sold 500 of his maps at 5s each, less the expenses of designing, engraving, and printing the map, he stood to make a £75 profit. To cover their engraving expenses of 600 livres the Dépôt de la Marine needed to sell 600–700 copies of the *Plan de la Baie de Narraganset.* The distribution began within the French marine. In March 1780, fifteen packets containing seventeen maps from the *Neptune Americo-septentrional* were sent to le Comte Hector, director of the port of Brest, from which the fleets left for North America.[86] Two collections of the *Neptune Americo-septentrional* printed on gilt-edged paper were sent to the Marquis de Chabert at Brest in January 1781, prior to his departure with Admiral de Grasse for the North American campaign.[87] Their voyage would end in the victorious battle at Yorktown in Chesapeake Bay. The atlas was also being used for strategic planning in Paris. Charles-Pierre Claret de Fleurieu, in charge of the daily management of the Dépôt de la Marine in Chabert's absence, was required to bring proof sheets of the *Neptune* to the king at Versailles, no matter what their state.[88]

The *Neptune* soon crossed the Channel, brought into England by the very man who had supplied the charts to the French navy. Faden ordered three copies of the *Neptune* in 1782; they may have been the first copies brought into England, destined for the homes of Alexander Dalrymple, hydrographer to the East India Company, the secretary of the admiralty, and King George III.[89] Besides selling it to the enemy, the Dépôt de la Marine also provided it to at least one American, sending an early but elegant copy to

Benjamin Franklin, one of the colonial representatives in Paris, and another to his host, Jacques-Donatien Le Ray de Chaumont.[90]

What the Public Saw

As noted above, French ships had experienced Narragansett Harbor well before the publication of the dépôt's *Plan de la Baie*. Their brief stay in August 1778 seems to have brought back no new information to the dépôt. The later, larger expedition of the army to Newport in 1781 under the command of General Rochambeau returned much material to the French archives, material that enriched later map productions. In 1779 and 1780, the dépôt could only rely on English charts, including the maps of the *Atlantic Neptune,* being issued by Des Barres. Requests from Buache de la Neuville for the "neptune atlantique de M Desbarres" are found in letters to William Faden in September 1782, when the dépôt was well into the production of the maps of the *Neptune Americo-septentrional.*[91]

Buache de la Neuville was first and foremost an entrepreneur for the dépôt, a mapmaker second. His role was to supply maps to the navy and to cover the dépôt's expenses while doing it; to accomplish this he had to create maps that would appeal to a greater public. His competition came not from English maps but from those produced by his compatriots, mapmakers like Le Rouge. Buache de la Neuville would count on selling the dépôt's one-sheet map of the Bay of Narragansett as easier to handle, and that it was entirely in French, including its longitude determined from the prime meridian of the Observatoire de Paris. Le Rouge's map was a mixture of French and English and retained Greenwich as the prime meridian. Both Le Rouge's *Port de Rhode Island* and the dépôt's *Plan de la Baie de Narraganset* met the criteria for a good map. They are clearly engraved and legible; the nautical soundings exuded an air of accuracy; they were based on observations made by a man certified as qualified by the London government. However, the dépôt's *Plan* manifests a richer range of engraving techniques for its topography, emulating the Faden map it copies, even without the silken undulations of the water. Finally, the dépôt *Plan* beat the Le Rouge *Port* for price: at 36 sous, it was cheaper than Le Rouge's two-sheet map, which probably sold at 3 livres. The dépôt's price was fixed, stamped clearly outside the border in the lower right corner.

[TOWARDS CONSUMERS AND CRITICS]

The publication of the *Neptune Americo-septentrional* raises the question of secrecy and security of cartographic information during wartime. William

Faden, shortly to become geographer to the king (1783), clearly felt no compunctions about selling maps to the French navy, even though some of the maps he supplied had been published with permission of the lords commissioners of the admiralty. It was only in 1782, nearly at the end of the war, that Buache de la Neuville's agent Famitte wrote to Faden. He was concerned about the difficulty of getting maps out of England but assured Faden that he would continue to have access to maps from the dépôt.[92] Thus did the commercial exchange of maps override national concerns, and the copying of maps fuel continued map publication.

The simple and cost-effective process of copying maps led inevitably to a proliferation of maps on the market. As we have seen, Blaskowitz's survey generated at least four printed maps for sale. (I say "at least" because it's possible that the copies may themselves have been copied.) Printed maps both created and responded to perceived markets. The Blaskowitz survey, as part of the larger Holland survey of North America, fulfilled one need—that of an administration to know the exact location and extent of territory, an important question for eighteenth-century mapmakers. When the raw data of Blaskowitz's survey went to print, it was transformed to serve different needs and purposes, but retained the authenticity of a survey taken on the spot and based on carefully measured observation, two of the basic requirements of a good map.

Des Barres' chart was printed at the behest of the Admiralty, to better serve the Royal Navy. We find the chart also in the possession of Sir Henry Clinton and Thomas Gage, British generals engaged in the civil war in North America. Work produced by their own engineers found its way back to London to engravers like William Faden, who made his chart of Narragansett Bay part of a larger commercial response to the situation in North America. That both these charts are dedicated to key figures in the North American conflict—Admiral Lord Howe and Hugh, Earl Percy—added to the cartographic claims to authority and authenticity. The charts revealed publicly for the first time a detailed picture of a significant, if small, area of North America that would be the scene of both British and French wartime activities. Yet neither of these printed works was considered classified material; the transfer of the information to the mapmakers of the Dépôt de la Marine only paralleled Des Barres' efforts: to serve the French navy and to sell to the larger public in order to finance publishing operations. Le Rouge's copy of Des Barres further emphasizes the commercial attraction of a map or chart of this part of world. All the map publishers here described were convinced that there was a sufficient demand for a map of Rhode Island to warrant the expense of engraving and printing a map to rival foreign or local competition.

Who were the customers for these maps and how did they choose among competing geographic images? The next chapter will explore how mapmakers and publishers perceived their customers and how the consumer made judgments about maps in a competitive market. Both these questions are rooted in the need of the eighteenth-century public for clear definitions of originality and accuracy, error and truth. Such knowledge bolstered consumer confidence in a mapmaker and protected his work from piracy. The very nature of mapmaking, reliant as it was on objective observation and compilation from a wide variety of sources, including maps of others, made determinations of error and truth difficult. Indeed, the use of error as an indicator of plagiarism or copying called into question the reliability of maps. What constituted a good map occupied producers and consumers, both public and private, throughout the eighteenth century.

PART THREE

EVALUATING MAPS

CHAPTER SIX

GIVING PLEASURE TO THE PUBLIC: TELLING GOOD FROM BAD

Le sieur de l'Isle, auteur des Globes & nouvelles Cartes . . . croiroit avoir fait quelque plaisir au public. —*Journal des Sçavans*[1]

The punishment for bad maps is that no one buys them.
—Nicolas Le Clerc[2]

The proliferation of printed maps at affordable prices left the potential consumer with only one question: which one to buy. Advice was plentiful but not always helpful. Advertisements in newspapers such as the *London Gazette* or the *Journal de Paris* announced the publication of maps, with title and price but little puff. Another segment of the periodical press in England—Edward Cave's *Gentleman's Magazine,* Thomas Astley's *London Magazine,* and John Hinton's *Universal Magazine*—published announcements of maps and regularly used maps as illustrations, most engraved by Thomas Jefferys, Emanuel Bowen, and Thomas Kitchin. The periodicals also published, albeit irregularly, critical comment on the merits of specific maps. The publishers saw the possibility of assembling the maps from their periodicals into atlases and competing for the map-buying market.[3] For fuller descriptions and some criticism in France, the potential buyer could turn to the *Journal des Sçavans,* the official publication for the Académie Royale des Sciences. The journal announced and described maps recently published and served as an outlet for geographers to expose in detail the sources and criteria used in their maps. The parallel publication of the Royal Society, the *Philosophical Transactions,* did not focus on maps; only two articles about maps appeared in it during the entire century, both concerning map projections.[4]

We can measure the interest of the scientific societies in cartography by a perusal of the minutes of the meetings. The *Journal Book* of the Royal Society and the *procès-verbaux* of the Académie Royale des Sciences reveal a steady stream of reports of surveys, presentations of maps, requests for approbation for maps, and solicitation for funds and intellectual help with maps.[5]

These discussions were often the result of international efforts. In London the Royal Society received new maps of Russia and the Caspian Sea from travelers who had been there; M. Delagrive sent his new map of Paris in solicitation of membership; the Homann firm sent globes; the French astronomer Godin sent reports from de La Condamine's expedition in South America; another report arrived from New York about measuring a degree of longitude. Similar reports were presented in Paris, and members of one society were often corresponding members of the other. But these meetings were not open to the general public. The map buyer not privy to this level of discussion turned to another source: the map seller.

Roch-Joseph Julien, who ran the first map store in Paris, opened his catalogue of 1763 with these remarks:

> The prodigious number of maps that inundate Europe makes it impossible for amateurs or even the foremost geographers to know which maps merit preference. The best means of acquiring such knowledge is, undoubtedly, to make a business out of it. But since such an enterprise demands considerable and heavy advances of money to begin, geographers are content to sell their own maps and those that they happen to have on hand from one moment to the next.[6]

From his long experience in the map business, Julien offered to help prospective buyers choose the maps that were right for them in order to make up an atlas.

> A good atlas is put together according *the use one wishes to make of it and how much one wishes to spend.* The best way to do this is to use my catalogue to choose the maps you would like and to consult me as to the choice of authors. My advice is much less suspect since none of the plates of the maps in my collection belong to me. Thus it is a matter of indifference to me whether I furnish one map or another. (emphasis added)[7]

Use and price, then, become the common denominators for defining a good map and predicting its success. We have already determined that a printed map should be priced as low as possible in order for a mapmaker to sell as many as possible. The map seller needed to know who used maps and what could they afford.

As with any consumer product that was not vital for daily life, the map was sold as much by persuasion as by necessity. The seduction began with the mapmaker's description of the potential readers, often found in the introduction to atlases or geographical treatises. For example, in 1681 Guillaume Sanson described the people who should pay attention to geography, which I translate here as "maps."

> Sovereigns and politicians cannot govern or sort out their interests with their neighbors without maps; generals and officers of the army need maps to make war successfully; churchmen need maps to govern their church's affairs; magistrates need maps to know the extent of their jurisdictions; financiers need maps to impose and collect taxes; men of commerce need maps to know the best routes; travelers could not succeed in their objectives without maps; no one with the least tincture of belles-lettres or employed in civil life can spend their leisure time in reading history and voyages without the aid of maps; poets, philosophers, and historians cannot neglect maps without falling into inexcusable errors.[8]

These sentiments changed little during the eighteenth century, though they gradually broadened to include, as Didier and Gilles Robert de Vaugondy described the subscription list to the *Atlas universel,* "a great number of studious people, those who read history as well as those who pay attention only to current events."[9] The fact that book and map dealers constituted the majority of subscribers, accounting for nearly half of the volumes sold by subscription, suggests an anticipation of a healthy retail sale of the atlas. They are followed by the aristocracy, nobility, magistrates and lawyers, and church dignitaries.[10]

In England, one encounters the description of an even broader range of map user: from the statesman to the rustic.

> [Geography] is for the soldier, the statesman or politician, for the education of young minds, for young Gentlemen (especially those preparing for the Grand Tour), for the rustic with time for self education, for the merchant or general trader. . . . Perusing these volumes will afford for maintaining useful and polite Conversation, [making one] qualified to confirm, to object, to explain, to correct, or to refute . . . and thereby keep up that spirit which is the Soul of good company. . . . Having good conversation allows one to avoid dangerous and destructive company . . . promotes the Good of Mankind in General . . . so great, so noble, so laudable an end as that of promoting a spirit of Science, which Freedom will ever attend.[11]

Maps then are promoted as having a powerful influence on young minds. The growing use of maps in education, the increasing need for maps to understand a colonial world, and the burgeoning desire for maps as part of the decorative motif in a visual culture quickened the market. To stimulate demand even further, mapmakers and publishers sought ways to make geography easy and maps accessible to a greater public. Books such as *Géographie pratique contenant les instructions suffisantes pour rendre une personne assez habile pour dresser lui-même des Cartes* made it possible to "learn in an easy way how to figure out distances between places, to survey a geographic plan, to measure a route on the sea, to observe longitudes and latitudes."[12] Maps, if not a necessity of life, were certainly one of its desirable extras.

[THE AFFORDABILITY OF MAPS IN FRANCE]

But could everyone from statesman to rustic afford maps? We have generalized the price of a map on a folio sheet to range roughly from 1 to 4 livres in France. Against this measure, we may place the wages of potential map users and the prices of other commodities (appendix 6). Wages of course varied from salaries to daily wages to piece rates, depending on the occupation of the wage earner. We can ask first whether the participants in the map trade itself could afford their own product. The *premier géographe du roi* received 1,200 livres a year, the other *géographes du roi* 600, the same range as the *ingénieurs-géographes,*[13] or army surveyors, who earned between 600 and 1,200 livres per year. For the surveyor or the geographer–map compiler, a complete atlas at 60 or 100 livres was very expensive, from around 5 to 17 percent of his annual pay. The map compilers (*dessinateurs*) in the Dépôt de la Marine were paid between 600 and 800 livres a year during the 1750s and 1760s; for them, as for surveyors, even a single map would constitute a large purchase. For the engraver who earned, as Guillaume Delahaye did in 1787, 4 livres 7 sous a day, a map cost between a quarter and the whole of his daily wage. Similarly, the copperplate printer who earned 3 or 5 livres per day would probably not quickly spend his daily wage on a map.

Turning from the map trade to other potential map users, we learn that a magistrate in Troyes could expect an income of 2,000–3,000 livres per annum,[14] increasing their ability to buy maps. Yet for another class of people, these figures would seem utterly paltry. The nobility and aristocracy could rely on income ranging from 4,000 to 50,000 livres a year. To live in the manner their class demanded—maintaining multiple homes, clothes for court, livery for servants, horses, and carriages, plus the costs of office—required in excess of 10,000 livres a year. A lesser income of between 4,000 and 10,000 livres per year only allowed for giving dinners a few times a month, keeping several servants, and maintaining five or six horses. The income for these classes came from the revenue from their properties, supplemented by pensions from the Crown if they were favored with Court connections. These pensions, including monies for ministers, diplomats, and others members of the government, could run from 8,000 to well above 20,000 livres a year. On the lowest level, a companion to one of Louis XV's daughters might well expect to receive 4,000 livres a year, notwithstanding other revenues she might enjoy. For many offices and titles connected with the Court and Crown represented royal borrowing: the office was purchased with the expectation of a return interest on the purchase price.[15]

The nobility, the aristocracy, the rising classes of the legal profession, and the upper reaches of the military were the book buyers, builders of libraries, acquirers of maps and atlases. They took a keen interest in the education of their children, which was begun early, at home, with a team of private tutors and masters, who taught history, chronology, geography, and mathematics, as well as dance and music. Girls might continue at home or in a convent school, while boys went on to a collège costing 700–1,500 livres a year, usually run by a religious order. From there they could enter the École militaire in Paris or the even more prestigious École des Pages du roi, which required 200 years of unsullied nobility to enter, though such degrees of nobility could be bought. A young noble could also attend an academy at 4,000 livres a year, where he was taught equitation, dance, surveying, languages, history, geography, and mathematics.[16] The rich nobility (as opposed to the poor nobility) who had the interest formed libraries in which works of geography took their small part, between 3 and 6 percent of their collections.[17]

Wages and income themselves need to be set against prices of commodities in eighteenth-century Paris (appendix 6). In 1785, journeymen in the building trades in Paris described their daily spending: 1 sou 6 deniers for a measure of spirits, 5 sous for a breakfast of a half setier of wine and 2 sous of cheese; 6 sous for a midday meal and 9 for the evening meal; 6 sous 3 deniers for two and a half pounds of bread; 7 sous for rent; 6 sous for wear and tear on clothes and tools.[18] Their total was 40 sous 9 deniers per day, or a little over 2 livres, the cost of two maps at the lower end of the price range.

The price of maps should also be compared to that of other printed goods. A book, with no illustrations, cost between 2 and 3 livres.[19] Graphic prints might sell for as little as 10 sous, but more typically between 1 and 12 livres, depending on the size, the complexity, and the fame of the engraver.[20] A map costing between 1 and 4 livres, though out of the ordinary range of the worker, was well within the budget of the leisured, reading classes. The duc de Croÿ, a map and voyage enthusiast, spent nearly 200 livres on books in one year (1767), and over a fourteen-year period (1762–76) his son spent an average of 500 livres a year on books, bindings, maps, and copies. The duc purchased maps not just for his own use but also to give away as gifts when he entertained.[21] The editor of the *Journal des Sçavans,* Jean-Jacques Dortous de Mairan, who did not command the annual income of a duke, spent 300 – 500 livres per year on books over the period 1720 to 1740.[22] For such buyers, works such as the *Atlas universel,* priced at 120 to 150 livres, were luxury items; they would cost nearly half the yearly book budget. But the atlas's subscription list accounts for 1,118 copies and its frequent presence in modern collections tells us it was not rare. Atlases, like globes, were becoming standard furniture in the private French library.

[THE AFFORDABILITY OF MAPS IN ENGLAND]

A similar exercise compares wages and prices in England (appendix 7) to the price of a map, which averaged from 6d to 2s. Turning first to the participants in the map trade itself, we see that a military surveyor might receive £50 per year[23] while a private surveyor, according to the *London Tradesman,* might expect 1 guinea per day.[24] The same manual promises that a journeyman engraver might receive 30s per week, though the best could expect as much as 10s 6d per day or £3 6s a week. The geographer to the king, however, could expect to receive nothing. The wages of an officer in the army, a likely map buyer, ranged from the 10s per day earned by a captain in the Foot on full pay to the £2 1s per day earned by the colonel in the Horseguards.[25] The map at 1s or 2s was within reach, even for the engraver, for whom it might constitute a fifth of his daily wage.

As in France, these wages for do not accurately depict the income of that class of people who would regularly be buying books, prints, and maps. Among the merchant classes, one could expect estates valued at £50,000 to £500,000, so great were the profits to be made in the slave trade, tobacco, sugar, government contracting, and investments in stocks and land.[26] The nobility could command small fortunes from their rent rolls. The earl of Kildare could count on £15,000 per year from his estates, in addition to the £10,000 that his wife, Emily Lennox, daughter of the duke of Richmond, brought as her dowry.[27] The king was not the richest man in the realm, but George III's budget for book and map purchases began at around £1,000 per year, growing to £1,500 per annum after 1770, continuing to increase to £4,500 towards the end of his life, roughly one-fifth of his annual income.[28]

Dr. Johnson had advised the young king's librarian on the development of the royal library. He urged him to include maps, especially those made on the spot, and not to be fooled by scale.

> It will be of great use to collect in every place Maps of the adjacent country, and Plans of towns, buildings, and gardens. By this care you will form a more valuable body of Geography than otherwise can be had. Many countries have been exactly surveyed, but it must not be expected that the exactness of actual mensuration will be preserved, when the Maps are reduced by a contracted scale and incorporated into a general system. The King of Sardinia's Italian dominions are not large, yet the maps made of them in the reign of Victor, fill two Atlantic folios. This part of your design will deserve particular regard, because in this, your success will always be proportionate to your diligence.[29]

The middle class, rapidly rising both socially and politically by dint of their accumulation of personal wealth, joined the gentry and the nobility as builders of private libraries, readers of books, and users of maps. Their instinct for self-education combined with a desire for self-improvement whetted their appetite for intellectual stimulation. Furthermore, they had more disposable income than their counterparts on the Continent, placing maps even more easily within their reach.[30]

The price of maps compared to other commodities confirms that maps were more affordable in England than in France. For foodstuffs, in 1734 a "middling sort of family" expected to live on £232 per year, which allowed 10s per week for fruit and toys. Tea cost 12s a pound and best French brandy, 2s 6d a quart.[31] The journal of the artist Arthur Pond records spending 2s 6d for a bushel of apples and 5s for a pound of chocolate.[32] On the entertainment side of the ledger, novels cost 3s, bound or 2s 6d in their paper wraps.[33] Playing cards cost 1s to 2s per deck, as did the entrance to the Ranelagh pleasure gardens: 1s for daytime, 2s 6d if tea and coffee were included, 5s at night with fireworks.[34]

Decorative prints, probably the map's greatest competitor, were offered at a wide range of prices from crudely executed engravings on thin paper at 3d or 6d to framed mezzotints at 3s 6d. Sets of prints, like multisheet maps, could cost 1 to 2 guineas. A catalogue like that of Robert Sayer and John Bennett, which listed prints and maps together, shows similar price ranges for prints and maps.[35]

Thus, the printed map in England was not a luxury item for the buyers of books and builders of libraries. Even for the man or woman who engraved it, the purchase of a map cost roughly a fifth of a day's wage. It was not quite as cheap as a brandy, but no more expensive than a novel.

Maps were affordable, and they made geographical information accessible. Their commercial success encouraged the imitators, plagiarists, and copyists. Guillaume Sanson noted, "This great facility for knowing something about geography has nurtured the idea in many people of setting themselves up in geography, which seems to them so easy, that by adding or changing something no one will detect their larceny. But their inadequacy causes them to make some strange mistakes."[36] In a word, mapmakers can get it wrong. How was a consumer to know if a map was right, if it was correct, if it was good?

[GOOD MAPS: THE GEOGRAPHER'S VIEW]

Just as they were prepared to define the map consumer, so were mapmakers ready to define the good map. The first thing a good map needed was an explanation. Claude Delisle warned that one should not take maps on faith; they must be accompanied by "instructions and rationales. . . . It takes more than maps to establish a geographic truth."[37] As described earlier in this book, in their lawsuit against Nolin, Delisle and his son Guillaume used the pages of the *Journal des Sçavans,* the publication of the Académie Royale des Sciences, to supply exactly those "instructions and rationales" for their cartographic publications. This geographic mémoire, which accompanied a map, analyzed all the sources used in constructing the map, explaining why some were accepted and others rejected. It was the forum in which the cartographer could justify his choices and show how he had arrived at his conclusions. He could explain his choice of projection, demonstrate the possibilities of a new projection, and establish the means by which he had determined longitude and latitude for the places on the map. The mémoire also allowed him to discuss uncertainties and describe the intellectual process he had used in drawing his conclusions. D'Anville expressed the mapmaker's obligation to be definite and the difficulties such boldness entailed:

> To commit oneself to the labor of preparing maps is a condition even more difficult than writing about geography. It requires you to take a stand, at the risk of being mistaken, in the infinite detail over which it would be easy to glide in writing, or to not even notice the problems because of not understanding them, since you have enough information to write about. Supposing that there is plenty of material, gathered in quantity: What chaos must be sorted out then? What depth of intellect does one have to acquire to be able to judge it all?[38]

The mémoire allowed the geographer to explain his positions and to alert the reader about those areas on the map that were not to be accepted as geographic reality and about which the mapmaker felt uncertain. The letters of Claude Delisle to the *Journal des Sçavans* concerned the geography of North America as it was found on the globes his son had prepared for Chancellor Boucherat and on the world map published by J. B. Nolin, discussed above. The series of letters, with Nolin's responses, form a geographic mémoire whose audience included not only the members of the Académie Royale des Sciences but also the general reader of the *Journal,* who was certainly a potential buyer of maps.

One of the geographic uncertainties of the early eighteenth century was the shape of California. Part of Claude Delisle's charge of plagiarism against Nolin was that Nolin had copied Delisle's California in the shape of a peninsula, a break from its late seventeenth-century insular shape. In order to show why he abandoned California as an island, Delisle reviewed sources from the sixteenth and seventeenth centuries, cartographic as well as textual. He attributed California's first appearance as an island to the Dutch map publisher Jansson, who had based the shape on the evidence of a Spanish map seized by the Dutch. But Delisle questioned the value of the Spanish map. Was it made on "good and faithful memoirs?"[39] Delisle noted that if the island of California were a sure and constant thing among the Spanish, all their maps would uniformly show California as an island, but they did not. Asserting that reports of travelers could be mistranslated, he insisted that *all* accounts must be read in their original languages to understand what they mean. The best sources were the words of those who had been to a place and whose words one could trust. What we know now, he said in May 1700, is that we do not know; there are discoveries still to be made. Delisle described his portrayal of the peninsular coast of California on both the Pacific and the Gulf of California sides as cut up and disconnected, "like tooth stones in an interrupted work."[40] This graphic device to show uncertainty was one of the features copied exactly by M. Nolin. Delisle asserted that Nolin would be unable to explain *why* he had rendered the coastline in this way.

Delisle treated his placement of the mouth of the Mississippi River and his depiction of Japan as an island in the same way, reviewing the printed sources, the published maps, and the words of travelers who had been there.[41] What marks his testimony is his insistence upon agreement: literary sources must agree, astronomical observations must agree, travelers' accounts should agree. Without concurrence, the cartographic equivalent of replication of an event, the duplication of evidence by experiment, a fact was not a fact. Delisle admitted the dangers of being tricked by malice or ignorance or lack of diligence on the part of mapmakers, explorers, and other writers. In the end, it was only "by reasoning, by conjecture, by estimate, and by relation with neighboring lands" that he could draw any conclusions at all.[42] He also maintained that these conclusions always awaited further clarification.

A mapmaker content to live with doubt understood the importance of explaining how his map was made. Delisle's profession of uncertainty was emulated by Jacques-Nicolas Bellin in the many mémoires he wrote in support of the charts he made during his tenure in the Dépôt de la Marine.[43] These brief but dense works were published both in pamphlet form and on folio sheets printed as introductions for the *Hydrographie Française* (1756–73), the Dépôt de la Marine's collection of sea charts of the world. Bellin used the same approach in each of his mémoires: a discussion of the history of discovery of a place, its physical features, a history of the cartography of the place, the process of establishing latitude, bearings, and distances for the place, the problems of determining longitude. He accounted for those areas for which knowledge was uncertain and explained how he had arrived at his cartographic conclusions. Bellin extended this justificatory writing to the work he supplied for others, for example the map of the Great Lakes, which he prepared for le père Charlevoix' *Histoire de la Nouvelle France.*

Bellin's explanations of the construction of his maps forms the introduction to the third volume of Charlevoix' history. Bellin described the errors that he tried to correct on these maps, dwelling in some detail upon those areas where knowledge remained uncertain. He admitted he had not used the maps of the most well-known geographers, such as Sanson, Coronelli, and Delisle, but had instead relied, understandably, on documents in the Dépôt de la Marine. He also used travelers' reports, particularly those of Charlevoix himself. "Everywhere, compass in hand, he made observations about latitude and longitude whenever he could, and the distances from one place to another, in sum, neglecting nothing that could be useful to the understanding of this region."[44]

The Jesuit father's observations gave new shapes to four of the five lakes and added to the number and names of rivers known to flow into the great waters. Yet in spite of Charlevoix' care, Bellin was still not satisfied with his own cartographic work for the south coast of Lake Erie and the north coast of Lake Huron, but claimed it was not possible to do any better.[45] He expressed similar sentiments about his rendering of Lake Superior. "The mémoires in the dépôt have given me the means to represent it a little more faithfully than what one has seen up to now. But I think we still have to wait for more information, for all these regions appear to me to be unequally determined."[46]

Bellin was so honest about his map that he alludes to his concession to political pressure when he explains why there are no depth soundings along the coast on his map of Canada. "One will be surprised to find no soundings on my plans, that is, the number of fathoms or feet of water; I know such details are extremely useful, and it would have been easy for me to fill them in with exactitude. But particular reasons, which have nothing to do with geography, stopped me from doing so. I have put them on the plans of ports which do not belong to France."[47]

Cartographic mémoires of French geographers repeatedly concentrated on three areas: the accumulation of verifiable astronomical observations of longitude and latitude that allowed the cartographer to create a projection and a base map; a clear understanding of distances and how they were measured in order to determine the map's scale; and the naming of places and their correct orthography. The recurring theme in the discussion was the need for agreement, for the resolution of conflicting information. There is a pervasive reluctance to let "old" information go if it could not be replaced with verifiable "new" information. One voyager's word was not enough; one astronomer's measurement was insufficient.

As important as mémoires were for buttressing a map's intellectual value, they were not best-sellers. Didier Robert de Vaugondy, admitting that "these types of literary work find very few readers," promised that his own list of sources for his maps would sacrifice erudition to brevity.[48] Bellin agreed, allowing that such mémoires could be a dry and weary read.[49] Thus, they did not enjoy wide circulation. When William Faden asked for copies of d'Anville's mémoires, he was told that the ninety-five mémoires had been published originally by the Académie des Belles-Lettres in nine or ten of their annual volumes, but not separately by d'Anville himself.[50] However, such justificatory pieces were much to be desired, even if it meant fewer maps. As one critic maintained, "one might then at least hope that [the maps] would be more exact and less often copied; the public would be better able to judge them and render to the author the justice he deserves."[51]

For the French, a good map was an analysed map, one accompanied by a printed justificatory essay that invited the map user to become part of the compilation process. This aspect of map publication is, in general, a much less frequent feature in the English map trade in the first half of the century. Introductory essays or accompanying mémoires either have disappeared or were never published, a situation explained by the fact that many mapmakers were engravers, not geographers. Some mapmakers placed blocks of prose explaining sources directly on the map itself, as on Herman Moll's maps in *The World Described* or John Mitchell's map of the British and French dominions of North America of 1755. For example, Moll explained the use of the Mercator projection on his map of the world and credited Henry Pratt with the improved depiction of roads on the map of Ireland. On the map of Denmark and Sweden, Moll marked the route "King Charles ye XI and his Mathematicians took Anno 1694 in making their observations which are carefully inserted here with many other Remarks left to ye Curious to observe."[52] Similarly, Mitchell added large blocks of text to the second edition of his eight-sheet map of North America to answer criticisms that he had "produced no Vouchers to support his Performance."[53] Perhaps in general response to this sort of criticism and to growing competition in the market for maps, by the last half of the century, surveyors and publishers of English county surveys were publishing introductions to their maps that explained methods of triangulation and the purposes of the survey. By the second half of the century in colonial North America, British administrators like Thomas Pownall and surveyors such as Lewis Evans, William De Brahm, Bernard Romans, and J. F. W. Des Barres published their sources and described their methods of survey, their analyses of data, their principles of projection.

Though abused for being long-winded and little read, the mémoire nonetheless helped the consumer develop a cartographic sensibility and discernment for the characteristics of a good map. While the mémoires focused on a map's content, there were aesthetic considerations as well. The geographer d'Anville presented his strict specifications for size, scale, projections, layout, and orthography. For example, d'Anville advised that every general map should be on at least two sheets, large enough so that the information is not squeezed. A world map should have a diameter of at least twenty inches for each hemisphere. When determining the scale at which a map was to be prepared, one should bear in mind that the paper would shrink after printing, reducing the scale slightly.[54] He warned that a map projection could easily fool the viewer with its distortions ("false ideas and monstrous forms").[55] He counseled making the general map *after* the detailed maps were complete, so that the general emerged from the specific, not the other way around. Maps in a collection should be consistent in their orientation. Mémoires should be published with maps to show the geographer's thinking.

D'Anville had particularly strong views regarding orthography and lettering. For place-names with more than one spelling, he recommended using the version that was as close as possible to the historical origins of a name, not the vulgate. For example, he preferred *Bourdeaux,* not *Bordeaux,* since the additional *u* derived from the Latin *Burdigala; Alemagne* from the Latin *Alemannia,* with one *l,* not two; *Daufiné* with an *f,* not a *ph,* from the ancient *Delfinus.* D'Anville did not limit his sights to French place-names. In the twelve pages he devoted to nomenclature, he used examples from England (e.g., the importance of keeping the place-name ending *chester* though often suppressed in local pronunciation), from Ottoman Greece (removing the prefix *is-* to a place-name, derived from the Greek preposition *eis,* or "to"), and from the Middle East (*Sourie* instead of *Syria,* because the Greek upsilon was pronounced *ou*).[56]

The distribution of place-names drew d'Anville's attention too; he advised the mapmaker to think carefully about their arrangement, making sure they are well centered in the area named. He reminded the would-be cartographer to ensure that river names actually ran along rivers. Yet he warned against being overfastidious in writing all the names horizontally, which becomes monotonous, and even fatiguing to the eye: "It does not give the sense of variety that Nature has placed in her work."[57] (Figure 36 gives an example of d'Anville's concerns.) Along with the disposition of place-names, d'Anville was concerned with the hierarchy of size of place-names on his maps. Similar concern with lettering may be found in the printed "Model of Forms and Heights of Characters to be used in the execution of *La Carte de la Guyenne*" by Pierre de Belleyme, one of Cassini de Thury's engineers. It clearly sets out the scale and style of lettering for the draftsmen and engravers.[58]

While d'Anville's principles of map design were not new, they had rarely been so clearly formulated. In the late seventeenth century the scientist Robert Hooke outlined his design ideas for the maps that would make up Moses Pitt's *English Atlas,* a noble venture and one of many that failed after one volume for lack of capital.[59] Hooke wanted the maps of the atlas to be on similar scales, to avoid using a variety of projections, to eliminate the confusion of place-names, to use different lettering for different features, and to unify the positions in labeling the maps. "I shall endeavor to avoyd all those inconveniences which I have observed in the work of diverse others. Very eminent men have been encumbered with disproportion of the parts, one to the other . . . [with] distortion and making out of shape and confusion by mixing things not at all proper for that place where they are placed."[60] Hooke was reaching for an aesthetic of map layout and style, of conventional signs, wanting a unified look for the maps of the *English Atlas.* Similar desires animated Herman Moll, who set out his goals on the title

FIGURE 36. Pierre Moullart-Sanson's reworking of a map by Nicolas Sanson, *Celto-Galatia, seu Celto Gallia* (c. 1700). It shows an example of the kind of lettering that geographers tried to avoid: upside down, overly curved, difficult to read. Map 48 in *Atlas [de la] géographie ancienne,* an atlas factice of proof states of maps, mostly by the Sansons, Moullart-Sanson, and Gilles Robert de Vaugondy. WLCL, Atlas W-3-D. (William L. Clements Library, University of Michigan.)

page of *The World Described:* "Each map is neatly engraved on Copper and printed on two sheets of elephant paper; so that the scale is large enough to shew the chief cities and towns, as well as provinces, without appearing in the least confus'd."

Didier Robert de Vaugondy described the *Atlas universel* (1757) as "manageable, complete, uniform, and consistent" ("commode, complet, uniforme et suivi"). Manageability entailed that maps and atlases not be so big as to be difficult to handle, yet not so small that they cannot contain all the necessary information. Completeness implied a sufficient number of maps, large- and small-scale, for the study of both modern and ancient history. The maps should be uniform in their orientation and consistent in scale, so that maps of neighboring regions easily overlap, and the user need not change his or her visual scale. An atlas should use the same symbols throughout in order to be easily understood.[61] These sentiments were echoed in the review of the *Atlas universel pour l'étude de la Géographie et de l'Histoire ancienne et moderne,* by Philippe de Prétot, praised as one of the most "commodes, étendus, complets" atlases available at a modest price (48 livres). The reviewer noted especially that the atlas consisted of maps separately engraved, not just one map colored differently for each leaf of the atlas.[62]

Geographers were the sharpest critics of other geographers. Philippe Buache admonished the abbé Jean Delagrive, who prepared a series of maps of the River Seine in 1738. Buache complained that they lacked sequential order, geodetic orientation, and uniform scale; in other words, they could not be fit together to make a cohesive whole.[63] The *ingénieur-géographe* Brion de la Tour used criticism to defend himself against a charge of plagiarism upon the publication of his *Atlas général, civil, ecclésiastique, et militaire* (Paris: Desnos, 1766–86). Accused of copying the *Atlas moderne* published by Jean Lattré in 1762, he retorted that an examination of both atlases would reveal many faults in the Lattré atlas, faults that Brion would not be so ignorant as to commit. The maps in the Lattré atlas were badly colored, too big to be manageable, in the wrong order, crowded, and with lettering so fine a magnifying glass was necessary to read them. The atlas should be retitled *Atlas moins moderne.* "All the faults are concealed under a beautiful veil; it is the elegance of the burin in certain parts . . . that makes the maps pleasing to the eye but a pity to the spirit."[64] Brion particularly singled out Rizzi-Zannoni's maps. His map of Switzerland "is seductive in the way in which the mountains are treated, although they are without proportion, but the excessive fineness of the lettering is off-putting" (figure 37).[65] He saves his harshest words for Rizzi-Zannoni's map of Italy, in which many of the local boundaries are wrongly drawn and colored, showing incorrect political allegiances,

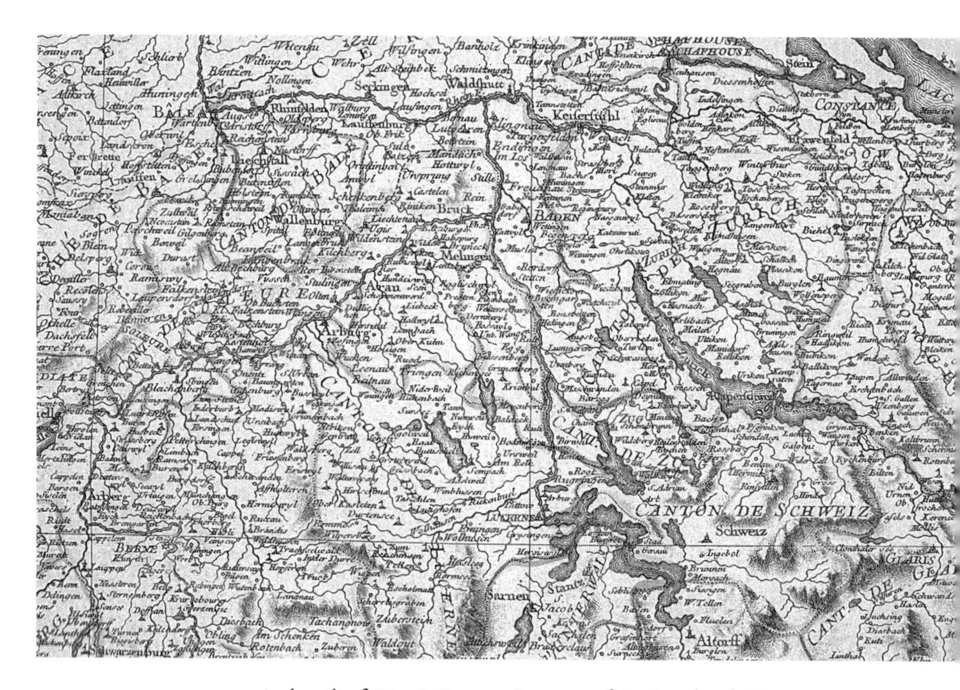

FIGURE 37. A detail of Rizzi-Zannoni's map of Switzerland. From Jean Lattré, *Atlas moderne* (Paris, 1762). Brion de la Tour objected to the fineness of the lettering and the rendering of the mountains. WLCL, Atlas E-3-B, map 9. (William L. Clements Library, University of Michigan.)

an amazing display of this Italian geographer's ignorance of his native country.[66] Perhaps Lattré took note of Brion de la Tour's words, or perhaps the publisher-engraver fell out with Rizzi-Zannoni for some other reason; in later editions of the *Atlas moderne,* Lattré replaced Rizzi-Zannoni's map of Switzerland with one by Rigobert Bonne, the only one of Lattré's various authors whom Brion de la Tour considered a true savant.

[GOOD MAPS: THE CUSTOMER'S VIEW]

Concerns about suitable projections, appropriate scales, historicity of spelling, and uniformity of coverage were not always uppermost in the buyer's mind as he or she contemplated the choice of maps. A consumer sometimes had simpler tastes, unsullied by the technicalities of production. Louis Renard, correspondent of Guillaume Delisle and briefly his business partner in Amsterdam, expressed some of the public's desiderata in maps. That maps should be offered at the lowest price possible had already been

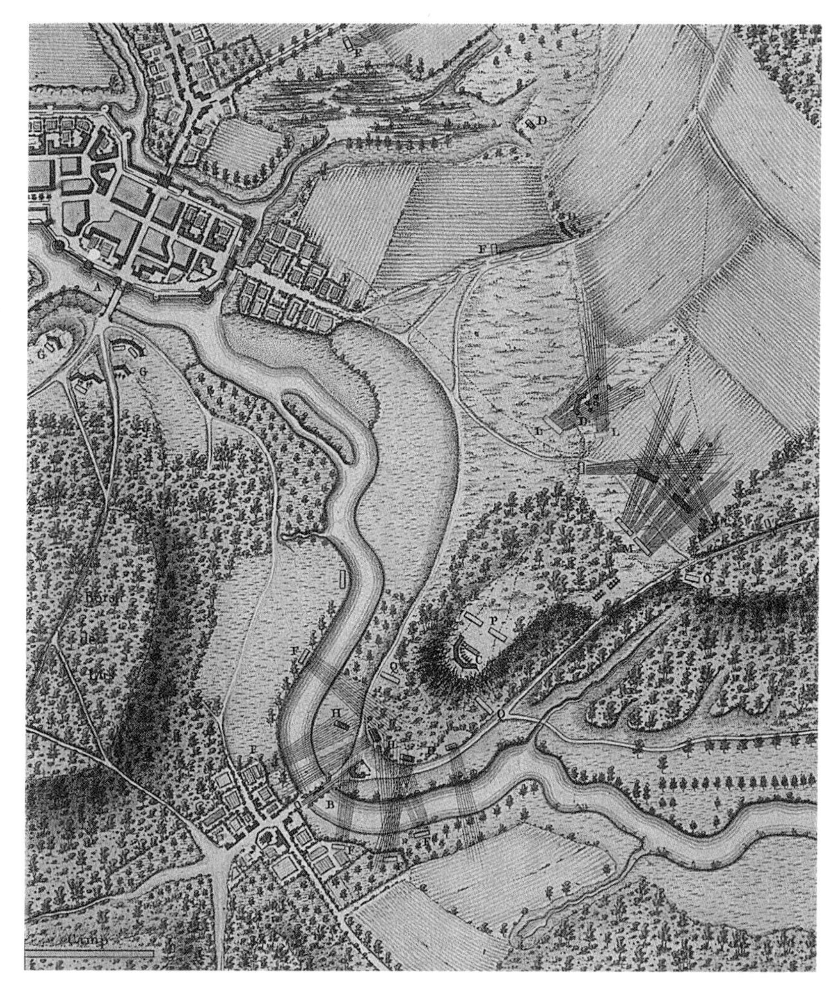

PLATE I. Plate II from Charles Louis François Fossé, *Idées d'un militaire* (Paris, 1783). Engraved by Louis-Marin Bonnet, using a combination of aquatint and stipple engraving and printed in five colors. The plates in this volume represent examples of the author's views on appropriate and nuanced coloring of plans by young military officers. Special Collections UG 443 .F75. (Rare Book Room, Harlan Hatcher Graduate Library, University of Michigan.)

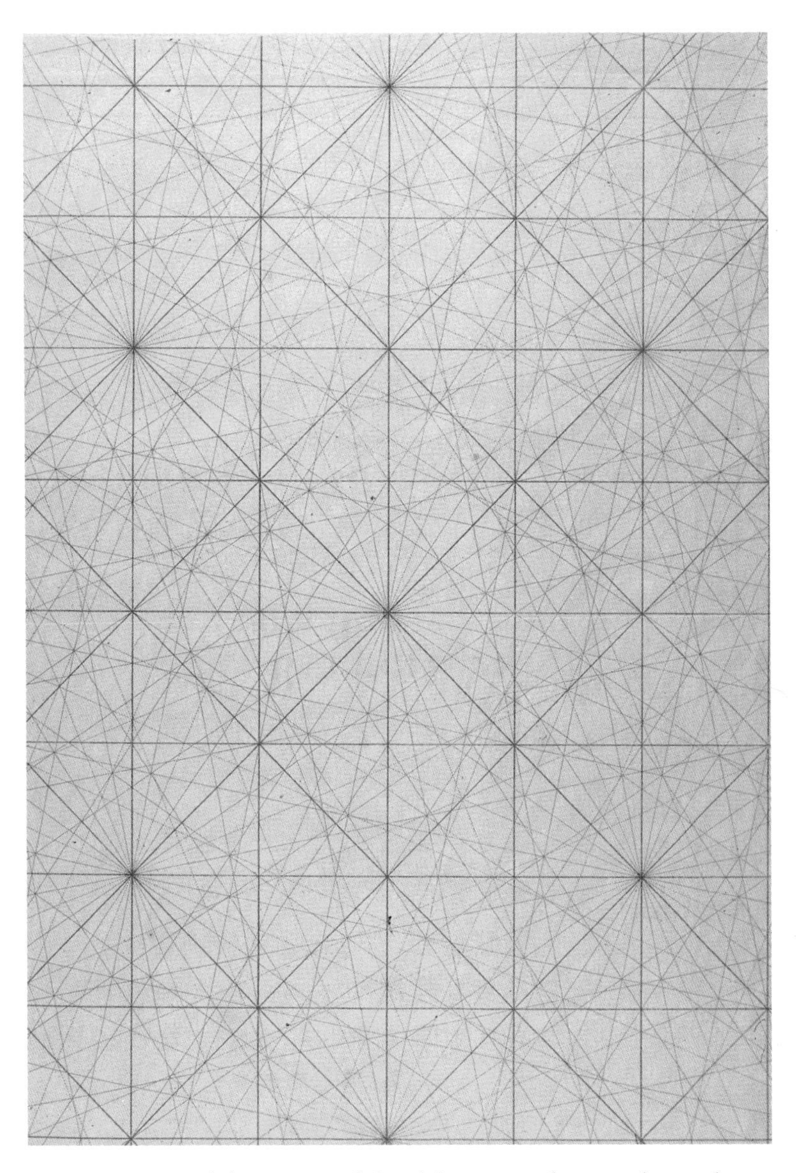

PLATE 2. Dépôt de la Marine, red rhumb lines printed separately on a sheet that would later be overprinted with a chart. The cost of engraving such a plate was assessed at 600 livres, about one-half the cost of engraving a chart. The cost of printing plates of various sizes of rhumb lines with red ink was assessed at 45 livres, over a three-year period. See appendix 1. (Private collection)

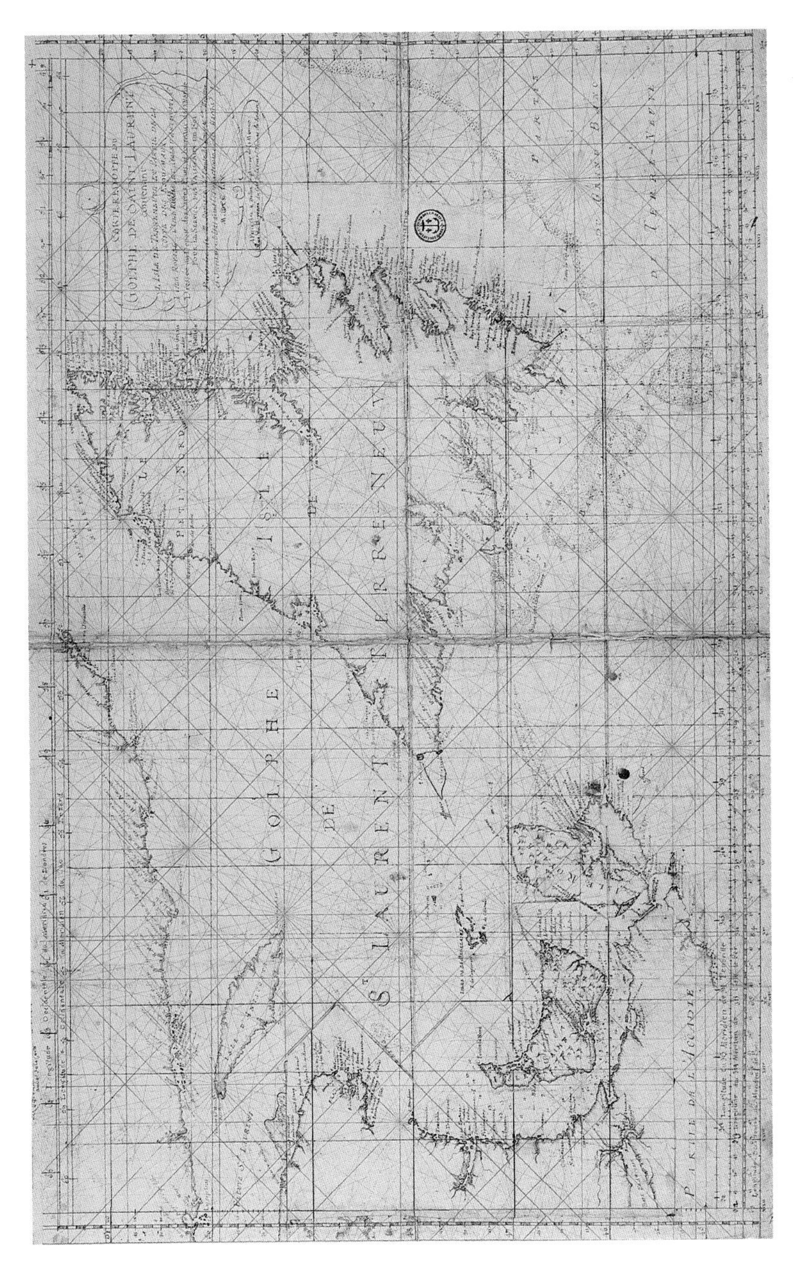

PLATE 3. J.-N. Bellin, manuscript chart of the mouth of the St Lawrence. Corrections have been made to the Ile de St Jean (Prince Edward Island) and part of the Ile Royale (Cape Breton Island). Compare to fig. 5. The map has been drawn on a sheet with the rhumb lines already printed in red, from a plate similar to that in color plate 2. CP MAR 6/JJ/76 C ou III, no. 247A. (Archives nationales, Paris.)

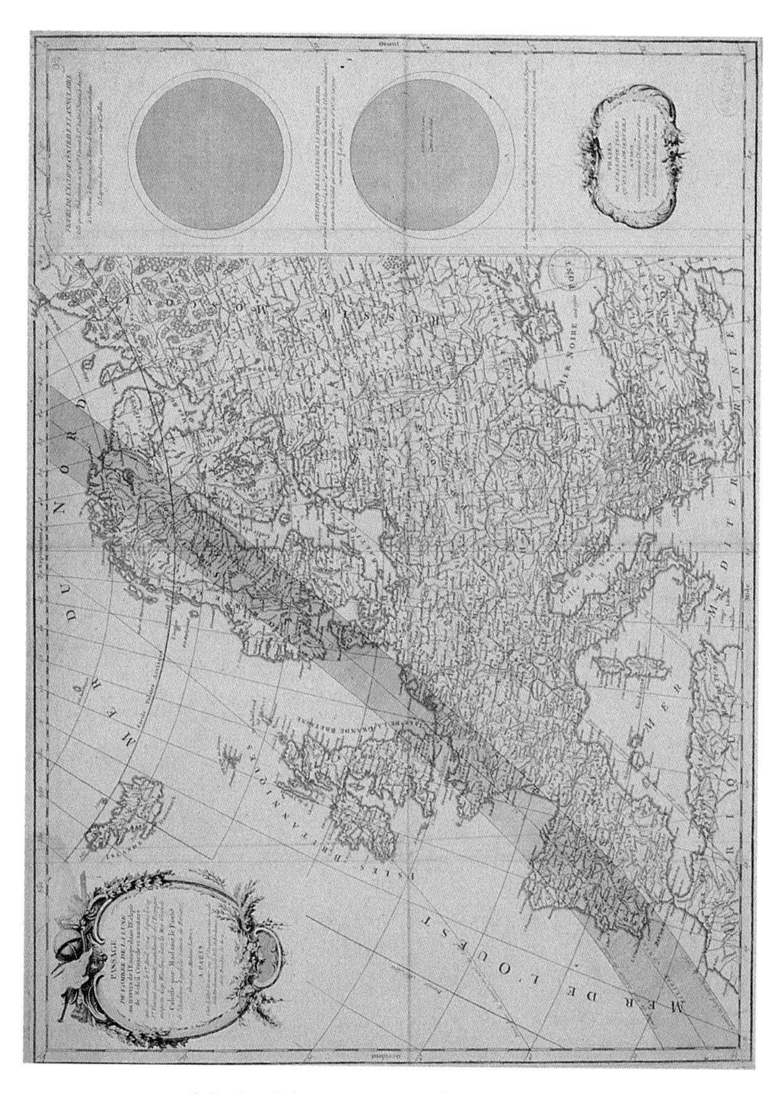

PLATE 4. *Passage de l'ombre de la Lune au travers de l'Europe dans l'Eclipse de Soleil Centrale et annulaire qui s'observera le 1er Avril 1764 . . .* (Paris: Jean Lattré, [1762]). The map was based on calculations made by Mlle Le Paute, who worked with the astronomers Clairaut and de La Lande. Mme Lattré, wife of the engraver and publisher Jean Lattré, engraved this map; her colleague, Mme Tardieu, engraved the cartouche. The latter was perhaps Marie-Anne Rousselet (1733–1826), daughter of Gilles Rousselet, an engraver, and wife of Pierre-François Tardieu, one of a well-known family of engravers and copperplate dealers. The map is printed in two colors: black and red (or sepia). 480 mm × 660 mm. BNF, Rés Ge. C. 8811. (Bibliothèque nationale de France, Paris.)

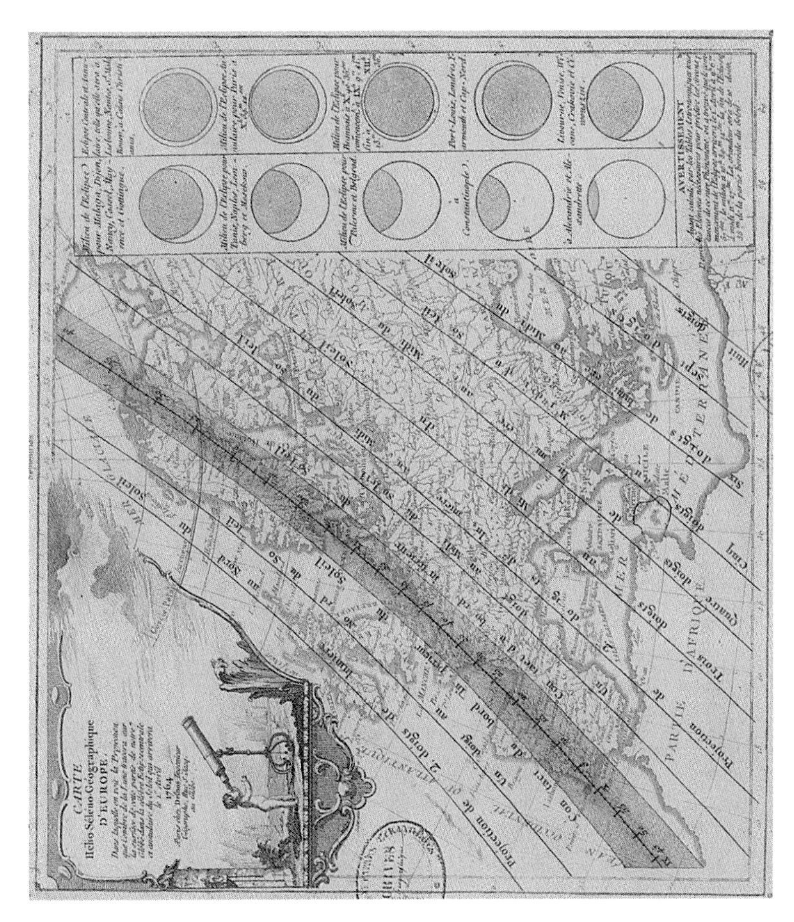

PLATE 5. *Carte Hélio-Séléno-Géographique d'Europe dans laquelle on voit la Projection que l'ombre de la Lune tracera sur la surface de cette partie de notre Globe dans la célèbre Eclipse central et annulaire du Soleil qui arrivera le 1er Avril 1764* (Paris: [Louis-Charles] Desnos, 1763). Lattré accused the geographer Rizzi-Zannoni of copying the map prepared by Mlle Le Paute for his 1762 production on Desnos' behalf. Acting on Lattré's complaint, the Paris police seized the plate, but Desnos pleaded ignorance of Rizzi-Zannoni's activities. Desnos' map is considerably smaller than the Lattré map, 232 mm × 263 mm. BNF, Ge. DD. 2987 (51), Collection d'Anville. (Bibliothèque nationale de France, Paris.)

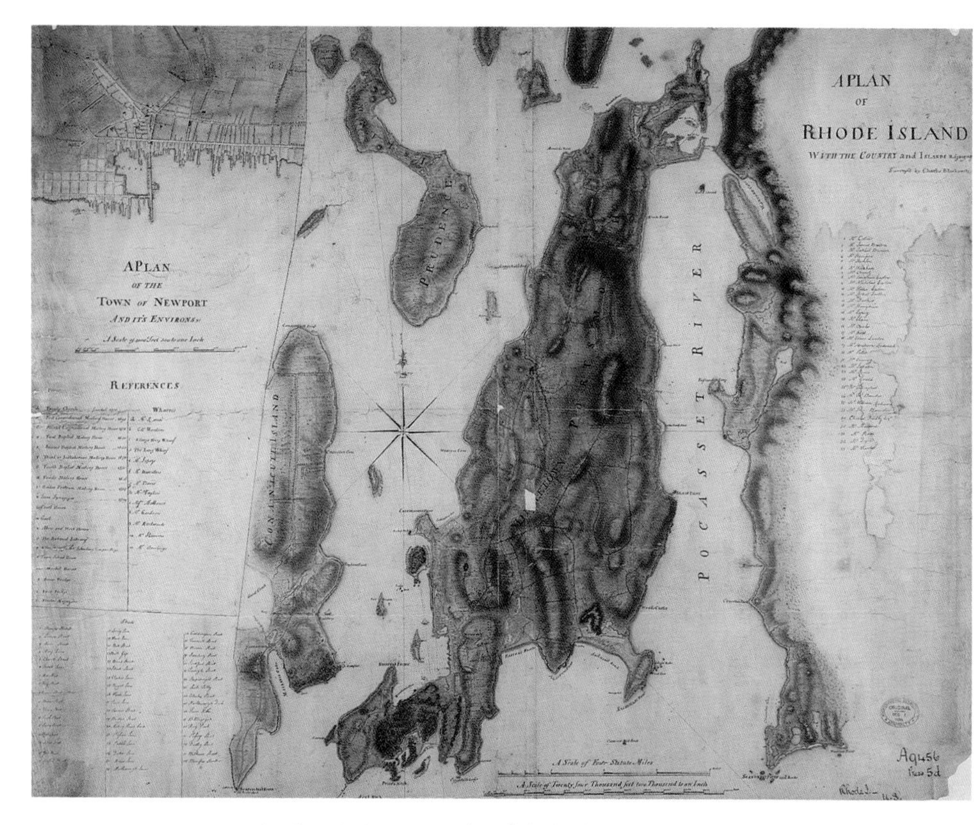

PLATE 6. Charles Blaskowitz, *A plan of Rhode Island with the Country and Islands adjacent.* The manuscript map may be dated after 1770, for that is the date of the foundation of the Fourth Baptist Meeting House, marked G on the inset plan of Newport. This may be one of the manuscripts resulting from the survey of 1774, performed by Blaskowitz and Wheeler. The large manuscript may also have served Des Barres and/or Faden for their printed versions of Narragansett Bay. 1205 mm × 970 mm. Hydrographic Office Library, Taunton A9456 Shelf 5d. (© British Crown Copyright 2003. Published by permission of the Controller of Her Majesty's Stationery Office and the UK Hydrographic Office [www.ukho.gov.uk].)

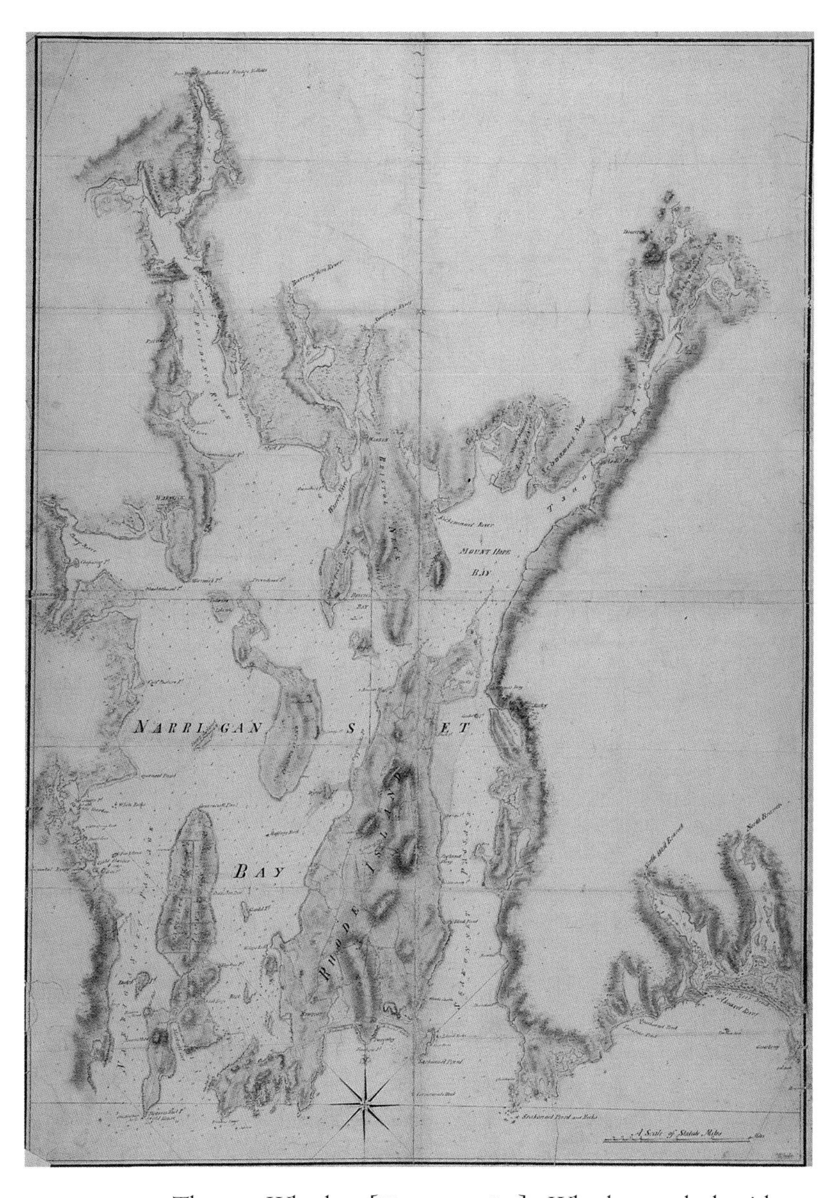

PLATE 7. Thomas Wheeler, [*Narraganset Bay*]. Wheeler worked with Blaskowitz on the survey of Narragansett Bay in 1774, under the general supervision of Samuel Holland. This manuscript was in the collection of Sir Henry Clinton, who wintered in Newport and environs during the winter of 1777. 1014 mm × 725 mm. WLCL, Map 3-J-7, Clinton ms. 58. (William L. Clements Library, University of Michigan.)

PLATE 8. [Charles Blaskowitz], *A map of the Baye of Narragansett with the Islands therein and part of the country adjacent* [1777]. This manuscript is attributed to Blaskowitz because several of its features are found on William Faden's printed version of the map, which credits Blaskowitz in the title. However, the calligraphy is cruder than Blaskowitz's signed work, as seen in plate 6. G3772 .N3 1777 B49 (Faden 87) (Library of Congress, Washington.)

made clear to Delisle. In addition, they should be on big, beautiful ("beau et grand") paper, not small paper like those Delisle had sent before. The cartouche should be "of such exquisite beauty that it will be the mark of the map's exactitude. You know that maps are a source to which everyone has a right, and good works such as yours ought to be sought after by the ignorant for their external beauty, as well as by the scholars because of their intrinsic worth."[67]

However, Renard's advice about paper and cartouches drew a sharp retort from Delisle:

> It is not true that I have always sold my maps in Paris on beautiful, large paper. On the contrary, they have always been sold on the paper I sent to you; it was only a little while ago that some folks wanted to have them with bigger margins. I had them printed especially for them and made them pay 2 sous more. As far as the advice you have given me, it is really useless to talk to me about the beauty of cartouches when the maps are already engraved, and even when they are not, I would not have them done any other way. A geographer who knows his business ought to know what's important in his work.[68]

Along with big and beautiful paper, some customers simply thought of maps in terms of size of paper or number of sheets. William Faden's correspondents, many of them map producers, showed a complete disregard for the rhetoric of map titles or advertising. They did not order "the most recent" maps, or those "based on latest observations," or the "most beautifully engraved." Instead, if they did not order a specific title from one of Faden's published catalogues, they ordered maps by the number of sheets: a North America in two sheets instead of eight, for example.[69]

The later editions of William Camden's *Britannia,* published in 1695 under the editorship of Dr. Edmund Gibson, contained maps prepared by Robert Morden that were larger than the original. The maps were designed according to the "newest surveys, very correct, and contained all the towns mentioned in Camden's but were not thought large and comprehensive enough by some judicious and ingenious Gentlemen that assist in the work, who think it Proper and necessary to have the Maps of every County full as large as this sheet will admit and to comprehend every town, village, etc through out."[70]

The abbé Nicolas Lenglet Du Fresnoy, whose *Méthode pour étudier la géographie* became a kind of consumer guide for map buying, concurred: "Our maps, whether general or particular, are all on different scales. This causes a kind of confusion and mental difficulty. If one is supposed to examine five or six maps by the same author for five or six neighboring regions, why should one have to change the scale from one map to another, thereby losing the mental image one just used?"[71]

[A MAP BUYER'S GUIDE BOOK]

The *Méthode pour étudier la géographie* included a catalogue raisonné of maps regarded by Lenglet Du Fresnoy as the best available. The *Méthode* was published several times throughout the century (1715, 1736, 1741, 1768); the map catalogue gives us a broad picture of the maps available to the reading public of France at midcentury. The author did not list all maps published for fear that this would "throw the curious and the amateurs into a labyrinth." His recommendations are models of brevity, wasting no time on analysis of sources. He hails some maps as "good," "original," "useful," or "curious, very well made" and dismisses others as "bad" or "of little use."[72] He does not limit his choice of the best maps to the French, though the works of Sanson, Jaillot, Delisle, and the Robert de Vaugondys dominate. His list is organized geographically, listing the recent maps available of a particular area. Thus for Turkey, maps by French authors are cited alongside those of Van der Aa, Ottens, De Witt, Allard, Visscher, Homann, Laurenberg, de Rossi, Coronelli, and Cantelli. Cantelli's work, though made sixty years earlier, he called a "good map made according to the latest observations."[73] He singled out maps made on the spot in the field, such as the map of the Pyrenees by Roussel and La Blottière, in nine sheets; he called it "an excellent map, which has been surveyed on the spot; the only thing against it is that it is badly oriented."[74]

Lenglet Du Fresnoy also emphasized nomenclature and language. He praises Nolin's map of Hungary in four sheets for showing the names of the principal cities in several modern languages along with their ancient names and the dates of their seizure by Christians and infidels. The map also included a small map (probably an inset) showing the extent of the Hungarian kingdom prior to the Turkish conquest.[75]

Size and scale were also a concern for Lenglet Du Fresnoy, especially when they met the demands of a particular subject. In the case of Henri Fricx's twenty-four-sheet map of the Catholic Low Countries, he finds that "these maps, so beautiful and detailed, are very necessary for the encampments and marches of armies. They are all on the same scale and may be assembled into a single map."[76] But he finds that Delisle's map of France of 1703 would have been better if it were "in a greater volume."[77]

Throughout his catalogue, the quality of engraving counts for much. The Sanson mappemonde, published by Jaillot in 1674, is "well engraved"; the more recent Guillaume Delisle *Globe terrestre gravée pour l'usage du roi* (1720), "very badly engraved."[78] The author admits the limits of his emphasis on beautiful engraving in his description of the *Carte Universelle de tout le monde* by Dankaerts and Tavernier of 1628: "This map is beautiful and well engraved, but what of the discoveries made during the 120 years since it has appeared?"[79] The problem with poor engraving may be found, for example, in Bailleul's map of the Election and Archbishopric of Trèves in four sheets: "although good, it is harshly and heavily engraved, which only spreads confusion."[80]

LEGIBILITY

Bad engraving made a map difficult to read. It was no wonder that legibility found a place near the top of the list of qualities of a good map. The Robert de Vaugondys refused payment to their engravers because of the crowded, squeezed italic lettering of certain maps in the *Atlas universel.* An aversion to cramped lettering and symbols permeates map criticism; map buyers simply did not like too much detail squeezed into a map. The local officials of Brittany complained to Jean-Dominique Cassini (also known in the literature as Cassini IV), son of Cassini de Thury, that the map of Brittany produced as part of the *Carte de France* seemed badly engraved when compared to that of Normandy (figure 38). The Brittany map was too crowded with names and difficult to read. Cassini defended his engravers, explaining that the nature of the landscape and the man-made features in it determined the look of the map. The "more or less agreeable aspect" so admirable in the maps of Normandy was achieved "by a happy disposition of opposites, the white making the black stand out." This was because Normandy's natural and man-made features were less numerous and more evenly distributed than those of Brittany, which was densely packed with woods, hills, moors, and little settlements. It required a detailed nomenclature that at such a small scale could inevitably cause confusion. Cassini allowed that the dense detail obscured the background of the map and did not allow for that black-and-white opposition and its "piquant effect" on the eye. One could only blame nature herself, not the engravers.[81]

J.-N. Delisle clearly understood the aesthetic aspect of the map. Having dropped Philippe Buache as his map designer, he had invited J.-N. Bellin to redesign his map of the putative discoveries of Admiral de la Fuente. He explained that "It is not enough that maps be curious or new or at the very least interesting and useful to the progress of geography. . . . It is even more ad-

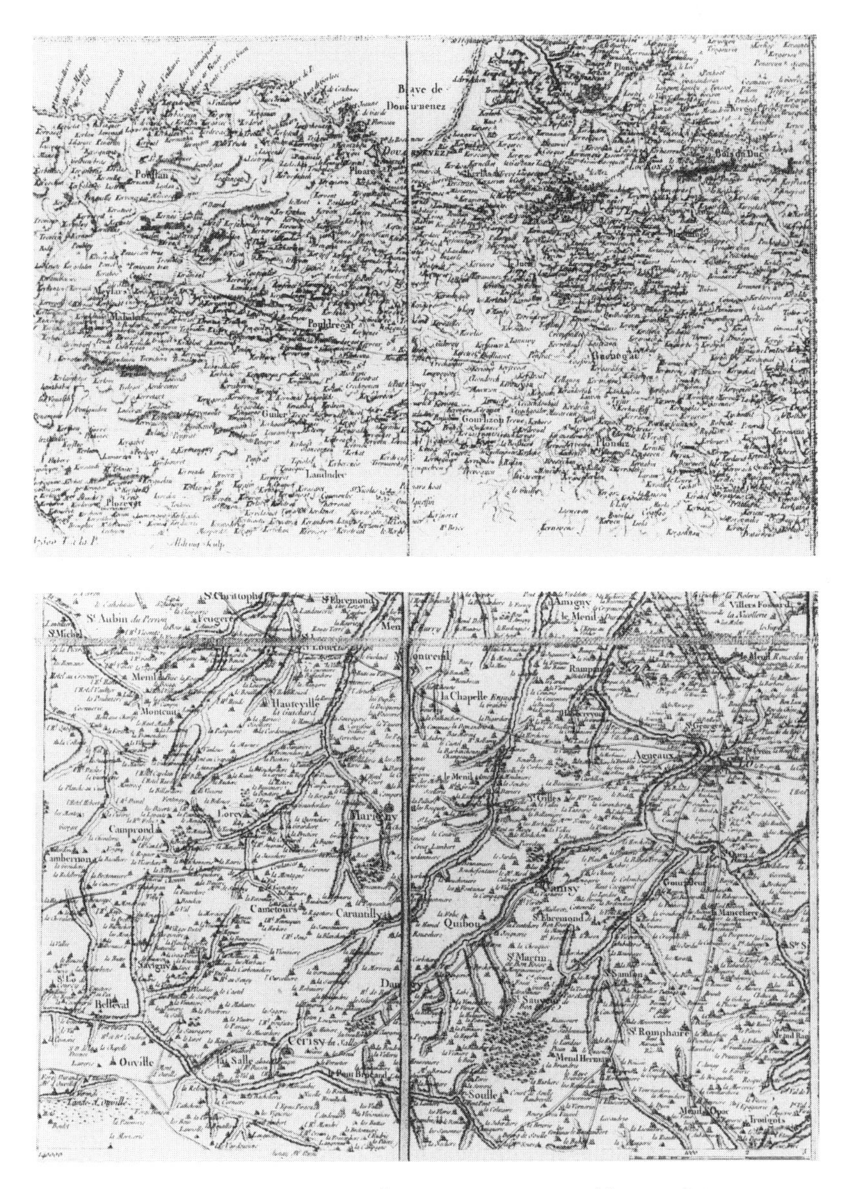

FIGURE 38. From Cassini, *Carte de France*. Top. Brittany (sheet 174). Bottom. Normandy (sheet 91). Note the contrast between the messy look of the Brittany map, to which the governing officers of the region objected, with the "effet piquant" of the Normandy sheet, with its contrasts of shade and light areas and place names more evenly spaced. (Map Library, Harlan Hatcher Graduate Library, University of Michigan.)

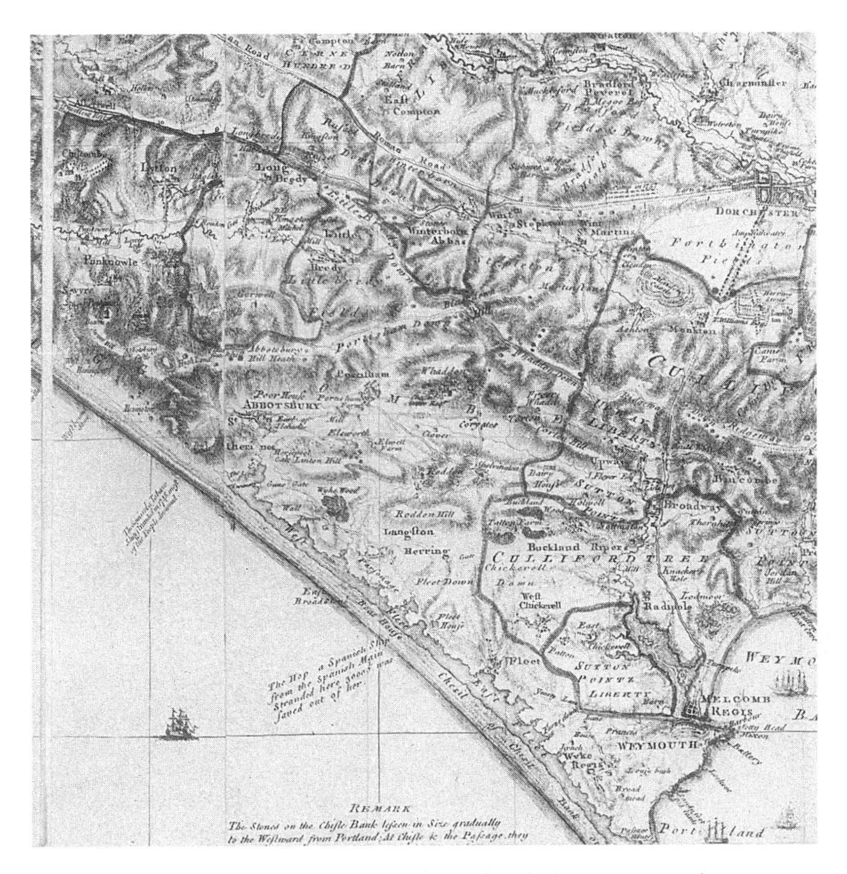

FIGURE 39. Isaac Taylor's map of Dorset (1765) demonstrates what some critics regarded as the untidy, gray engraving of a self-taught amateur. BL, Maps 2153 (3). (Map Library, British Library, London.)

vantageous that they be pleasing to the eyes and engraved with taste and neatness."[82]

The struggle between beautiful engraving and content could be the determining factor in a map's success. Isaac Taylor produced a map of Dorset in 1765 (figure 39) in pursuit of the Society of Arts award for best trigonometrically surveyed map. One critic judged it a "capital survey." But someone who had helped Taylor perform the survey described his previous troubles with his engravers. Their "indolence, want of punctuality, and caprice" had forced Taylor to engrave his map himself, in a manner his friends found "clumsy, gloomy, and unequal." So unappealing was Taylor's effort with burin and etching needle that, in spite of the map's accuracy, the Society of Arts prize was awarded to Benjamin Donn for his map of Devonshire.[83]

A similar incident occurred in North America. The General Court of Massachusetts found John Norman's engraving "inelegant and unacceptable" for Osgood Carleton's map of Maine and Massachusetts. This must have been particularly hurtful since Norman, an Englishman by birth, had been an apprentice to the printer William Faden, father of the map engraver and seller, also named William Faden, who was particularly well known for his elegant engraving. Norman was given seven months "to take out the many accidental strokes in the plate, and also, that they make the margins of the rivers, ponds, and seacoasts neater, and that the whole plate be better polished."[84]

A quick-witted publisher needed a sharp response against charges of illegibility and untidiness. When the duke of York suggested to Moses Pitt that the lettering of the maps in the *English Atlas* was too close together and that the maps themselves were rather small, Pitt replied, "Ah, yes, but they have a fair margin."[85]

BLANK SPACES

Modern historians of cartography are sometimes inclined to see the blank space on a map as an intentional "silence," indicating the suppression of information. This interpretation of cartographic design has led some researchers to look for powerful forces encouraging the manipulation and removal of critical geographic or topographic information.[86] The critical response in the eighteenth century to the blank spaces on maps was more subdued and often couched in the aesthetic discussion of map design. If a map could be criticized for being too crowded, could a map also be too empty? In contrast to their northern compatriots in Brittany, whose map was too detailed, the local worthies of Provence were alarmed at the amount of blank space on the map of their province. Pressed for a response, Cassini IV assured them that it was possible to omit things on maps by negligence, but sometimes one omitted by design, "in order not to multiply its features too much and to avoid confusion when a region is too filled up."[87] In 1762 Seguin, one of Cassini's *ingénieurs,* asked the Electors of Burgundy how they wished to fill the spaces that were left by the sinuous contours of the borders of the province on the map's outline rectangle. He was instructed to place cities and larger towns in the space and to prolong the roads beyond the provincial borders, but to add no detail of mountains, woods, or fields; this smattering of information along the edges of the map would "render [the blank spaces] less shocking."[88]

Didier Robert de Vaugondy alerted readers that the map of Canada in the *Atlas universel* displayed none of the recent discoveries of lakes and rivers west of Lake Superior, for these were not part of Canada but of French Louisiana. To see them, readers would have to turn to d'Anville's map of the Mississippi.[89]

SYMBOLS

Another area in which consumer taste played a part was in the conventional signs or symbols used on maps. The map user was not always aware of the meanings of symbols. For some, the entire ensemble of the map was incomprehensible.[90] The seventeenth-century cleric Augustin Lubin published the *Mercure géographique ou le Guide du curieux des cartes géographiques* (Paris, 1678) to aid the reader in understanding the map; it offered a dictionary of symbols to the map user.[91] Didier Robert de Vaugondy dealt with the problem more succinctly by offering his readers a map entitled *Introduction à la coñoissance et à l'usage des Cartes,* found in the *Atlas portatif universel et militaire* of 1748 (figure 40). It is a practice map, showing all the typical features and lettering one might find on a geographic map.

Many printed maps from the late seventeenth century included either an explanatory table or a key to the symbols used. Symbols used on English maps remained essentially the same since the publication of maps of the Elizabethan surveyor Saxton; though symbols were not standardized, there were few innovations and little variety.[92] In France, an attempt at standardization is found in the "table of comparative place names on topographic and chorographic maps" in Buchotte's *Les regles du dessein et du lavis* (Paris, 1721). At the same time, new symbols were applauded. An anonymous reviewer in the *Journal des Sçavans* noted that Nicolas de Fer, publisher of a new map of the environs of Paris, "has invented characters to distinguish cities, towns, parishes, villages, chateaux, abbeys, priories, windmills, crosses, and trees. He has not omitted parks, paths, waterways, routes, and roads, and has added many names of streams and their sources, the divisions of the suburbs of Paris and of the Election of Paris."[93] Ever alert to the new and the topical, de Fer popularized the work of the French engineer, maréchal de Vauban by showing towns in fortified plan rather than in elevation, as on the maps in his *Forces de l'Europe ou l'Introduction à la Fortification,* published in 1695.[94] This style was soon copied in Amsterdam by Pierre Mortier (*Atlas royal,* 1695) and C. Allard (*Théâtre de la guerre du pais bas catholique,* 1696). The new urban symbol caused Delisle's correspondent Louis Renard to be quite specific in his request that Delisle draw cities and fortified places in plan rather than in elevation, for "this style is beginning to be introduced and is very well thought

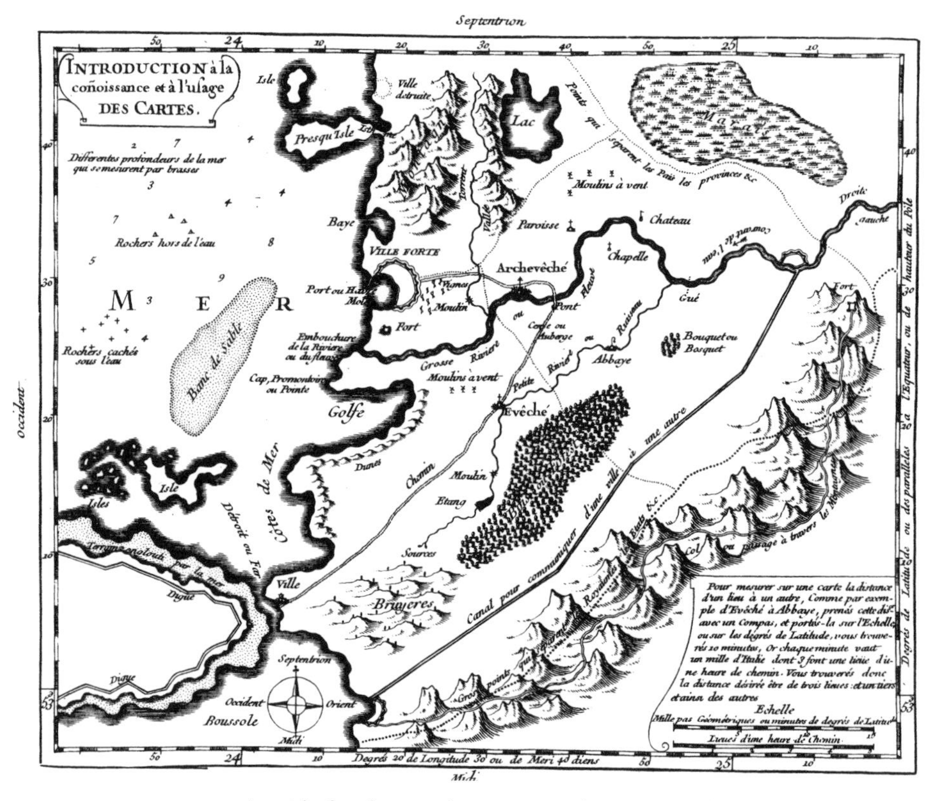

FIGURE 40. A guide for the neophyte map reader. From Robert de Vaugondy, *Atlas portatif universel et militaire* (Paris, 1748). Greenlee 4891 .R65 1784. (Newberry Library, Chicago.)

of these days."[95] It was recommended in Buchotte's *Regles du dessein* and used on the sheets of the *Carte de France.*

LANGUAGE

Because the range of symbols used could vary widely from map to map, from mapmaker to mapmaker, a key or explanation was almost always essential. The language of the key, and indeed, the language of the title, dedication, authority, place-names, and other written material in the map, demanded another decision. What language to use—that of the region being mapped, or the tongue of the cartographer, or the language of the consumer? Louis Renard advised Delisle to place two titles on all maps sent to Amsterdam: one in Latin "for the benefit of foreigners" ("pour la commodité des étrangers"), another in French or German or English.[96] Fifty years later, Didier Robert de Vaugondy evoked the advantages of Latin titles on maps, for the ancient tongue, like conventional signs, provided a universal language, especially on German maps. "If they [the Homann heirs] would publish their maps in Latin, many scholars who are strangers to German could profit from their enlightenment."[97] Georges-Louis Le Rouge, on the other hand, understood the need for multilingual marketing, as may be seen from the French and German used in the title cartouche of the second edition of his copy of John Mitchell's map of the British and French dominions in North America.

D'Anville's strict instructions eschewing vulgar spellings and promoting a more historical approach did not meet with Didier Robert de Vaugondy's approval. He criticized d'Anville's use of native orthography in his maps of Indochina, commenting that "these maps, being made for Europeans rather than for Asiatics, ought to have represented the names as they are known here."[98] Such remarks failed to set any trends. Latin did not become the universal language of maps and was reserved almost exclusively for maps of ancient geography. The language of a map tended to reflect the country of origin of the publisher, not necessarily the market for which it was intended.

COLOR

The fact that maps were offered to the public in both colored and uncolored versions suggests that each consumer exercised a distinctly individual taste as to map coloring. Renard asked Delisle to send his maps plain, without color, since his customers liked to have them colored to their own taste. Renard reminded the French geographer that in Holland one liked "a rich illumination and of the utmost delicacy, for which a great many connoisseurs spend big sums."[99]

[ACCURACY AND ERROR]

FINDING FAULT

Perhaps the greatest desideratum of a map and the most difficult to achieve was its accuracy. It also was the most difficult attribute to define. There was no way of ascertaining with certainty whether a region or area foreign to the user was represented with any degree of truth. Short of visiting the region him- or herself, the map user's only recourse was to compare maps of the same area and make choices based on the features described above: legibility, attractiveness, language, and coloring. He or she had to take on faith those assertions in the title that a map was made according to the latest observations or from surveys made on the spot. As we have seen, these words were easily added. Yet consumers became one of many sources used by mapmakers for ensuring the accuracy of maps. Consumer reaction to error was a source of correction.

Geographers and mapmakers readily admitted that errors were an inevitable part of cartography. As one declared, "And as for an Error here and there; whoever considers how difficult it is to hit the exact Bearings, and how the differences in miles in the several parts of the Kingdom perplex the whole; may possibly have occasion to wonder there should be so few."[100]

While observations of longitude and latitude were continually made throughout the century, not all mapmakers had equal access to these data. Nor did observations made on the spot always guarantee accuracy. The père de Bonnecamps wrote from Canada to J.-N. Delisle in Paris regarding the latitude of Quebec. It had been observed by a local military engineer but was too far north by seven or eight minutes. De Bonnecamps told Delisle that such an excess would not surprise him if he had seen, as de Bonnecamps had, the instrument used by the engineer. It was "a planchette of wood, measuring around 8⅔ inches in diameter, equipped with a copper limb divided into 360°, each of which measures ⅚ of a line [a line = 1/12 inch]."[101] In other words, it was much too small to give a reading of great accuracy.

Even a map drawn from actual surveys could disappoint. Thomas Pownall, governor of Massachusetts Bay and of South Carolina and lieutenant governor of New Jersey before the American Revolution and himself a topographic sketcher and amateur surveyor, found many faults in *A new and accurate Map of Virginia* surveyed by John Henry, published in 1770 by Thomas Jefferys. He called it "a very inaccurate compilation; defective in Topography, and not very attentive even to Geography; the Draughtsman or the Engraver has totally omitted the South Branch of Potomack River; Nor is that curious and interesting Piece of Information, the communication between the waters of Virginia and the waters of the Ohio, which were known when this was published, marked in it." Pownall grudgingly added, "This map of Mr Henry has indeed the Division lines of the counties of the province drawn on it, and if they are rightly drawn, it is certainly an improvement."[102] Yet this did not stop Henry's map of Virginia from being a map of choice for display. Lord Botetourt, one of Henry's subscribers, hung it in his Williamsburg dining room, where it still hangs today.[103]

Warnings to users about what to expect from a map could deflect criticism in advance. The *Gentleman's Magazine* alerted readers regarding the map of America that appeared in the 1755 issue: "The reader is not to expect the name of every hamlet, redout, or rivulet; let it suffice that the face of the country is delineated and the principal cities in every province are set down." Robert Plot preempted the captious user of his map of Oxfordshire: "Yet this map, though it contains near five times as much as any other of the county made before . . . is not so perfect, I confess, as I wish it were, there being . . . perhaps here and there a village over-looked."[104] He appealed to the gentry to inform him of any omissions of villages near them containing ten houses and to bring him the information "with the particular bearing and distances of their houses and villages from the most noted place near them" in order for an appendix to the map to be compiled and printed.[105]

As Plot suggests, the best way to ensure accuracy was to invite verification. That meant that more than one report had to substantiate either an observation or a depiction or configuration of an area. This process was easier to perform in places closer to the map's market, more difficult in the case of the faraway. Geographers and mapmakers relied on a wide network of consumers, friends, strangers, amateurs, and cartographic enthusiasts to check their maps. Corrections and suggestions for Guillaume Delisle's maps from friends in France, Switzerland, Holland, Spain, and North America (the abbé Bobé wrote from New Orleans) are preserved in his correspondence.[106] M. de Tancarville wrote from Montreuil regarding the map of the Boulonnais, commenting on the political divisions, the kind of lettering to use for towns, the locations of villages, the precise distances between places on the map; the abbé Bignon suggested the use of different line gradations on Delisle's map of the Caspian Sea; M. Violier wrote from Geneva with corrections for Delisle's map of Greece, sent copies of Scheuzer's new map of Switzerland, and questioned the longitude of Geneva.[107]

Didier Robert de Vaugondy also actively sought corrections and describes his correspondence with local scholars in his summary analyses of the maps of the *Atlas universel.* For example, he and his father Gilles Robert de Vaugondy sent their map of Burgundy to the savants of the region, receiving many corrections that, they claimed, contributed to its perfection.[108] In England, proofreading and checking maps also required community participation. For the 1695 edition of William Camden's *Britannia,* Dr. Edmund Gibson sent copies of proofs of the maps to "the most knowing Gentlemen in each county, with a request to supply defects, rectify positions, and correct false spellings."[109] Drafts of maps were hung up in public places as invitations for comment and correction. John Rocque's large twenty-four-sheet plan of London, published by John Pine, was displayed in the shops of Rocque, Pine, and their partner, John Tinney, at the Court of Aldermen of London, and at the Royal Society. "Proper persons" were sent with map in hand to all parts of London to check the topography.[110] The steady stream of county maps published in pursuit of the Society of Arts' prize for the best trigonometrical survey were subjected to public scrutiny prior to publication and even more rigorous checking for accuracy after publication.[111]

In France drafts of provincial maps were hung in every diocesan office for the members of the parish to study. To help him with the corrections and revisions of his map of Languedoc, Philippe Buache created a questionnaire for church officials, asking about the parish, covering all aspects of its geography, from its landscape to its cartobibliography.[112] From the seventeenth century, local clergy had been important suppliers of geographical information, especially for orthography and place-names, and often initiated map projects in France.[113] There were occasions, however, when the local priest was not always so obliging. Some refused Cassini's engineers entrance to church towers to make observations for the *Carte de France.*[114]

These prepublication critiques focused on location and spelling, rarely on topographical features or landforms. Location might extend to the coordinates of latitude and less often of longitude, though of course the latter became increasingly a subject of map correction as the century delivered more precise longitudinal information. More often location meant on what side of the road or river a place was located, or whether it had been placed too far north or south in relation to other features. Spelling of place-names was always a bone of contention, as was the place-name itself, especially when local pronunciation made it difficult to spell. This was one of the ways in which the local guide or *indicateur* was indispensable. Even when the area was not in the least remote or foreign, uncertainty could permeate the nomenclature, as on the map of Kent surveyed by Andrew Dury, John Andrews, and William Herbert published in 1769 (figure 41). The dual place-names they incorporate on their map suggest the problem faced by mapmakers encountering local dialects with no standardized orthography. In one small section of the map one finds "Nicknash or Dignash," "West Wellecon or Westwell Leacon," "Pating Fostal or Lenham Fostal," "Wingfield or Stalisfield Valley," "Stobbington or Stuppington."[115] Cassini warned of similar problems when his surveyors were working in Provence: "it is so easy to err when one is in a region whose language and pronunciation are not familiar."[116]

With errors so rife, proofreading was essential. Cassini paid *vérificateurs* 300 livres per plate not only to proofread but also to check on the spot the maps of the *Carte de France.*[117] Thomas Jefferson, who made his own map of Virginia to accompany his *Notes on the State of Virginia,* found no fewer than 172 errors on the map after it had been engraved the first time. Though he was resident in Paris, he had sent the map to England for engraving, fearing

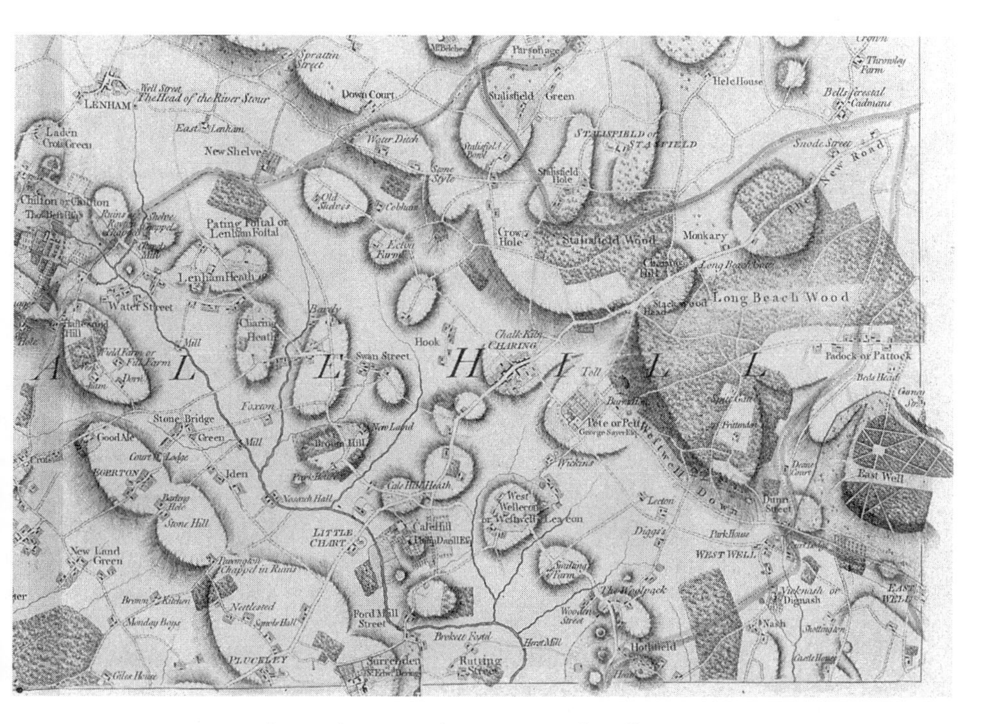

FIGURE 41. John Andrews, Andrew Dury, and William Herbert, *A Topographical Map of the County of Kent* (London, 1769). The detail from plate 13 shows place-names with alternative spellings, reflecting various pronunciations. BL, Maps 1 TAB.21. (Map Library, British Library, London.)

that a Frenchman would not be sufficiently familiar with the English language to cut the plate well.[118] That was his first mistake. His second was to go with the lowest bidder. Judging William Faden's estimate of 50 guineas for engraving the plate as too high, he sent the manuscript to Samuel Neele, who had offered to do it for about half Faden's price (c. £25), though his final bill came to £28 16s 9d. The low bid produced predictably low results. Neele's engraving errors were so numerous that Jefferson was forced to hire a Paris engraver, Guillaume Delahaye, to make corrections on the plate. Delahaye's charges for corrections and printing 250 copies came to nearly 220 livres (c. £10 sterling); Jefferson added to this the cost of transporting of the plate between Paris and London (11 livres 5 sous for the plate; 240 livres for the servant taking it) for final publication. Jefferson probably would have been better off with Faden's original high price.[119]

The criticisms of maps by eighteenth-century producers and consumers, centered as they are on issues of paper size, legibility, and spelling, may not seem subtle or sophisticated to our modern or postmodern eyes and ears. But those concerns should not overshadow the century's awareness of the ease with which maps erred and their inherently prejudicial nature. Certainly the French and English boundary commissioners, negotiating the borders of Nova Scotia (Acadia) to the terms of the Treaty of Aix-la-Chapelle in 1750, made it clear that maps were no reliable source of boundary information:

> For maps are from the Nature of them a very slight evidence. Geographers often lay them down upon incorrect Surveys, copying the Mistakes of one another, and if the Surveys be correct, the Maps taken from them, tho' they may shew the true position of a Country, the Situation of Islands and Towns, and the Course of Rivers, yet can never determine the Limits of a Territory, which depend entirely upon authentic Proof; and the Proofs in that Case, upon which the Maps should be founded to give them any Weight, would be themselves a better evidence.[120]

In a rare moment of cross-Channel alliance, the French commissioners did agree that maps could not be the decisive authority. Their authors were "more concerned with giving an air of system and truth to their maps as well as an appearance of science and research than with fixing the rights of Princes and the true boundaries of a region."[121]

Map titles of the eighteenth century display a consistent rhetoric. English maps are "new and accurate," "new and complete," "new and exact," "new and improved." French maps are "adjusted to the latest observations," "showing the most recent discoveries." Such language leads one to surmise market pressure for novelty and change. But some geographers tried to resist. Didier Robert de Vaugondy made no apology for not always including the latest information, maintaining that it was better to stick with a received idea than to forge a chimera just to provide something new.[122] Guillaume Delisle's eulogist commented on his practice of not wishing to shock the public with the new; instead, he published new information slowly, by alterations to his plates. The new states and editions would lead the public gradually into acceptance of new information.[123]

Yet the new and the recently observed were perceived as the factors that sold maps. Poor sales hampered the success of Didier Robert de Vaugondy's *Hémisphère austral ou antarctique,* published in 1773 under the auspices of the duc de Croÿ to support the duke's pet project, the Kerguelen expedition to the Pacific. It was sold at what the duke considered a low price of 14 livres, but which was in fact rather expensive by market standards. When it became clear that only few copies had been sold, Robert de Vaugondy wrote to the duke that he hoped the additions and corrections that the duke would be making on the map would "renew and pique the curiosity of the connoisseurs."[124] One hundred years earlier, the publisher Jaillot had inserted a specific clause in his agreement with Guillaume Sanson regarding the possibility of new borders created after new treaties and the correction of maps to show them.[125]

By contrast, William Faden's plans of battles of the American Revolution, made from on-the-spot observations and rapidly published after the fact, failed to capture a market in Paris, where interest in the war itself was otherwise high. In 1777 Jean Lattré advertised in the *Journal de Paris* that he had just received "a complete assortment of original English maps, especially relevant to the present war in North America."[126] But a year later Lattré was returning battle plans he had ordered to Faden for lack of sale, instructing the London mapmaker to send no more: "they make little impact here."[127]

In the end what sold maps was price. A copy or a counterfeit was as good as the real thing to the consumer, who could be adamantly undiscerning about maps. Louis Renard, committed to selling Delisle's original maps in Amsterdam, faced stiff competition from Pieter Mortier, who copied Delisle's maps and sold them for well under Renard's price, sometimes as much as 50 percent cheaper. Even worse, according to Renard, sometimes Mortier did not even copy a Delisle map, but simply put Delisle's name on another map in order to sell it. Renard explained that the public was uninterested in the chicanery, even when it was exposed. Once the counterfeits are known, "one gets used to them . . .; the testimony of experts does little good" when one can have copies at half the price of the original.[128]

Only among the mapmakers themselves did copies receive their well-deserved criticism, though even then economics shadowed geographic concerns. Thomas Pownall, about to publish his corrected and improved version of Lewis Evans's map of the Middle Colonies of North America, was distressed by a copy of Evans's map "pirated, in a most audacious manner" by Thomas Jefferys in 1758, "under a false pretense of improvement."[129] He wrote in the preface of his *Topographical Description* that "the Engraver was so totally ignorant of the Principles on which the Original was formed . . . it can scarce be called a Copy." The copyist ignored Evans's careful depiction of mountains, which Pownall claimed showed mountain ridges running in specific directions, sometimes ending in peaks, sometimes in knobs and bluff points, interlocking and having gaps, and sometimes spreading into hilly land. Pownall claimed that the copyist had placed mountains where there were none, had converted swamps into mountains, and in some places omitted marks of high ground, not understanding what the marks meant. "This thing of Jefferys might as well be a Map of the Face of the Moon."[130]

Pownall's criticism was published in abbreviated form in the *London Evening Post.* Soon Jefferys's son Thomas responded in the same paper. Young Jefferys maintained that his father's publication of the Evans map was not a piracy, for the plate had been in the possession of Emanuel Bowen, its engraver, from whom Jefferys had purchased it. "The scarcity of the original impression [i.e., Evans's map], and the estimation in which the work was held were sufficient inducements with my father to re-publish the map in England. He added, it is true, the line of forts at the back settlements, but he varied not a tittle from the geographic representation as it stood in the original. He made no pretence of improvement, and therefore the charge of having falsified his pretensions falls to the ground." Young Jefferys then shifted his defence. "I do not, Sir, mean to insinuate, that your affections, like your patriotism, ought to be suspected; but this I will affirm, that the man who can charge another with STEALING a work which he honestly PURCHASED, is so little careful to distinguish between truth and falsehood, that great caution should be used in believing what he says."[131]

The Jefferys plate had passed to the engraver and print seller Robert Sayer. Pownall had clearly reproached Sayer for continuing to publish such a faulty piece of plagiarism. According to Pownall, Sayer answered with his own economic argument. "He very honourably told me, that if the Plate stood as a single Article in his Shop, he would destroy it directly; but that it made Part of an Atlas already published by him; and was also part of another very soon to be published by him, which cost many thousand Pounds; and that he did not know how to take it out of these Collections." Pownall added sourly, "I can only say, it will disgrace any Collection in which it stands, and that I am sorry it is to disgrace any coming from a Shop in which there are so many valuable maps and charts." Pownall's words had some effect, as he noted in the preface to the third edition (1784) of his *Topographical Description* that Sayers had "since corrected from my Map for the purpose of making part of a Collection in a Pocket Atlas of North America."[132]

MAKING BETTER MAPS

The criticisms, the stockpiles of unsold maps, the problems of plagiarism all spurred geographers and mapmakers of the eighteenth century to suggest ways to improve cartography. They fall under three rubrics: systemized training; better pay and more government support; and a regulatory body to vet the quality of productions and to enforce privileges and copyrights in order to ensure that the market was not swamped with cheap imitations.

Training

The first thing a mapmaker needed was training sufficient for making good judgments about maps. When Rizzi-Zannoni wrote to William Faden to order maps from London, he was ready to defer to Faden's "taste and discernment," confident that Faden would be able to tell "at a glance a geometric map from those that are not."[133] The development of such taste and discernment was a leitmotif of eighteenth-century map criticism.

Early in the century Bradock Mead called for a "Reformation in Geography, by inciting Travellers and Geographers to Care and Diligence in making and collecting their geographical observations." The first thing a mapmaker must do, according to Mead, was

> not take every map that comes out, upon Trust, or conclude that the newest is still the best, but . . . be at the Pains to examine them by the Observations of the best Travellers, that so he may know their Goodness or Defects, and where he finds an Error, correct it in his own: But this not without producing his Authority for every such Alteration in some corner of the Map; and indeed every Alteration should be look'd upon as an Imposition, that was not so confirm'd, because all Copiers are for altering, to make a shew, as if they had corrected or improv'd the Original.[134]

Mead appealed to the mapmakers' commercial instinct by reminding them that "if they rightly consider'd, the Publick Advantage is their private, and where they now sell one Map, they would sell many, if they were correct; for then they would be bought by those that delight in them, for Use, as well as Ornament."[135]

A later critic of maps summed up more succinctly what was required of a mapmaker: "exact Geographical knowledge cannot be obtained by sitting at home."[136] The Rev. Jeremy Belknap thus chided John Norman and his map of New England, admonishing Norman for taking on faith the opinions and surveys of others. Mapmakers should check all the information themselves in the field. By midcentury Philippe Buache was suggesting that young geographers be trained by more experienced geographers and that they be required to present their work to the Académie Royale des Sciences for advice and approbation. He pointed out that this practice was already established for the invention of machines and other scientific projects in order to assure the public that they could have confidence in product. "Geography must not be turned over to those who traffic in imagery."[137] Buache complained that many printed maps were nothing more than "a confused and rude assemblage of pieces brought together without taste or order or based on any scientific principle," a sorry state caused by the "greed of some craftsmen."[138] Such bad works inundating the public made it impossible for *amateurs des cartes* to distinguish good sources of maps. Buache undertook to train young geographers who would be able to defend their works in front of the Académie and continue the struggle against the "commerce d'imagerie."

Government Support

The institutionalization of cartography would have many advantages, among them better recompense for mapmakers. In 1743 Gilles Robert de Vaugondy had argued that claimants to the title of "geographer" should undergo a rigorous examination before they be allowed to publish anything. This would reduce the number of geographers and the number of maps brought out by plagiarists and copyists. Thus the public would not be tricked, and France would receive more honor abroad, where such bad works also go. Good scholarly geographers would not be confused with ignorant ones and the good geographers would be better recompensed for their work, which had cost them dearly to produce.[139]

Philippe Buache and Robert de Vaugondy agreed that the establishment of a "cosmographical society" like those in Nuremburg and Stockholm would be a check against bad work.[140] Robert de Vaugondy further suggested that one, two, or even three more geographical appointments be added to the Académie Royale des Sciences as a true means of supporting geography, which was "very much on the decline in France."[141] The funding for these salaries and institutions would come not from the consumer but from the state.

In England, Bradock Mead made similar pleas for a centralized mapping agency in the first quarter of the century to enhance the nation's honor. "For what could redound more to the British Fame, of improving Arts and Sciences, than the setting so noble a Project on Foot? . . . Geography seems to merit the National Concern, more than it has hitherto experience'd."[142] Emanuel Bowen reiterated the cry at midcentury by calling for a national survey of the coasts and harbors of England "for safety and benefit of navigation." He felt that it was the duty of the "Legislature of this Kingdom" to "raise a sufficient sum by Act of Parliament and cause persons properly qualified to be employed in making a complete survey of the whole coast, and to publish Draughts thereof for the good of the publick."[143] Bowen's suggestions were not completely acted on until the establishment of the Hydrography Office in the Admiralty in 1795.

Governments in both countries, as we have seen, were reluctant to get into the map publishing business; they relied heavily on the private map trade to provide for the needs of both army and navy, both through irregular subsidy of private enterprises and also through access to government materials. In France, the branches of the government responsible for the army, the navy, and the infrastructure of the roads each supported a cartographic service. While the results of their efforts usually remained in manuscript, the employees were not prevented from publishing their maps, and the repositories were even offered as resources to the private mapmaker. The Dépôt de la Marine was open to non-navy personnel who used, and sometimes borrowed, cartographic materials. Didier Robert de Vaugondy frequently mentions using its archives for sources of the maps in the *Atlas universel.*[144] This situation was not always satisfactory to the personnel in charge. The marquis de Chabert, who was head of the dépôt when he was not at sea, expressed particular concern to the minister of the marine over materials being lost because of the ease of access. He requested that material be lent only to naval officers or to "a well-known scholar whose work could be useful, and with my agreement." He wanted pieces to be lent only with some kind of receipt and to be returned within a set period of time, and he asked that dépôt material showing fortifications, port cities, and journals of expeditions be locked up and not shown.[145] Such directives were not issued de vacuo; they were usually in response to actual situations.

Regulatory Agencies

In the last quarter of the century, the officers in charge of the dépôt became more aware that the unregulated use of sources in the dépôt's archive could lead to a proliferation of bad maps. In addition, many French mariners, for lack of maps produced by the dépôt, turned to foreign sources, particularly Dutch charts. An *arrêt du conseil du roi* was issued 5 October 1773 in an attempt to establish the Dépôt de la Marine as the responsible institution for chart production. It decreed that

> all marine charts, portulans, and instructions necessary for the conduct of vessels—warships as well as merchant ships—be composed, prepared, and published exclusively at the dépôt of His Majesty by persons capable of doing the job well and that these works always be accompanied by printed analyses, with an indication of which authorities support the work. [This is done] not only in order to inspire in navigators a just confidence by exposing

> to them in truth the degree of exactitude or doubt contained in these maps in each of their parts, but also in order to protect them from the dangerous uncertainty into which the heap of maps published without such particulars would throw them, maps that, although denuded of materials sufficient for their construction, are announced with fatuous titles, exaggerated in order to promote sales.[146]

The royal decision also made it clear that no individual—no scholar, geographer, hydrographer, Naval officer, or harbor pilot—could publish such nautical works without an express commission.

These new restrictions and redefinition of the role of the Dépôt de la Marine came shortly after the death of the chief hydrographer Bellin, who had been both responsible for and a beneficiary of the sale of charts published by the dépôt. With Bellin's death ended the system of semiprivate enterprise in this division of government mapping. After the death of Louis XV in 1774, a reorganization of the dépôt clamped down on the free and easy access to materials in the dépôt by "outsiders" as Chabert had requested some eighteen years earlier. The royal decision of 1773 allowed the marquis de Chabert and his new young associate Claret de Fleurieu to rethink the old methods of compiling charts, exemplified for over forty years by Bellin, who stayed in his *cabinet* in Paris, never once going to sea. Chabert and Fleurieu wished to change this desk job into a seafaring activity. Both were experienced surveyors, Chabert having traveled and surveyed in the 1740s and 1750s around Newfoundland and the coast of Acadia, Fleurieu on his voyage around the world in the 1760s. Their approach to chart making, manifested finally in the person of Beautemps-Beaupré, would revolutionize and revitalize hydrography in France at the end of the century and into the next. It would also follow the pattern set in the second half of the century by other agencies within the government, such as the École des Ponts et Chaussées, that were pursuing a more active role in training and supporting cartographers.[147]

The government in France also responded to the plight of the good geographer who was being undersold by the "plagiarist and copyist." One result of the plagiarism lawsuit between Delisle and Nolin was an *arrêt du conseil* (16 December 1704) obliging geographers to submit their new maps to a commission of experts, in order to avoid counterfeiting suits such as this in the future. However, to judge from ensuing lawsuits, the regulation appears to have been rarely enforced and little used until eighty years later, when it was reissued and refined. The new version aimed to prevent French geographers from copying foreign maps, so that the errors perpetrated by other nations concerning boundaries would not be perpetuated and that imperfect maps not be used for shipping, endangering commerce and communication. The *arrêt* enjoins:

> all geographers, engravers, and anyone else wishing to engrave, publish, and sell geographical maps of any sort or plans of cities, ports, harbors, bays, coasts, frontiers, or elsewhere to obtain permission from the chancellor or the garde des sceaux. To do this, the manuscript design or the proofs of the maps along with their foundations or proofs in support of how they were made must be presented and examined by the departments that might have an interest in such maps before they may receive permission for publication, under penalty of 600 livres and the seizure and confiscation of the printed maps and plans, as well as proofs and plates.[148]

The reissue of this *arrêt* probably resulted from the contretemps surrounding the publication of the *Atlas du Commerce* by Nicolas Le Clerc in 1785. Le Clerc, a retired doctor, had served many years in Russia; during this time, he became interested in producing this atlas of the Baltic Sea and the Gulf of Finland. He copied from Swedish maps of the area by André Akerman and Jonas Hahn. The underminister of the navy, Claret de Fleurieu, thought the copies were bad enough to make them dangerous to ships. He asked the minister of the navy to stop their publication. Le Clerc replied that the maps of Akerman and Hahn, published originally by order of the Swedish government, were the best available and that his copies only made them more widely known. He further asserted that he had alerted the Dépôt de la Marine to his intentions to publish these works and that the dépôt made no effort to stop their appearance until after all the arrangements had been made. But Fleurieu, on behalf of the navy, pointed out that charts used both by the merchant marine and the navy needed to undergo some kind of critical scrutiny before being published. This decisive moment when a department of state butted heads with freewheeling private enterprise pushed Fleurieu and his successor Beautemps-Beaupré to encourage the navy to take a more active role in producing charts made by survey, not compilation.[149]

Such interventions by the state were only ad hoc measures. Not until government-supported mapping agencies emerged at the very end of the century were the issues of training and consistent standards of accuracy more fully addressed. These developments did not ring the death knell for the private publisher. On the contrary, they merely provided him with a better map to copy.

Some mapmakers objected to the state locking up maps and materials, complaining that the action reduced their sources of information. The "Discours préliminaire" of Julien's edition of Jefferys's *Atlas des Indes Occidentales* (London and Paris, 1777) urged the governments of Europe to remove the "mysterious veil" with which they enveloped their geographical collections. "The fruit of your puerile preoccupations will thus be to keep in ignorance all the men who would make honest use of the knowledge forbidden to them. . . . The modern definition of the word 'homeland' justifies, it is said, this exclusion and many others, but when the uselessness of monopoly is demonstrated by long experience, one would do well to renounce it."[150] In the immediate context of this atlas of the West Indies, these words seemed to refer to Spain and Portugal, whose secretive policies about maps were legendary. They could equally apply to restrictions on access to geographic information promoted by the French government.

But more effective than government control or royal edict were economic forces in changing the nature of the map trade. The mapmakers and publishers of both London and Paris in the late eighteenth century saw their stocks bought up and consolidated into fewer and fewer hands. William Faden used this technique to establish himself as one of the most successful map producers and sellers at the end of the century in London.[151] In addition, he managed to become the sole distributor of Ordnance Survey maps in the infant days of that institution, to the consternation of his fellow map sellers. As Dezauche had done with the Dépôt de la Marine prior to the Revolution, Faden had negotiated advantageous terms with the Ordnance—a 17 percent discount and six months' credit. John Cary and other London map sellers objected to the exclusive arrangement and in a commercial retaliation began to make their own reduced copies of the OS maps as they appeared, in spite of the fact that such actions would infringe the Ordnance Survey's copyright. However, the Board of Ordnance and its chief surveyor, Lt. Col. Mudge, realized that to antagonize the private map sellers would be counterproductive, and they did not wish "to force the Trade into a combination among themselves to injure the sale of the Ordnance Maps."[152] Selling maps paid for maps, and the complexities of distribution and retail work were not part of the institutionalization of cartography.

In Paris, as geographers and mapmakers died, their inventory passed into the hands of the just a few private map publishers. Desnos bought the stocks of de Fer and Jaillot, to which he added work by Brion de la Tour,

Rizzi-Zannoni, and others. Dezauche combined Delisle, Jaillot, and some of the Sanson stock. Delamarche bought the Robert de Vaugondy stock, which included Sanson work, to which he added the Lattré stock.[153] The consolidation of the production end of the market was matched by growing demands for maps from rapidly developing educational markets in both England and France. As the nineteenth century approached, these trends would be translated into the industrialization of map production by larger and larger map firms and the development of geography as an intellectual discipline.[154] Despite these developments, no regulatory agencies for the map marketplace ever emerged. The consumer was still left to be his or her own judge of quality and the market remained open to anyone who wanted to make a map.

CONCLUSIONS

This book has studied many of the economic and consumer-driven forces that affected the printed map trade in France and England during the long eighteenth century. My premise has been that printed maps result from a complex interplay of social, technological, and economic activities. The effect of the printed map on the user is shaped by the choices and limitations imposed upon its maker by the resources available. The printed map traces its source to a manuscript map that incorporated observations of space; these observations resulted from physical effort, such as a survey, or intellectual effort, such as a compiled map, or a combination of both. The manuscript map may be copied, enlarged, or reduced, its information removed, altered, or expanded several times over before a version of it (or several different versions of it) is engraved on copperplate, ready for printing. Thus the information embodied by a printed map has already been distilled several times before the engraving, a process that itself alters the content of the map. Nor does the printed map mark the end of this information highway, for it too may be copied and reprinted for different purposes, for different audiences, by different authorities, further transforming its contents, with different results for its users.

Even though a printed map may have a certain afterlife in the hands and minds of its users, to appreciate fully its effect as a purveyor of information, it is important to understand the various factors that have contributed to its existence and defined its purpose. It is sometimes tempting to overload a map with meaning before we know its complete history and can verify the claims of its title or the story behind its contents. Without pretending to write a history of eighteenth-century cartography, I have tried to highlight those factors in the production, distribution, and consumption of printed maps that may be considered when assessing a printed map's role. By contrasting printed map production in France and England, I have shown that the route from survey to printed map, while essentially the same, has sufficient national variation in its twists and turns to warrant a comparison between the two.

[THE DEVELOPMENT OF A MANUSCRIPT MAP]

French map production benefited from the centralized, institutional support and training provided for surveyors in the army (the *ingénieurs-géographes),* the engineers of the Ponts et Chaussées, the map compilers in the Dépôt de la Marine, the engineers of the *Carte de France,* and the scholarly compilers, the *géographes du roi.* Three kings (Louis XIV, XV, and XVI), a minister (Colbert), and a regent (Philippe d'Orléans), as well as numerous lesser ministers and a state-supported Académie Royale des Sciences, created an environment of cartographic appreciation. In such a context, a map trade could flourish and a national survey could be accomplished.

England could not claim such centralized support for cartographic projects. A parliamentary government reduced the influence of even the most cartographically inclined monarch. Though engineers in the army were trained in some survey techniques and the Admiralty offered erratic support for some local chart making, Parliament did not recognize the role of the scholar-geographer as warranting special status nor give a national survey any priority. Surveying remained a privately funded activity, and only in the last third of the century did the army and navy begin to emphasize survey and reconnaissance as worth the expense, though both institutions took advantage of talent trained elsewhere in Europe.

The difference in support for cartography between the two nations meant that English map publishers were heavily reliant on French and other European maps for cartographic material, except for local surveys.

[THE TRANSITION FROM MANUSCRIPT TO PRINT]

In France, the print trade was free, unregulated in numbers or entry, though most engravers had entered the trade through apprenticeship in the *communautés,* the system of guilds. This provided a deeper labor pool in which a mapmaker could find an engraver to translate his manuscript onto copperplate. It did not necessarily mean a more competitive pool, since the requirements of map engraving were quite specific as opposed to those of print engraving, and engravers specialized as *graveurs en géographie.* While engravers themselves could and did publish maps as well as prints, as the century progressed, mapmakers, whose original work had already been supported institutionally, increasingly dominated the market as publishers.

In England, entry to the craft of the engraver was restricted to members of the companies or guilds that in turn controlled the trade in prints. Furthermore, the number of capable engravers in England, London specifically, was neither large nor particularly skilled. The French Huguenot community in London played a large role in the engraving trade, and French engravers working in London in the 1730s and 1740s were to exert a particular influence on the improvement of English engraving skills. Mapmakers, themselves already few in number and without institutional support, faced a more restricted pool of labor; thus, few mapmakers published their own work. More often engravers published maps, finding manuscript maps in the work of engineers or surveyors who approached them with original material, or hiring unnamed talent to compile material from other printed maps, or merely copying maps from those already published in England or abroad. The number of engravers specializing in maps was limited, and relatively few engraved maps exclusively. Engraver-publishers generally mixed their maps with a more varied inventory of prints in order to survive commercially.

In both countries, the costs of engraving and printing were borne by individual publishers, rarely by an institution or government agency. By the end of the century, this situation was changing in both places. The French Revolution effected the establishment and centralization of government agencies to administer the young republic and the establishment of a national school system. These institutions required the products of the national mapping agencies that already existed, in the army, the navy, and the civil engineering corps of the Ponts et Chaussées. In England, the establishment of Admiralty Hydrographic Office and the Ordnance Survey in the last decade of the century brought disparate mapping efforts under the control of Parliament. By the end of the century, both nations could boast national mapping agencies.

[THE PUBLICATION AND DISTRIBUTION OF MAPS]

The map trade in both France and England sought financial support for publication of maps in similar ways: through solicitation from the government, through subscriptions, through advance advertising, and through borrowing. Maps were sold in a similar manner too: by subscription and delivery, by purchase at the shop, and by correspondence on a national and international level.

As maps entered the public domain in a similar fashion in both countries, so were they similarly exposed to scrutiny. Articles in newspapers and other periodicals and in scientific publications and the analysis of maps in diplomatic treaties and lawsuits, as well as the commentary on maps in geography texts, were manifestations of the public discourse surrounding maps. Where England differed critically from France was in the dearth of mémoires or explanatory texts published to accompany maps, a specialty of French cartography of this period.

From the sources presented here we can conclude that any lines drawn between groups involved in the production of maps are often arbitrary and faulty. The overlapping skills of surveyor, compiler, engraver, and publisher blur the boundaries separating them. Their shared economic interests, crucial to survival, created a mutually dependent world, defined by many interlocking personal relationships. Individuals, rather than corporate groups, loom large in this analysis, whether a particular government minister, a surveyor, an engraver, or a bookseller. Every participant played a significant role in the production of a map, since the labor of each had an economic value and no labor was extraneous to the final production. The consumer exerted an influence on the production of maps not only through sheer buying power, but also through public critique and the private comments made directly to the producer about maps. The world of the printed map trade was small and personal.

The gathering and compiling of data was the most expensive aspect in the production of printed maps. This amount so greatly outweighed the costs of copper and engraving that copying maps that were already published was a more economically viable approach to production, whatever its intellectual and scientific drawbacks. Once a map was engraved, whether from previously printed material or from original surveys and observations, the owner of the plate could manipulate it in many ways. Stocks of copperplates as well as stocks of maps thus had their own economic value, even if their intellectual value was reduced with time and changing information. Yet the value of old stock was realized only when it was sold.

Selling maps became the chief means of recouping the costs of production, even for government-sponsored mapping efforts. However, it was difficult to increase the volume of sales to the point that all the costs of any particular map could be recovered. Map publishers learned to increase their sales stock with products other than maps: with books, prints, and scientific instruments. The burden of inventory and the importance of moving stock highlight the role of criticism of maps, both public and private. The small world of the map trade made all criticism ultimately personal, and the vigorous competition for the market added a sharp edge to public comment. A map producer relied on an intuitive perception of public desire to estimate market demand; he could only guess how effective his prepublication advertising would be in generating that demand. The map business benefited from the participation of publishers, geographers, and surveyors in the public arena of map criticism, the scientific debate on issues such as longitude and measurement, the publication of geographic mémoires, and private correspondence with others involved in aspects of the trade. The map publisher, whether an engraver or geographer or surveyor, was a key figure in creating a market and refining the public's taste for maps.

There is no doubt that in the course of the century the growing role of "science"—the understanding of the importance of the verifiability of phenomena—influenced map production. French and English printed maps from the eighteenth century often reflect new understandings that had been observed: places and peoples heretofore unrecorded in Western Europe. The innovative work of eighteenth-century geographers focused on the problem of determining the exact location of places and peoples, particularly their longitude. The corollary of the longitude investigations was the determination of shapes of landforms and coastlines, as well as the shape of the earth itself. Without the accuracy of location achieved by the large-scale surveying efforts of the eighteenth century, aided by the development and refinement of highly calibrated instruments, the development of thematic maps in the nineteenth century would not have been possible. The eighteenth-century printed map tended to mirror work done rather than show the way to new questions about natural or man-made phenomena.

But this is not to say that the use of the printed map as a tool for understanding and analyzing the distribution of phenomena—observed and hypothetical—did not exist. We need only consider Edmond Halley's maps of trade winds (1686) and magnetic variation (1701)[1] or Philippe Buache's maps showing putative global mountain chains.[2] Maps showing the route of the visibility of the eclipse of the sun across Europe, illustrated in the color plates, also exemplify a printed cartographic response to scientific inquiry. Benjamin Franklin's efforts at mapping the Atlantic Gulf Stream similarly fall into this category.[3] These examples, however, were not common. Except for the eclipse maps, they were generally printed for publication in the journals of scientific societies or for private distribution. Nonetheless, the eclipse maps demonstrate a perception on the part of the map trade that there was a market for the interplay of science and cartography. This market grew during the close of the eighteenth century; by the early decades of the nineteenth it could support the production of a wide range of thematic maps.[4]

The general public did not hunger for printed maps that elucidated scientific hypotheses; they wanted maps that merely looked scientific or made claim in the title to be scientific. With no way to determine the accuracy of maps from their own observations, the consuming public relied on the look and the price of a map. They wanted maps that were printed clearly, intelligible, beautifully engraved, and reasonably priced. Recognizing that Terra Incognita was indeed incognita, map producers like the Delisle family shifted public taste by eliminating the fantastic and the apocryphal from the content areas of maps. No longer did monstrous ships and outsize galleons fill the seas or *indigeni* of odd attire and peculiar dining habits populate unknown lands. Such people and objects were relegated to the margins and cartouches, decorative forms that remained on maps until the end of the century. What became important was that the map claim the authority of science by adding to the title "based on the latest observations" or "surveyed on the spot." Whether true or not, these claims demonstrate a pervasive understanding that a good map was one that showed an area verified by a witness: it was a fact. Disputes over a map's accuracy were disputes over these claims to fact: Who were the observers? What were their instruments? What were their measurements? How was their testimony to be interpreted? These questions of authority were the questions of science, and in that way the rhetoric of mapmaking emulated science, even if the maps themselves, in the never-ending effort to cut costs, might present information that was years out of date.

By assembling a wide range of data in order to study the map trade from the perspectives of the participants, I hope to have highlighted the problems facing the commercial mapmaker of the eighteenth century. Concerns about money and the labor pool, the logistics of anticipating and meeting demand, the difficulties of distribution, and the ongoing debate about quality engrossed even the most derivative of mapmakers and publishers. Their problems become ours when we look at a printed map and begin to draw conclusions from it.

Maps are pictures. When they purport to portray facts, they are among the most deceptive forms of evidence used by historians. Mark Monmonier has asserted that it is "not only . . . easy to lie with maps, it is essential."[5] The lie is rooted in the very nature of mapmaking, as Monmonier has pointed out—the rendering of three-dimensional space onto a two-dimensional surface at a highly reduced scale. The lie blossoms in the economic and social hothouse surrounding the map's production. This is not to deny the printed map its place among important human artifacts; it is only to remind us that it is such an artifact.

APPENDIX I

COSTS OF MAP PRODUCTION IN FRANCE

Year	*Cost (in livres)*	*Comment*
	SURVEYING	
1721	32,000/year	estimated for 3-yr. survey of Brittany (total 96,000)[1]
1729	10,450/year	estimated for 20-yr. survey of Languedoc (total 215,000)[2]
1739/40	34,645	survey of the meridian of Paris (eleven months)[3]
1742	14,098	survey of the line perpendicular to the meridian[4]
1746	40,000/year	survey, *Carte de France*[5]
1756	80,000/year	survey, *Carte de France* (56,000 for personnel, 24,000 equipment)[6]
[1763]	6,000/year	survey meridian in Bavaria, (total 12,000 for two years)[7]

continued

Year	*Cost (in livres)*	*Comment*
	SURVEYING	
Surveyor's wage		
1721	2,000/year	estimated for surveyor in Brittany[8]
	12,000/year	J. N. Delisle to supervise survey of Brittany[9]
1728	26/day	fodder and upkeep for 6 horses, 2 mules[10]
1729	2,825/year	surveyor in Languedoc[11]
1739/40	600–1,200	astronomer's wage[12]
1768	1,400	survey of *seigneurie* of Cobrieux, near Lille[13]
1773	5 Dutch florins per diem	survey of *commanderie* of La Braque in Brabant[14]
1777	1,800/year	*ingénieur-géographe* in army[15]
	30 sous/day	guides (*indicateurs*), army rate
1778	4 sous 6 deniers per arpent	cadastral survey[16]
1779	6 sous/*mencaudée*	1 *mencaudée* = 80–100 sq. rods[17]
	MAP DESIGN AND COMPILATION	
17th cent.	~30/design[18]	
1670	4,000	18 maps by Sanson[19]
1672	1,150	4 maps, a globe, a sphere, and a 6-sheet map of France[20]
	1,800	4 6-sheet wall maps
		(75/plate; 450/map)[21]
	2,700	for 29 maps[22]
1673	176	2-sheet map
	88	1-sheet map[23]

Year	*Cost (in livres)*	*Comment*
MAP DESIGN AND COMPILATION		
1717	60	a map of Italy by Delisle for the duc d'Orléans[24]
1728	600	3 large, 15 small maps of China by d'Anville for Du Halde[25]
1729	800	increased cost for maps of China by d'Anville[26]
1750–53	96	large manuscript map of Côtes de Poitou, environs de Marseilles, and others
	9	Cours de la rivière d'Orne
	72	Isle d'Oleron[27]
1751–69	600–800/year	salary of *dessinateur* (draftsman)[28]
1756	1,200	30 maps by Buache[29]
1757–58	144	100 manuscript maps and plans and a folio volume (very rare) on magnetic needle[30]
1759	750	for maps of Gaul by d'Anville, text and folio map (reduced from 1,200)[31]
1760	1,500	towards a 6- or 8-sheet map of Germany by Rizzi-Zannoni[32]
1763	6/day	draftsman for Rizzi-Zannoni working on map of Germany[33]
1764	240 livres 17 sous	map of Spain, paid to Rizzi-Zannoni[34]
1767	48	calculation for projection of *Carta geografica del Regno di Napoli* (in 4 sheets)
	2,800	layout and corrections of same[35]
1772	4–5/day	for a good *dessinateur*
	8–10/sq. ft.	at 2–3 days/sq. ft.
	9/sq. ft.	compromise price for work on 6×6 ft. map

continued

Year	*Cost (in livres)*	*Comment*
	MAP DESIGN AND COMPILATION	
	24	to reduce 6×6 ft. map to small map
	324–432	for a 6×6 ft. map[36]
1773	400	reduction of Dutch maps and coastal views[37] (six months' work)
1774	2,138	Rizzi-Zannoni's charges for a map of Mediterranean: 600 for 2 *dessinateurs* at 3/day × 100 days 400 to prepare longitude & latitude grid [i.e., projection] 400 to prepare cartographic materials and sketches 570 for application of new information to map 168 for cartouche design[38]
1778	27,600	14-sheet map of Provence, including 100 printed copies, and the reduction to a 2-sheet map[39]
an 2 [1793–94]	5,000 francs	14 maps[40]
1813	800 francs plus 75 copies	for map and commentary, Barbié du Bocage's edition of Sallust[41]
1815	20 livres	6 maps[42]
1817	2,250 francs/year	*dessinateur,* Dépôt de la Guerre[43]
	COPPER	
Unengraved		
17th cent.	10[44]	*per plate*
18th cent.	1 écu/pound[45]	
1765	6	average/plate, various sizes (Bellin, *Petit atlas maritime*)[46]
1767	77	per folio plate, 308 for 4 plates[47]

Year	Cost (in livres)	Comment
	COPPER	
Unengraved		
1771	35	per plate (c. 56 sous/pound)[48]
1779	2 livres 17 sous	per pound (average/plate, 68 livres 8 sous)[49]
1780	50	per folio plate[50]
1795/96	65	per pound[51]
Already engraved or used		
1730	c. 15 sous	per pound, used plate (1 sou/quarto plate red copper; 4 livres 10 shillings/used folio plate)[52]
1744	30	used copper
	70	red copper (with mappemonde already engraved)[53]
1780	8	per plate, old copper[54]
1780	400	for engraved plates[55]
1783	200	for 4 engraved plates[56]
1791	5,000	for 13 folio plates[57]
Copper and engraving (per folio sheet unless otherwise specified)		
1644	33	400 for 12 plates, map of France[58]
1721	500	estimated, for map of Brittany[59]
1724	600	G. Delisle, for map of America[60]
1737	650	for west coasts of Europe and part of Africa
	750	for each plate, 3-sheet map, Mediterranean

continued

Year	*Cost (in livres)*	*Comment*
	COPPER	
Copper and engraving (per folio sheet unless otherwise specified)		
	500	for each plate, 2-sheet map, St Lawrence River, N. America
	600	for each plate, 2-sheet map, Gulf of Mexico
		(prices include rhumb lines and writing)[61]
1748	1,200	for design and engraving map of the Languedoc[62]
1752	220	per plate for maps for the *Atlas universel*[63]
1750–53	900–1,100	per plate for charts, Dépôt de la Marine[64]
1753–55	1,100–1,200	per plate for charts, Dépôt de la Marine[65]
1755–56	1,050–1,350	per plate for charts, Dépôt de la Marine[66]
[1756]	400	per plate[67]
1757–58	1,000–1,200	per plate for charts, Dépôt de la Marine[68]
1759	400	for maps of Gaul by d'Anville; reduced from 600[69]
1766	550–1,320	per plate for charts, Dépôt de la Marine[70]
1767	660–1,050	per plate for charts, Dépôt de la Marine[71]
1768	2,000	per plate, 8,000 for 4 sheets, for map of the diocese of Narbonne by Philippe Buache[72]
1769	323 livres 10 sous	for 2 plates, including paper and printing, for the *Plan du camp de Verberie,* engraved by Delahaye[73]

Year	*Cost (in livres)*	*Comment*
	COPPER	
Copper and engraving (per folio sheet unless otherwise specified)		
1770	1,150	per plate for charts, Dépôt de la Marine[74]
1777	200	map in two plates, from the *Carte de France*[75]
1778	1,000	general map, *Neptune Americo-septentrional*
	600	detailed map, ibid.[76]
1807	1,000 francs	map of Poland[77]
1817	110 francs	per plate, for engraving the *Carte de France*[78]
Copper and engraving/quarto sheet		
1746	50	mineralogical maps of France[79]
1756	70	includes printing[80]
1762	37 livres 10 sous	3,000 livres for 80 plates, for maps prepared by Bellin for the *Petit atlas maritime*[81]
c. 1763	36	reengraving the plan[82]
1765	48	per plate[83]
1769	68	per plate[84]
1782	72	2 small supplements[85]
ENGRAVING (NOTE THAT MANY PRICES *MAY* INCLUDE PRICE OF COPPER)		
Per plate		
1672	84	per plate (includes sea hachures, plans of fortified towns, all place names)[86]

continued

Year	Cost (in livres)	Comment
ENGRAVING (NOTE THAT MANY PRICES *MAY* INCLUDE PRICE OF COPPER)		
Per plate		
1767	15	degrees of latitude, longitude
	72	design of map
	100	line work
	85–90	undulations of water
	250	lettering
	350–520	mountains[87]
1768	700	plan only[88]
1780	800	lettering only[89]
1787	313	lettering, plate 1
	600	plan, plate 2
	417	lettering, plate 2
	719	plan and lettering, plate 3
	400	plan, plate 4
	525	lettering, plate 4[90]
Per 100 words		
late 17th c.	3	italic
	4	roman capitals[91]
c. 1763	6	italic
	24	roman capitals
	3	to retouch italic[92]
1765	6	italic
	24	roman capitals[93]
1768	6	italic
	24	roman capitals[94]
1771	16	
	12 sous	for the title in capitals[95]

Year	Cost (in livres)	Comment
ENGRAVING (NOTE THAT MANY PRICES MAY INCLUDE PRICE OF COPPER)		
Shading hachures		
1756–57	160	map of Archipelago, Dépôt de la Marine
	200	per sheet for 3-sheet map of Mediterranean[96]
Rhumb lines for charts, Dépôt de la Marine		
1755–56	150	to extend lines on plate[97]
1755–56	450	to reengrave a plate of rhumb lines[98]
1757–58	600	a new plate engraved with rhumb lines[99]
Per day		
1780	6	corrections
	8	new engraving[100]
1785/6	12	supervisory[101]
1787	4 livres 8 sous	110 for 25 days[102]
an IX (1800/1801)	8 francs[103]	
1807	10 francs[104]	
Per year		
1672	550	Jaillot hires Caumartin every day, 6 a.m. to 8 p.m. (with meal time)[105]
1770	c. 2,640	at 100 florins/month, 1,200 florins/year[106]
1780	3,000	Delahaye, supervising engraving[107]
1817	2,000 francs	Dépôt de la Guerre[108]

continued

Year	*Cost (in livres)*	*Comment*
ENGRAVING (NOTE THAT MANY PRICES *MAY* INCLUDE PRICE OF COPPER)		
Corrections		
1752	40	per folio plate, Delahaye[109]
1750–53	60	per chart, to engrave longitudes[110]
1753–55	100	chart, New Guinea, new observations
	300	chart, St Dominique, new observations[111]
1755–56	200	to efface about one-half chart of St Dominique[112]
	700	to redo the *Carte de la Mer du Sud* with new observations and rhumb lines
	450	to reengrave separate rhumb line plate with the burin[113]
	300	to engrave different meridians for chart of New Guinea
	150	to prolong rhumb lines on same chart[114]
1763	36	reengraving[115]
1767	75	average/plate, Dépôt de la Marine[116]
1767	48	to retouch mountains
	134	to retouch lettering, etching of mountains
	303	for corrections, reengraving, smoothing plate[117]
1778	2 livres 10 sous	to retouch 100 words
	18	to retouch circles and squares[118]
1780	150	average/chart, Dépôt de la Marine[119]
1782	6	per day, for various plates[120]

Year	Cost (in livres)	Comment
ENGRAVING (NOTE THAT MANY PRICES *MAY* INCLUDE PRICE OF COPPER)		
Cartouches and other noncartographic elements		
1755–56	1,000	frontispiece, *Hydrographie Française*
	1,350	*Tableau des Pavillons, Hydrographie Française*[121]
1756–57	1,900	37 vignettes: plans, ports, views[122]
1757–58	600	frontispiece and 11 vignettes[123]
c. 1763	30 sous	to engrave a stamp[124]
	15	retouch a cartouche
1765	45	frontispiece[125]
1773	100	each view on a quarto plate[126]
1774	1,800	149 views on 25 plates[127]
1787	550	cartouche for a map of Aquitaine[128]
Erasing a plate		
n.d.	48	to efface a map of Gaul[129]
1778	12	to efface the lines from a plate
	12	to efface the words from a plate[130]
PAPER		
Per sheet (grand aigle or chapelet)		
1706	5 sous[131]	
1721	2 sous[132]	
1724	1 sou 5 deniers[133]	
1759	2 sous (including coloring)[134]	
1807	6 sous[135]	

continued

Year	*Cost (in livres)*	*Comment*
	PAPER	
Per quire		
1728	9	large paper for maps & plans[136]
1743	9	grand aigle (975 × 665 mm)
	6	grand colombier (845 × 570 mm)
	3	nom de Jésus (690 × 525 mm)
	2 livres 5 sous	grand raisin (650 × 480 mm)
	1	petit raisin ou carré (585 × 445 mm)
	10 sous	serpente, pas battu
	12 sous	serpente, battu
	1 livre 15 sous	serpent de Hollande
	2	serpente de Hollande, battu
	3 livres 10 sous	serpente verni, pour calquer
	1 livre 5 sous	serpente huilé[137]
Per ream		
1724	35	grand aigle[138]
1756/57	21	quarto[139]
1759	11	carré fin de Montargis[140]
1769	130	grand aigle[141]
1773	125–130	grand aigle[142]
1774	120	grand aigle
	57 livres 15 sous	grand jesus
	18–26	grand raisin
	13 livres 10 sous – 15 livres	ecu en quarré
	9–9½	couronne[143]
1775	90–180	chapelet or grand aigle[144]

Year	Cost (in livres)	Comment
	PAPER	
Per ream		
1778	30	chapelet moyen
	31 livres 11 sous	chapelet fin
	25	chapelet bulle
	24	jesus fin
	21	jesus moyen
	20	grand raisin fin
	13	cavalier fin
	10	cavalier
	6	couronne moyenne[145]
1778	110	grand aigle
	240	grand aigle d'Hollande
	42	jésus
	19	carré fin double[146]
1787	120	grand aigle[147]
1807	130 francs[148]	
Vellum		
1728	2	per sheet[149]
	PRINTING (FOLIO SHEETS UNLESS OTHERWISE NOTED)	
Per 100		
1721	6	estimated[150]
1724	4 livres 10 sous[151]	

continued

Year	*Cost (in livres)*	*Comment*
PRINTING (FOLIO SHEETS UNLESS OTHERWISE NOTED)		
Per 100		
1756–57	30	includes overprinting the rhumb line sheet in green
	4	on quarto paper
	12 livres 10 sous	on grand carré[152]
1759	20[153]	
1767	c. 20[154]	
1769	25[155]	
1778	4	for half sheets
	15	for grand aigle[156]
1779	15[157]	
1786/89	20[158]	
1787	15[159]	
1787	36	grand aigle
	18	half sheets[160]
Paper and printing		
1737	4 sous	includes color after printing[161]
1755–56	10 sous[162]	
1759	6 sous	includes color[163]
1778	10 sous[164]	
1782	15 sous 2 deniers	at 76 livres/100[165]
1787	9 sous[166]	
1807	2 francs 15 centimes[167]	

Year	Cost (in livres)	Comment
PRINTING (FOLIO SHEETS UNLESS OTHERWISE NOTED)		
Printing in color		
1750–53	45	to print rhumb lines from various plate sizes, in red, budgeted over 3 years[168]
1755–56	60	per 100 sheets, from a rhumb line plate in green[169]
MISCELLANEOUS		
Gluing maps on linen		
1751–58	30–42	per year, gluing on linen[170]
1753–55	25	includes a map onto cardboard backing in order to be able to "pointer des routes" without damaging the map[171]
1756	36	includes a 6-sheet Scotland, a 6-sheet England plus two smaller maps[172]
1757	40	includes the Mitchell map of North America (8 sheets) and La Martinique (large paper)[173]
1772	30	to glue map 8 feet square[174]
Cleaning a plate		
1750–53	10	per plate[175]
1765	6	per quarto plate[176]
1779	12 sous	
	48 livres 10 sous	plates for 1,602 maps[177]

continued

Year	*Cost (in livres)*	*Comment*
	MISCELLANEOUS	
Cleaning, collation, coloring, assembling		
1767	2 sous	per map[178]
Coloring		
1708	6 deniers	per map[179]
1787	3 sous	per map[180]
Ink		
1756	19 sous	per pound German black[181]
1756	72	green ink for 600 rhumb line plates[182]
Roller press for copper plate printing		
1683	12	a press with two rollers[183]
1730	35[184]	
1773	400	2 presses for grand aigle paper of 29 inches
	60	4 extra rollers, at 15 livres each
	120	3 tables of grand aigle size, 2 ft. × 4½ ft., at 40 livres each
	250	a large press of iron for pressing charts
	20	a dampener in grand aigle size
	16	2 tables for wiping, 2 inkwells, 2 stretchers[185]
Instruments and other items necessary for mapmaking		
1728	1,600	quadrant of 3½ feet, C. Langlois

Year	*Cost (in livres)*	*Comment*
	MISCELLANEOUS	
Instruments and other items necessary for mapmaking		
	250	plane table, by Langlois
	50	beam compass of 2½ feet, Baradelle
	30	beam compass of 16 inches
	25	beam compass of 9 inches
	20	wing compass, 4 inches
	130	copper measuring stick, Jean Lefèvre
	600	2 clocks with second hands, Julien Le Roy
	400	2 pendulum clocks with second hands, Le Roy
	100	a telescope of 5 feet, covered in leather, Le Bas, le fils
	15	2 thermometers, by Amontans[186]
1755	16	2 rulers in walnut, 4 feet long, and 1 in ebony, 2½ feet, long
	45	3 copper alphabets for tracing and writing titles, plus 2 sets of copper numbers, all sizes
	48	large compass, with changing points for pencil or ink
	6	a case for the compass
	100	glass for tracing and design work
	8	for 1 portfolio, grand aigle, in which to store printed maps[187]
1773	1,920	achromatic telescope, 3½ feet long, from Dolland in London[188]
1778	3,300	quadrant of 2 feet by Bezout, requested by Chabert for Marine[189]

APPENDIX 2

COSTS OF MAP PRODUCTION IN ENGLAND

Year	*Cost*	*Comment*
	SURVEYING (MAY INCLUDE OTHER EXPENSES)	
1679	£1,441 5s	proposed survey of England and Wales for *Atlas Anglicanus,* Seller et al.[1]
1679	20s	town plan for *Atlas Anglicanus,* Seller et al.[2]
1703	£100	survey and 8 copper plates[3]
1722–29	3d/acre	over 100 acres
	6d/acre	enclosed land
	9d/acre	common land
	10s 6d/day	valuing, other work attached to survey[4]
1737	£444 17s 6d	estimate for the survey of the Welsh Coast by Lewis Morris: £182 10s/year for surveyor at 10s/day, working 365 days £100 for 5 surveyor's assistants at £20/year £30 for repairs to surveyor's vessel at 30s/ton £30 for a ship's mate £102 7s 6d for provisions for 7[5]

Year	*Cost*	*Comment*
	SURVEYING (MAY INCLUDE OTHER EXPENSES)	
1738	25 guineas	Survey of St Simon Island, Georgia by Samuel Augsberger[6]
c. 1761	£2,110 12s	reimbursement for survey of Jamaica[7]
1764	£1,117 12s	1 year's work on the survey of the Northern District in North America[8] £100/year for deputy surveyor 7s/day for assistant surveyors 5s/day for draftsman 6d/day for color men, chainmen, signalmen, private men to assist in camp £100/year for extraordinary expenses for horses and guides £416 15s for 2 sets of proper instruments for the surveyor in the first year
1765	£2,000	an "accurate survey" of Devon[9]
1766	£2,778 12s	proposed for first year of survey by William Roy for a military map of England[10]
	£2,700	same survey, for each of the following 6 or 7 years
1769	£350	a survey of Northumberland by Andrew Armstrong, includes equipment with a team of 1 surveyor, 1 assistant, 2 laborers, plus occasional laborers; 3 years preparation resulting in a 9-sheet map[11]
1773	£2,000	survey of the coasts of North America by J. F. W. Des Barres[12] 10s 6d/day for staff 13s 6d/year for food £150/year for equipment and instruments £40/year for an instrument maker, repairs to instruments

continued

Year	Cost	Comment
SURVEYING (MAY INCLUDE OTHER EXPENSES)		
		£80/year for housing £800/year for Samuel Holland and the land surveys 20s/day for Des Barres' salary (£356/year) 10 guineas/year for supplies
1777	£2,400	£400/year, 6 years to survey Sussex to produce an 8-sheet map at 2 miles:1 inch for surveying, drawing, and engraving[13]
1782	1,000 guineas	surveying, drawing, and engraving of map of Somerset, to William Day, "exclusive of his own time and labor"[14]
1787	2d/acre	survey and a fair plan of Sussex at 6 inches:1 mile[15]
1805	£2 2s/day	survey and valuation[16]
DESIGN AND COMPILATION		
1678	£10 6s 10d	a map of New England[17]
1679	£850	for designing county maps for the *Atlas Anglicanus* by using old surveys
	£100	for adding topographical detail to the maps of the *Atlas Anglicanus*[18]
1706	£10 15s	for drafting a plan of the battle of Ramillies[19]
1706	£32 5s	for a plan of Kensington House & Gardens[20]
1745	12 guineas	compilation of surveys for map of Georgia by John Thomas, Jr.[21]
1783	£2 12s 6d	map and valuation of Coveney, Cambridgeshire[22]

Year	Cost	Comment
	COPPER	
Per plate		
1674	20s	Camden's *Britannia*
	55s	6 plates for map of England[23]
1675	12s 9d	map of Cambridge 14 × 17 in., Robert Morden[24]
c. 1679	25s	for maps in Seller's *Atlas Anglicanus*[25]
1703	£100	for 8 plates and survey of Stepney (plate sizes vary: 615 × 247, 470 × 247, 615 × 310, 470 × 510 mm.)[26]
1745–48	3s 3d – 11s	plans and views in George, Lord Anson's *Voyage* (average plate size 22 × 45 cm.)[27]
1769	£1 12s	for each of 9 plates, 24 × 18 in.[28]
1774	£3 2s 6d	large copperplate, chart of Saybrook Bar, Connecticut[29]
1774	£2 2s 8d	B. Roman's map of East Florida
	£1	smoothing the plates for engraving[30]
Per pound		
1674	2s 8d[31]	
1680	3s	planished for etching, any thickness[32]
1700	18d[33]	
1742	2s 4d	elephant plate (27 × 22 in.)[34]
Copperplate and engraving		
1700	£10	2 plates for a globe, for W. Berry and R. Morden[35]

continued

Year	Cost	Comment
	COPPER	
Copperplate and engraving		
	£15	half of map of Greater London, Robert Morden[36]
1784	£5475	for 10 years' work on plates of *Atlantic Neptune*[37]
Value of engraved plate		
1674	5s/plate	from Dürer's drawing book
	4s/plate	from the *English Traveller*
	20s/plate	plates of Canaan, England, Ireland
	55s/plate	6 plates, large map of England[38]
1782	£20 for 9 plates	map of Northumberland[39]
1789	£95 for 6 plates	Yates's map of Staffordshire[40]
Engraving (per plate)		
1675	£8 10s	map of Cambridge, Robert Morden[41]
c. 1679	8s	estimated for maps of Seller's *Atlas Anglicanus*[42]
	£15	£30 for 2 large plates for J. Adam's map of England[43]
1700	5s	coat of arms in border, Gascoyne's map of Devon[44]
1742	£50	maps of South Wales, engraved by Emanuel Bowen
	£10 10s	close work on a map the size of elephant sheet (27 × 22 in.)
	£7 3s	1 sheet, including arms and decoration
	£3 10s	open work such as sea charts, elephant sheet[45]

Year	*Cost*	*Comment*
	COPPER	
Engraving (per plate)		
1745	£2 12s 6d	R. Seale, plan Juan Fernando (22.5 × 40 cm)
	£5 14s 6d	R. Seale, 2 plans on 1 plate (23 × 40 cm)
	£5 15s 6d	R. Seale, 2 maps and plan of Payta
	£4 8s 6d	Wood, view of Tenian (22.5 × 45 cm)
1747	£3 3s	Wood, islands of Lima (3 plans on 1 plate, 20 × 45 cm)
	£11 11s	R. Seale, tract of Acapulco & Victu Bay (23 × 40 cm)[46]
1765	£3 3s	Alesui (ancient Salisbury)[47]
1769	£10	for each of 9 plates (24 × 18 in.)[48]
1774	£18 8s	for Romans's map of east Florida[49]
1774	£15	2 large copper plates, chart of Saybrook Bar, Connecticut[50]
1775	35 guineas	for a plate of the *Atlantic Neptune,* Des Barres[51]
Engraving (per hour)		
c. 1660s	5s	observed by Overton and Garrett in Paris[52]
c. 1670s	12d	charged by Wenceslaus Hollar[53]
Engraving (other rates)		
1747	£1 10s/week	competent journeyman engraver
	10s 6d/day	the best engravers[54]

continued

Year	*Cost*	*Comment*
	COPPER	
Engraving: writing, corrections, and reengraving		
1748	£1 1s	for retouching plate for Anson's *Voyagers*[55]
1748	£1 2s	for engraving titles on 11 plates for Anson's *Voyages*[56]
1773	15s	corrections on the plate of ancient Alesui (Salisbury)[58]
	PAPER (PER REAM)	
1674	8s 6d	Royal (25 × 20 in.) [conjectured price][59]
1734	£3 5s	Genoa (16 × 13 in. = foolscap)[60]
1735	£14	double elephant (40 × 27 in.), at 14s/quire (25 sheets)
	£5 5s	Dutch Imperial (30 × 22 in.), at £2 12s 6d/10 quires
	£3 7s–£3 15s 6d	English Imperial[61]
1742	£2	elephant (27 × 22 in.)[62]
1746	£9	double elephant at 9s/quire[63]
1748	£124 13s 1d	total paper for plans and views for Anson's *Voyages*, 70 plates[64]
1769	£3	(24 × 18 in. = royal or medium)[65]
1773	15s	fine demy (c. 22 × 17 in.)[66]
	PRINTING	
1699	9d/100 sheets	Cambridge, James Child[67]
1705	8d/100 sheets	London, James Child[68]
1708	10d/100 sheets	for Thomas Hearne[69]
1722	10s/1,000[70]	[= 1s/100]

Year	*Cost*	*Comment*
	PRINTING	
1740s	£48 + beer	for 70 plates, number of copies uncertain[71]
1742	6s/100	elephant[72]
1748	£200	for 42 plans and views for Lord Anson's *Voyages*; number of copies uncertain[73]
1769	£20	for 500 9-sheet maps[74]
1773	£4 6s 4d	200 sets of *Alesui* (ancient Salisbury) [£2 3s 2d/100][75]
	COLORING	
1672–1714	6s	1 map[76]
1745	2s 6d	to add color to a print[77]
	MISCELLANEOUS	
1677	£1 6s 6d	pasting and mending maps of Jamaica and Newfoundland[78]
c. 1700	6d	pasting descriptions to maps[79]
c. 1700	3s 6d	mounting on linen with rollers[80]
Instruments		
1764	£21	an astronomical quadrant, survey of North America
	£30	a theodolite with a vertical arc and telescope divided to every minute
	£20	a large theodolite divided to every minute
	£1 10s	a pocket theodolite
	£4 14s 6d	1 azimuth compass

continued

Year	Cost	Comment
	PRINTING	
Instruments		
	£2 12s 6d	1 12-inch round protractor with index and nones to every minute
	£5 5s	a Mecrometer [*sic*]
	£4 14s 6d	1 Hadley's quadrant, 18-inch radius
	£5 5s	1 telescope and rule
	£36 15s	Mr Short's 24-inch reflecting telescope, Rockworth stand
	£40	a Shelton clock or timepiece for astronomical observations
	£6 6s	a pair of globes
	£2 2s	a copying glass
	£1 1s	a brass chain, 50 feet long[81]
1773	£60	an astronomical quadrant, coastal survey, North America
	£40	a reflecting telescope
	3 guineas	a 17-inch globe[82]
	ADVERTISING	
1668	10s/ad	in the *London Gazette*[83]
c. 1750	2s/ad	10–12 lines in the *Daily Advertiser*[84]
	RIGHTS TO ATLAS	
c. 1760	100 guineas	for rights to the *Atlas Methodique,* 1 vol. in duodecimo[85]

APPENDIX 3

COSTS OF MAP PRODUCTION IN NORTH AMERICA

SURVEYING		
1774	2s 3d	100 cleared acres
	4s 6d	100 uncleared acres[1]
1780	15s	per day, common surveyor
	20–40 s	per day, surveyor of ability
	$2–4	per day, surveyors
	$ .50	per day, chain bearers[2]

ENGRAVING		
1786	£1 4s	for engraving arms[3]

PRINTING	
1786	6s/100 sheets[4]
1791	6s/100 sheets[5]

continued

These are the expenses for Osgood Carleton, *An accurate map of the District of Maine; being part of the Commonwealth of Massachusetts . . . with An accurate map of the Commonwealth of Massachusetts Exclusive of the District of Maine* (Boston: Carleton and Norman, 1798). The eight-sheet set of maps was engraved by John Norman.[6]

Initial expenses	
$1,136	for compilation of map by Osgood Carleton
$170	for 8 copper plates
$1,100	for engraving the 8 plates
$105	estimated for paper for 400 sets of maps
$160	for printing 400 sets of maps
$100	estimated for coloring
$2,771	Total

For 100 additional sets of maps (800 sheets)	
$54	paper
$40	printing
$150	doing up on cloth
$25	coloring
$269	Total

APPENDIX 4

MAP AND PRINT PRICES IN FRANCE

MAPS				
Date	*Author*	*Description*	*Sheets*	*Price*
1644	Tavernier	France	12	40 sous[1]
1668	Sanson		1	10–20 sous[2]
1670	Sanson/Jaillot	various	1	6 sous[3]
1706	Delisle		1	10–15 sous[4]
1737	Buache	charts	1	20 sous printed on paper
				50 sous–3 livres printed on vellum[5]
1757	Robert de Vaugondy	folio atlas	108	120–50 livres[6]
1759	Robert de Vaugondy	various	1	1 livre 5 sous[7]
1762	Seguin	Burgundy	15	48 livres[8]
1763	various	various	1–4	1–6 livres
	Bellin	charts	½–2	2 livres/sheet
	various	atlas (folio)		30–150 livres
	Cassini	*Carte de France*	1	4 livres[9]
1766	Robert de Vaugondy	quarto atlas		21 livres[10]

continued

MAPS				
Date	*Author*	*Description*	*Sheets*	*Price*
1770	Cassini	*Carte de France*	1 sheet uncolored	4 livres
			1 sheet on linen	6 livres
			1 sheet colored	8 livres
			1 sheet colored on linen	10 livres
			1 glued on taffeta	14 livres
			1 glued on satin	15 livres[11]
1777	Robert de Vaugondy	general map	1 grand folio	3 livres
		polar map	1	1 livres 10 sous
		quarto atlas	36	21–36 livres[12]
1778	Dépôt de la Marine	charts	1–2	1 livres 10 sous[13]
1779	Dépôt de la Marine	charts from *Neptune Americo-septentrional*	1–2	1 livres 16 sous–3 livres[14]
1791	Dépôt de la Marine	charts	1	1 livres 4 sous[15]

PRINTS		
Date	*Description*	*Price*
1720–40	prints, architectural	18 sous, book of designs for cornices
		12 sous, set of goldsmith's designs
		8 sous, set of fireplace designs[16]
1750–90	prints, decorative	1 livres 4 sous–16 livres[17]
1762	prints, portraits	10 sous–1 livre 4 sous
	prints, flowers	1 livres 4 sous, set of 10[18]

APPENDIX 5

MAP AND PRINT PRICES IN ENGLAND

N.b. The source of the price is the map itself unless otherwise noted.

Date	*Author*	*Description*	*Price*
1674[1]	deWit	2 worlds on cloth	£1 4s
	Speed	100 county maps	18s
	Jenner	100 maps	8s
1676[2]	Seller	maritime coasts of America, in 16 sheets	£1 10s
		chart of West Indies	15s
		Mercator map of W. Indies	12s
1689	Ogilby & Morgan	map of England	1s[3]
1696	Moxon	large wall map of world in 21 sheets	£2[4]
1713	Price & Senex	*New Sett of Maps*	7d[5]
1718[6]	Senex	*English atlas,* 30 maps	£20
		two-sheet map	1s 6d
		one-sheet map	1s
[1720?]	Bowles	John Seller, charts	6d each
		Britannia Depicta, Ogilby improved	10s

continued

Date	*Author*	*Description*	*Price*
		Atlas Minor (62 maps)	12s plain. 15s colored
1732	Moll	single map from *The World Described*	5d plain, 8d colored
		a complete folio of maps from same	5s plain, 6s colored[7]
1745	Rocque	London and 10 miles around in 16 sheets	£2 2s
1748	Rocque	London, including Southwark and Westminster, in 24 sheets	£3 3s
1760–65	Donn	Devon in 12 sheets	£1 10s
		mounted on canvas	£2
1761	Rocque	Berkshire, in 18 sheets	£2 12s 6d [= 2½ guineas]
1767	Burdett	Derbyshire in six sheets	£1 1s by subscription, £2 2s for nonsubscribers
1774–75		1-sheet maps	1s–3s
		2-sheet maps	4s
		4-sheet maps	5s–10s
		multisheet maps	10s 6d–£3 3s
		atlases (quarto)	7s 6d–£1 1s
		atlases (folio)	1–3 guineas
		atlases (foreign)	1–8 guineas[8]
1777	Burdett	Cheshire in 4 sheets	£2 2s
1777	Armstrong	Northumberland in 9 sheets	£1 11s 6d by subscription, £2 2s for nonsubscribers

Date	Author	Description	Price
1777	Armstrong	Ayrshire in 6 sheets	£1 1s
1786	Yates	Lancashire in 8 sheets	£1 11s 6d
		mounted on canvas and colored	£2 7s
1796	Hamilton	West Indies in 4 sheets	10s 6d
1822	Faden	Hampshire in 6 sheets, surveyed by Taylor	£1 10s[9]

COMPARISON OF MAP PRICES WITH PRINT PRICES

In the first half of the eighteenth century, print prices ranged from 1 to 2s for portraits; 4s for two- or three-sheet views; and 2 to 3s for historical prints. By the latter half of the century, prices had risen generally to range from 5s to 10s 6d (half of a guinea) for an ordinary range of prints up to 1 guinea for large prints. Small satirical prints ranged from 6d in the earlier period to 1 – 6s in the later period.[10]

APPENDIX 6

WAGES AND EXPENSES IN FRANCE

WAGES		
Date	*Occupation*	*Wage*
c. 1730	tapestry worker	1–1 livres 10 sous/week: beginner
		2–3 livres/week: advanced apprentice
		4–6 livres/week: worker in flowers
		9–12 livres/week: draper, landscaper[1]
1750–60	astronomer (Dépôt de la Marine)	1,700 livres/year
	copyist (Dépôt de la Marine)	800 livres/year
	dessinateur (Dépôt de la Marine)	600 livres/year
	office boy (Dépôt de la Marine)	400 livres/year[2]
c. 1750	artisan	180–216 livres/year[3]
1755	*ingénieur-géographe* (army)	1,800 livres/year
	lieutenant (army)	1,200 livres/year[4]
1768	printer	13 livres 1 sou 6 deniers/week[5]
1770	printer	3 livres–3 livres 10 sous/day[6]
1771	compositor	4 livres/day
	printer	3 livres/day[7]

WAGES		
Date	*Occupation*	*Wage*
1785	mason	40–50 sous/day[8]
1787	master printer	100 sous/day
	printer's assistant	40 sous/day[9]

EXPENSES		
Date	*Occupation*	*Expense*
1750	English diplomat	£1,200/year to live in Paris (c. 48,000 livres)
		8 guineas/month lodging (c. 160 livres)
		15 guineas/month for coach and horses (c. 300 livres)
		30 shillings/month for his secretary[10]
1750–60	Dépôt de la Marine	c. 720 livres/year for paper, grand aigle
		96 livres/year for pencils
		180 livres/year for wood and matches
		c. 85 livres/year for candles
		9 livres/year to dust the books[11]
c. 1750	artisan	c. 150 livres/year[12]
1782	student	500–600 livres/year[13]
1785	builder	1 sou 6 deniers for a measure of spirits
		5 sous for morning meal
		9 sous for midday meal
		6 sous 3 deniers for 2½ pounds of bread
		6 sous for tools and clothes, wear and tear
		7 sous for rent/day[14]

APPENDIX 7

WAGES AND EXPENSES IN ENGLAND

WAGES		
Date	*Occupation*	*Wage (annual unless otherwise stated)*
1676	William Blathwayt, assistant to the Lords of Trades and Plantations	£150
	keepers of the chamber, supplying fire and candles	£72
	cleaning woman	10s[1]
1675	John Ogilby and William Morgan, cosmographical and geographical printers	£13 6s 8d[2]
1700	army, half pay	3s/day[3]
1707	stud master at Newmarket	£200
	groom at Newmarket	£57
	stable boy at Newmarket	£3
	Turkish girl	£40[4]
1707	Lord chamberlain	£1,200
	page of the backstairs	£80
	poet laureate (Nahum Tate)	£200
	royal trainer (Tregonnet Frampton)	£600[5]

WAGES		
Date	*Occupation*	*Wage (annual unless otherwise stated)*
	half pay, army	5s/day[6]
1776	captain in the infantry	10s/day
	captain in the footguards (household infantry)	16s 6d/day
	captain in Horseguard (or cavalry)	£1 1s 6d/day
	price of commission:	
	for captain, infantry	£1,500
	for lieutenant-colonel, infantry	£3,500[7]

EXPENSES		
Year	*Item of expense*	*Cost*
1674	lease of Thomase Jenner's shop, Royal Exchange	£30[8]
1700	drawing lessons as advertised in the *London Gazette*	5s entrance, 5s per month[9]
1731	Hogarth's *Harlot's Progress,* 6 plates	1 guinea per set
1743	rent for student in Edinburgh	£4
	food for student	4d/week[10]
1734–50	2 pounds chocolate	10s
	bushel of apples	2s 6d
	Overton's boy, a tip	1s
	hogshead of cyder	£2 2s
	year's lodging	£30
	2 bottles of wine for ketchup	3s 6d[11]
1710–50	fee for apprenticeship to map engraver	£22 average[12]
1750–1800	fee for apprenticeship to map engraver	£39 average[12]

continued

EXPENSES		
Year	*Item of expense*	*Cost*
1770s	novel, bound	3s
	novel, paper wraps	2s 6d[13]
1789	London Hackney fares	1s – 5s
	Thames watermen fares, London to Windsor	6d – 14s[14]
1789	courtisans, d'après Boswell	6d – 60 guineas/night
	Vauxhall Gardens, entrance (in 1792	1s 2s)
	musical evenings	1 guinea
	Ranelagh Gardens, entrance plus tea & coffee	2s 6d
	if fireworks	5s
	gardens only, daytime	1s[15]

NOTES

ABBREVIATIONS

AN	Archives nationales, Paris
BL	British Library, London
BNF	Bibliothèque nationale de France, Paris
RGS	Royal Geographical Society, London
WLCL	William L. Clements Library, University of Michigan, Ann Arbor

PREFACE

1. Mary Sponberg Pedley, ed., *The map trade in the late eighteenth century: letters to the London map sellers Jefferys and Faden* (Oxford: Voltaire Foundation, 2000).

INTRODUCTION

1. On consumerism in England and France in the eighteenth century, see Neil McKendrick, John Brewer, and J. N. Plumb, *The birth of the consumer society: the commercialization of eighteenth-century England* (Bloomington: Indiana University Press, 1982); and Daniel Roche, *A history of everyday things: the birth of consumption in France, 1600–1800,* trans. Brian Pearce (Cambridge: Cambridge University Press, 2000).

2. No one interested in the map trade in the Netherlands need look further than the consistently thorough work of Dutch scholars such as Cornelis Koeman, Günter Schilder, and Peter van der Krogt. For example, Cornelis Koeman, *Atlantes Neerlandici,* 6 vols. (Amsterdam: Theatrum Orbis Terrarum, 1967–85), currently being revised by van der Krogt as *Koeman's Atlantes Neerlandici,* 3 vols. ('t Goy-Houten: HES Publishers, 1997–2003); Günter Schilder, *Monumenta cartographica Neerlandica,* 7 vols. (Alphen aan den Rijn: Uitgeverij Canaletto, 1986–2003); Peter van der Krogt, *Globi Neerlandici* (Utrecht: HES, 1993). For one of the most prolific eighteenth-century map publish-

ing houses in Amsterdam, see Marco van Egmond, "The secrets of long life: the Dutch firm of Covens & Mortier (1685–1866) and their copper plates," *Imago Mundi* 54 (2002): 67–85; for a study of an important Nuremberg map publisher, see Markus Heinz, Michael Diefenbacher, and Ruth Bach-Damaskinos, *"Auserlesene und allerneueste Landkarten": Der Verlag Homann in Nürnberg 1702–1848* (Nuremberg: W. Tümmels, 2002).

3. For the English print trade, see Timothy Clayton, *The English print: 1688–1802* (New Haven: Yale University Press, 1997). The Paris print trade has been the subject of two theses, which remain unfortunately unpublished: Corinne Le Bitouzé, "Le commerce de l'estampe à Paris dans la première moitié du XVIIIe siécle," unpublished thesis, Ecole nationale des Chartes, Paris, 1986; and Pierre Casselle, "Le commerce des estampes à Paris dans la seconde moitié du 18ème siècle," unpublished thesis, Ecole nationale des Chartes, Paris, 1976.

4. Aspects of the London map trade are covered in Catherine Delano-Smith and Roger J. P. Kain, *English maps: a history* (London: British Library, 1999); and David Woodward, "English cartography, 1650–1750," in *The compleat plattmaker: essays on chart, map, and globe making in England in the seventeenth and eighteenth centuries,* ed. Norman J. W. Thrower (Berkeley: University of California Press, 1978). Sarah Tyacke's works on seventeenth-century map sellers are listed in the bibliography. See also Donald Hodson, *County atlases of the British Isles published after 1703,* 3 vols. (Welwyn: Tewin Press, 1984–97). In France, the works of Mireille Pastoureau and Monique Pelletier, cited in the bibliography, have taken into account aspects of the map trade. The ongoing work of Catherine Hofmann in preparation for her study of eighteenth-century French atlases and that of Donald Hodson and Laurence Worms on London engravers and mapmakers will add considerably to our detailed knowledge of both markets; both have generously added to this study.

5. The reasons for the choice of copperplate are not entirely clear, even to scholars as distinguished as Antony Griffiths, *Prints and printmaking* (London: British Museum, 1980), 20. While one may argue that intaglio reproduction allows for greater accuracy of detail than woodcut, the work of such eminent figures as Dürer and the wood engravers of the nineteenth century belies this hypothesis. Some other reasons are possible. The intaglio metal plate could produce larger numbers of prints (up to two thousand), thus offering greater marketing potential to the producer. Copperplates were more durable than wood blocks, less susceptible to damp, and more easily repaired, altered, and reengraved. The line work of the burin, the drypoint, stipple, and crayon manner, and mezzotint and aquatint all offered a wider tonal range to reproduction from metal plate than wood could offer.

6. René Taton, ed., *Enseignement et diffusion des sciences en France au XVIIIe siècle* (Paris: Hermann, 1986). See especially articles by Roger Hahn on scientific teaching in the military and naval academies, by Gaston Serbos on teaching in the École des Ponts et Chaussées for civil engineers, and by François de Dainville, S.J., on the training of "géographes" and "géomètres."

7. Douglas W. Marshall, "Military maps of the eighteenth century and the Tower of London drawing room," *Imago Mundi* 32 (1980).

8. Sarah Bendall, *Dictionary of land surveyors and local map-makers of Great Britain and Ireland 1530–1850,* 2nd ed. (London: British Library, 1997); J. B. Harley, "The re-mapping of England, 1750–1800," *Imago Mundi* 19 (1965).

9. The library of James West, president of the Royal Society, was sold at auction over thirteen days and included forty-four lots of maps, which included both antiquarian and contemporary maps, some manuscript, many printed. Langford Auctioneers, "Sale of Library of James West, Esq. deceased, late President of the Royal Society, 19 January 1773 and 12 following days," in *Sales Catalogues* (London: Langford, 1773). Other library sales in this same collection of catalogues include many lots of maps. For more on auction catalogues, see J. B. Harley and Gwyn Walters, "English map collecting 1790–1840: a pilot survey of the evidence in Sotheby sale catalogues," *Imago Mundi* 30 (1978). For France, see Michel Marion, *Bibliothèques privées à Paris au milieu du XVIIIe siècle* (Paris: Bibliothèque nationale, 1978).

10. Robert Walton, 1655, quoted by Sarah Tyacke, "Map-sellers and the London map trade c. 1650–1710," in *My head is a map: essays and memoirs in honour of R. V. Tooley,* ed. Helen Wallis and Sarah Tyacke (London: Francis Edwards and Carta Press, 1973), 66 n. 16. A page of Walton's catalogue is illustrated in Sarah Tyacke, *London map-sellers 1660–1720* (Tring: Map Collector Publications, 1978), 108.

11. Delano-Smith and Kain, *English maps,* 123.

12. Clayton, *English print,* 23.

13. Katie Scott, *The rococo interior* (New Haven: Yale University Press, 1995), 191.

14. Colin Jones, *Madame de Pompadour: images of a mistress* (London: National Gallery, 2002), 132–34.

15. See David C. Jolly, *Maps in British periodicals, pt. 1: Major monthlies before 1800* (Brookline, Mass.: David C. Jolly, 1990).

16. Casselle, "Le commerce des estampes"; Le Bitouzé, "Commerce de l'estampe à Paris."

17. Clayton, *English print.*

18. Harley and Walters, "English map collecting." Further sales catalogues from auctions of private collections and the collections of map sellers may be studied in both the Print Room of the British Museum and the Département d'Estampes of the Bibliothèque nationale de France. These are a rich and as yet untapped resource ready for the statistical analysis that would reveal much about collecting habits and price structures for the eighteenth century.

19. Mary Sponberg Pedley, "The subscription list of the 1757 *Atlas universel:* a study in cartographic dissemination," *Imago Mundi* 31 (1979). J. B. Harley also analyzed subscribers to John Senex's *A new general atlas* (London: D. Browne, 1721) and to John Cary's *New and Correct English Atlas* (1787). J. B. Harley, "Power and legitimation in the English geographical atlases of the eighteenth century," in *Images of the world: the atlas through history,* ed. John A. Walters and Ronald E. Grim (Washington: Library of Congress, 1997), 168–73. More recent work was presented by A. D. M. Phillips, University College, London, at the twentieth International Conference on the History of Cartography in Boston and Portland, June 2003: "The local market for the late seventeenth century local printed county map in England: the subscribers to Gregory

King's Map of Staffordshire, 1679 – 81." It remains a field in which much remains to be done, especially regarding France.

20. Diaries and travel memoirs that are published and indexed quickly reveal that maps play a role in the collections and daily lives of their authors. See, for example, Joseph-Jérôme Le Français de La Lande, *Journal d'un voyage en Angleterre, 1763* (Oxford: Voltaire Foundation, 1980), 102, 106, who describes shopping in London for maps; la Marquise de La Tour du Pin, *Journal d'une femme de cinquante ans, 1778 – 1815* (Paris: Librairie Chapelot, 1920), 361, wherein she notes the perspicacious placement of Moscow on a map of Germany in 1812. The Lennox sisters made use of geography jigsaw puzzles for their children's lessons. Stella Tillyard, *The aristocrats: Caroline, Emily, Louisa, and Sarah Lennox 1740 – 1832* (New York: Farrar, Straus and Giroux, 1994), 213. Joachim Rees, "Supplemente der Erinnerung zum Gebrauch populärer Druckgraphik in Reisetagebüchern des 18. Jahrhunderts," presented at the conference Interkulturelle Kommunikation in der Europäischen Druckgraphik vom 18 zum 19 Jahrhundert, Monte Verità, Ascona, Switzerland, April 2002, notes the purchase of maps at auction in preparation for a journey.

21. Jean Chatelus, "Thèmes picturaux dans les appartements de marchands et artisans Parisiens au XVIIIe siècle," *Dix-huitième siècle* 6 (1974): 309 – 10. For the period 1726 – 1759 Chatelus carefully analyses the various pictorial themes found in the inventories, but does not provide a fixed number or percentage for maps. He finds that the lists of prints often included maps or, more rarely, plans.

22. Peter Stent, catalogue of 1700, quoted in Tyacke, "Map-sellers," 65.

23. The map was published by Peter Scalé and William Richards in 1763. J. H. Andrews, "The French school of Dublin land surveyors," *Irish Geography* 5, no. 4 (1967): 284.

24. Laurence Sterne, *The Life and Opinions of Tristram Shandy, Gentleman,* ed. James Aiken Work (London: R. and J. Dodsley, 1760; reprint, New York: Odyssey, 1940), 88. For the publication of the prints and maps of the duke of Marlborough's campaigns, see Clayton, *English print,* 79.

25. Walter W. Ristow, *American maps and mapmakers: commercial cartography in the nineteenth century* (Detroit: Wayne State University Press, 1985), 19, defines it as "maps . . . produced by private individuals and small commercial publishers."

26. Catherine Delano-Smith, "The map as commodity," in *Plantejaments I Objectius d'una Història universal de la Cartografia,* ed. David Woodward, Catherine Delano-Smith, and Cordell D. K. Yee (Barcelona: Institut Cartogràfic de Catalunya, 2001), 94 – 100. Delano-Smith places most printed maps into the categories of "middling" or "little" maps in terms of their perceived intellectual value. The author is quick to point out the importance of the midden heap of "little" maps towards understanding the place of maps in society. Ibid, 108.

27. Norman J. W. Thrower, *Maps and civilization,* 2nd ed. (Chicago: University of Chicago Press, 1996), 125, examines commercial and government publishing as separate entities. Lloyd A. Brown, *The story of maps* (New York: Bonanza Books, 1949), 174, uses the term "commercial maps" to imply those maps made for public customers by map publishers.

28. J. B. Harley and David Woodward, *The history of cartography*, vol. 1, *Cartography in prehistoric, ancient, and medieval Europe and the Mediterranean* (Chicago: University of Chicago Press, 1987), xvi.

29. Ibid., 2–3.

30. Among others, Harley, "Power and legitimation in the English geographical atlases of the eighteenth century"; Matthew Edney, "Reconsidering Enlightenment geography and mapmaking: reconnaissance, mapping, archive," in *Geography and Enlightenment*, ed. Charles Withers (Chicago: University of Chicago Press, 1999); J. B. Harley, "Maps, knowledge, and power," in *The iconography of landscape: essays on the symbolic representation, design, and use of past environments*, ed. Denis Cosgrove and Stephen Daniels (Cambridge: Cambridge University Press, 1988); Christian Jacob, *L'empire des cartes: approche théorique de la cartographie à travers l'histoire* (Paris: Albin Michel, 1992); Mark Monmonier, *How to lie with maps* (Chicago: University of Chicago Press, 1991); Denis Wood, *The power of maps* (New York: Guilford Press, 1992).

31. Johannes Blaeu, *Atlas Novus* (1654), translated by I. Cunningham, quoted in Charles W. J. Withers, *Geography, science, and national identity: Scotland since 1520* (Cambridge: Cambridge University Press, 2001), [vii].

32. J. B. Harley, *The new nature of maps: essays in the history of cartography*, ed. Paul Laxton (Baltimore: John Hopkins University Press, 2001); Wood, *The power of maps*.

33. See particularly studies in English cartography, such as Tyacke, *London map-sellers;* Tyacke, "Map-sellers"; Laurence Worms, "Location in the London map trade: a talk given to members of IMCOS on 3rd June 2000," *International Map Collectors' Society Journal* 82 (autumn 2000); Laurence Worms, "Thomas Kitchin's 'journey of life': hydrographer to George III, mapmaker, and engraver," *Map Collector* 62 (1993). For French cartography, see, Mireille Pastoureau, "Confection et commerce des cartes à Paris aux XVIe et XVIIe siècles," in *La carte manuscrite et imprimée du XVIe au XIXe siècle: Journée d'étude sur l'histoire du livre et des documents graphiques*, ed. Frédéric Barbier (Munich: Saur, 1983); Mireille Pastoureau, *Les atlas français XVIe–XVIIe siècles* (Paris: Bibliothèque nationale, 1984); Monique Pelletier, *La carte de Cassini: l'extraordinaire aventure de la Carte de France* (Paris: Presses de l'École des Ponts et Chaussées, 1990).

34. While George Vertue was a central figure in attracting French engravers to London, other print sellers associated with the map trade, such as Henry Overton and Thomas Bowles, hired French talent. Clayton, *English print*, 25–48, 55–57.

35. Andrews, "French school."

36. John Green [Bradock Mead], *Remarks in support of the New Chart of North America in six Sheets* (London: Thomas Jefferys, 1753).

37. Pierre Gasnault, *Érudition Mauriste du XVIIIe siècle* (Paris: Institut d'études Augustiniennes, 1999).

38. See, for example, J. B. Harley, "John Strachey of Somerset: an antiquarian cartographer of the early eighteenth century," *Cartographic Journal* (June 1966): 2–7.

39. Thomas Pownall, *A topographical description of the dominions of the United States of America [being a revised and enlarged edition of] a Topographical description of such parts of North America as are contained in the (annexed) map of the Middle British Colonies etc. in North America*, ed. Lois Mulkearn (1775; reprint, Pittsburgh: University of Pittsburgh Press, 1949).

40. Even the bastion of state cartography in France, the Dépôt de la Marine, relied heavily on the private sector for many aspects of production and distribution of its charts, even though many of the maps in the stock of the private map publisher were old or out of date. Olivier Chapuis, *À la mer comme au ciel: Beautemps-Beaupré et la naissance de l'hydrographie moderne (1700–1850)* (Paris: Presses de l'Université de Paris-Sorbonne, 1999), 311.

41. Monmonier, *How to lie with maps,* 1. Speaking of modern maps, Monmonier notes "Yet cartographers are not licensed, and many mapmakers competent in commercial art or the use of computer workstations have never studied cartography. Map users seldom, if ever, question these authorities, and they often fail to appreciate the map's power as a tool of deliberate falsification or subtle propaganda."

42. David Woodward, *Maps as prints in the Italian Renaissance: makers, distributors, and consumers* (London: British Library, 1996), provides a model for this close range survey; readers will find many parallels between the map trade in Italy in the sixteenth century and that in France and England in the eighteenth.

CHAPTER I

1. "Rien de plus commun & de plus facile, que de faire des Cartes; rien de si difficile, que d'en faire de passables. Un bon Géographe est d'autant plus rare, qu'il faut que la nature & l'art se réunissent pour le former." Jacques-Nicolas Bellin, "Remarques de M. Bellin . . . sur les Cartes et les Plans, qu'il a été chargé de dresser . . . ," preface to vol. 3 of le révérend père de Charlevoix, *Journal d'un voyage fait par ordre du Roi dans l'Amérique Septentrionale* (Paris: chez Didot, libraire, 1744), 3:xii.

2. The word "cartographe" did not appear in French until 1877, according to François de Dainville, S.J., *Le langage des géographes: termes, signes, couleurs des cartes anciennes 1500–1800* (Paris: Editions A. et J. Picard, 1964), ix. In English, "cartography" occurs first in 1859, "cartographer" in 1863. *Oxford universal English dictionary,* ed. William Little, H. W. Fowler, and J. Coulson (Oxford: Oxford University Press, 1937), 2:269. Another analysis of the first appearance of the word "cartography" may be found in the preface of Norman J. W. Thrower, ed., *The compleat plattmaker: essays on chart, map, and globe making in England in the seventeenth and eighteenth centuries* (Berkeley: University of California Press, 1978), xv.

3. David Buisseret, ed., *Rural images: estate maps in the Old and New Worlds* (Chicago: University of Chicago Press, 1996); P. D. A. Harvey, *The history of topographical maps: symbols, pictures, and surveys* (London: Thames & Hudson, 1980).

4. Sarah Bendall, "Estate maps of an English county," in *Rural images: the estate plan in the Old and New worlds,* ed. David Buisseret (Chicago: University of Chicago Press, 1996).

5. R. Campbell, *The London Tradesman: Being a compendious View of all the Trades, Professions, Arts, both Liberal and Mechanic, now practised in the Cities of London and Westminster: Calculated for the information of Parents, and instruction of youth in their Choice of Business* (London: T. Gardner, 1747), 274–75.

6. *Freeman's Journal,* 13 January 1770, cited in J. H. Andrews, *Plantation acres: An histori-*

cal study of the Irish land surveyor and his maps (Belfast: Ulster Historical Foundation, 1985), 285.

7. Roger Desreumaux, "Relations entre les arpenteurs et leurs employeurs au XVIIIe siècle dans la région lilloise," *Revue du Nord, histoire et archéologie: Nord de la France, Belgique et Pays Bas,* no. 66 (April–September 1984).

8. Jean Meyer, *La noblesse bretonne au xviiie siècle* (Paris: S.E.V.P.E.N., 1966), 1084–85. The tenant's refusal to cooperate with surveyors is attested from the sixteenth to the eighteenth century. For example, the well-known Elizabethan surveyor Christopher Saxton, in his first survey work after completing the national survey of England and Wales, found that reputation and a permit from the Crown were not guarantees of success. He was seen off the premises of Broughton Hall in Yorkshire by the sitting tenant Henry Tempest with his two sons and a spaniel dog in May 1587. Ifor M. Evans and Heather Lawrence, *Christopher Saxton: Elizabethan map-maker* (Wakefield: Wakefield Historical Publications and the Holland Press, 1979), 80.

9. A. H. W. Robinson, *Marine cartography in Britain* (Leicester: Leicester University Press, 1972), 74.

10. "Il a donc fallu former de nouveaux sujets, faire un choix de ceux qui, par la force de leur âge, par le désire de s'instruire, par les heureuses dispositions de la nature, annonçaient du zèle et de l'ambition pour se distinguer, il fallait les éprouver, suppléer à ce qui leur manquait de connaissances et souvent au défaut de fortune, leur indiquer les *nouvelles méthodes qu'ils devaient suivre, bien différentes de celles de l'arpentage,* les exercer longtemps dans la pratique des observations, dans le calcul, et surtout dans le dessin, pour établir une uniformité tant dans les caractères qui distinguiaient les objets, que dans la représentation du pays." Cassini de Thury described the role of the thirty-four geographical engineers recruited for the survey: two would survey the large triangles and establish the base map for the other engineers who would be charged with making the large scale maps. Four others would be responsible for verifying maps before publication by communicating them to local worthies and curates. Ludovic Drayperon, "Enquête à instituer sur l'éxécution de la grande carte topographique de France de Cassini de Thury," *Revue de géographie* 38 (1896). The information cited by Drayperon is found in the manuscript collection of Louis-Henri Charlet, one of the subscribers to the *Carte de France.*

11. "Nous ne nous sommes pas transportés dans chaque village, dans chaque hameau pour en lever le plan. . . . Nous n'avons pas visité chaque métairie, suivi et mesuré le cours de chaque rivière, . . . Un tel détail ne peut convenir qu'à des plans particuliers de quelque terre seigneuriale, la grandeur raisonnable à laquelle on doit fixer la carte d'un pays ne permet pas qu'on y puisse marquer tant de choses sans une extrême confusion. . . .

Nous avons préféré le sommet des montagnes et des clochers élevés pour découvrir la campagne. Là, avec le secours de plusieurs indicateurs pris dans les villages les plus voisins, nous nous faisions montrer et nommer tous les lieux qui s'offroient à notre vue, nous prenions même les alignemens des lieux enfoncés dans les vallons, qui ne pouvoient être aperçus d'aucun endroit, lorsque l'indicateur se reconnoissoit assez bien pour ne pas le manquer; et nous dressions, sur les lieux mêmes, la carte de ce canton au moyen des angles que nous avions pris et sur l'estime des éloignemens respectifs en y joignant le cours de rivières, des chemins, etc. dont nous prenions, indépendamment de cela, des mémoires avec des dessins dans tous les lieux où nous pouvions." Archives départemental, Hérault, C.États 138, cited in François de Dainville, S.J., "La levée d'une carte en Languedoc à l'entour de 1730," in *La cartographie reflet de l'histoire* (Geneve: Slatkine, 1986), 369–70. Also in Monique Pelletier, *Les cartes des Cassini: la science au service de l'état et des régions* (Paris: Comité des travaux historiques et scientifiques, 2002), 182.

12. Matthew Edney, "Reconsidering Enlightenment geography and mapmaking: reconnaissance, mapping, archive," in *Geography and Enlightenment,* ed. Charles Withers (Chicago: University of Chicago Press, 1999), 170–75.

13. François de Dainville, S.J., "Enseignement des 'géographes' et des 'géomètres,'" in *Enseignement et diffusion des sciences en France au XVIIIe siècle,* ed. René Taton (Paris: Herman, 1986).

14. Douglas W. Marshall, "Military maps of the eighteenth century and the Tower of London drawing room," *Imago Mundi* 32 (1980), 24.

15. Sarah Bendall, *Dictionary of land surveyors and local map-makers of Great Britain and Ireland 1530–1850,* 2nd ed. (London: British Library, 1997), 39–40.

16. Benjamin Donn, *An epitome of Natural and Experimental Philosophy including Geography and the Uses* (London: Law & Kearsley; London: Heath and Wing; Bristol: Donn; [1769]), endpaper.

17. For example, the seventeenth-century Down Survey of Ireland remained in manuscript. Copies were made on demand, for a price. Andrews, *Plantation acres,* 85.

18. "Un bon Géographe est d'autant plus rare, qu'il faut que la nature & l'art se réunissent pour le former. Il doit tenir de la première la mémoire, l'amour pour le travail, la patience & un esprit d'ordre & d'arrangement; de l'autre des connoissances suffisantes dans la Géométrie & dans l'Astronomie, après lesquelles viennent l'étude longue & sterile des Voyageurs, la discussion critique de leurs Relations & de leurs Journaux, sources continuelles d'incertitudes & d'erreurs, que souvent le travail le plus assidu ne sçauroit vaincre: joignez à cela quelque intelligence des Langues Etrangères." Bellin, "Remarques," 3:xii.

19. Jacques-Nicolas Bellin, *Observations sur la carte de la Manche, dressée au Dépôt des cartes, plans & journaux de la marine, pour le service des vaisseaux du roi . . . en 1749* (Paris: l'Imprimerie de Didot, 1750), 3.

20. Jean-Baptiste Bourguignon d'Anville, *Considérations générales sur l'étude et les Connoissances que demande la composition des Ouvrages de géographie* (Paris: Lambert, 1777), 53.

21. "Il faut passer des tems considérable à se préparer et à rassembler les connaissances nécessaires, et souvent avec le travail le plus assidu, à peine peut-on se flatter de pouvoir vaincre les difficultés qui se présentent." Bellin, *Observations,* 3.

22. François de Dainville, S.J., *La géographie des humanistes: les Jésuites et l'éducation de la société française* (Paris: Beauchesne, 1940); François de Dainville, S.J., *L'éducation des Jésuites (XVIe–XVIIIe siècles)* (Paris: Editions de Minuit, 1978). See especially in the latter, "L'enseignement de l'histoire et de la géographie et la 'ratio studiorum,'" 427–54.

23. De Dainville, *L'éducation des Jésuites,* 25–42.

24. "Les hommes ne sont pas néz pour mésurer des lignes, pour examiner le raport des angles, et pour emploier tout leur temps à considérer les divers mouvements de la matière. Leur esprit est trop grand, la vie trop courte, leur temps trop précieux pour l'occuper à des si petits objets." Quoted ibid., 332. Jean Bouhier (1673–1746) was born in Dijon and educated by the Jesuits of that city, though his teachers' interests in science and math seem not to have taken hold of their pupil. He was an exceptional linguist, member of the Dijon parlement and of the Académie Française. His works covered a wide range of jurisprudence, ancient history, and literature.

25. Mireille Pastoureau, *Nicolas Sanson d'Abbeville: atlas du monde 1665* (Paris: Sand & Conti, 1988), 14–15.

26. Jean Mabillon, *Traité des études monastiques* (Paris, 1691), 384, quoted in Pierre Gasnault, *Érudition mauriste du XVIIIe siècle* (Paris: Institut d'études Augustiniennes, 1999), 32. "La verité qui est la partie la plus essentielle de l'Histoire et qui consiste dans la certitude des faits." Dom Rivet, *Histoire litteraire de la France,* 1:xxiii, quoted in Gasnault, *Érudition mauriste,* 32.

27. Gasnault, *Érudition mauriste,* 37.

28. Desnos, "Inventaire aprés décès de Mde Desnos," AN, MC, LXXV (745), 16 February 1778; Desnos, "Liquidation et partage de la communauté d'entre le S. Desnos et la dlle. Loy sa femme et de la succession de lad. Dame," AN, MC, LXXV (824), 10 February 1785.

29. Margaret Bradley, "The financial basis of French scientific education and scientific institutions in Paris, 1790–1815," *Annals of Science* 36 (1979): 451.

30. Mary Sponberg Pedley, *Bel et utile: the work of the Robert de Vaugondy family of mapmakers* (Tring: Map Collector Publications, 1992).

31. A genealogy of the Delisle stock of maps may be found in R. V. Tooley, *The French mapping of the Americas: The Delisle, Buache, Dezauche succession (1700–1830)* (London: Map Collectors' Circle, 1967). Tooley's summary is out of date and does not include many important aspects of the contributions of these geographers to the history of cartography. The sale of the Delisle-Buache stock to Jean Claude (not J. A.) Dezauche is recorded in AN, MC, XXVIII (481), 5 June 1780.

32. Wilfrid Prest, *Albion ascendant: English history, 1660–1815* (Oxford: Oxford University Press, 1998), 174–78.

33. Campbell, *The London Tradesman,* 19–20.

34. See particularly the work of Sarah Tyacke and Laurence Worms on this subject.

35. The exceptions appeared in the late seventeenth and early eighteenth centuries, in the persons of Joseph Moxon, who was expelled for not paying his dues, John Senex, and Henry Popple. Later geographers to the king, such as Thomas Jefferys and William Faden, were not members.

36. See chapter 5 for a further discussion of maps in the life of the Royal Society.

37. Peter Barber, "Maps and monarchs in Europe, 1550–1800," in *Royal and republican sovereignty in early modern Europe,* ed. Robert Oresko, G. C. Gibbs, and H. M. Scott (Cambridge: Cambridge University Press, 1996), 88, 95.

38. G. R. Crone, "Further notes on Bradock Mead, alias John Green, an

eighteenth century cartographer," *Imago Mundi* 8 (1951); G. R. Crone, "John Green: notes on a neglected eighteenth century geographer and cartographer," *Imago Mundi* 6 (1949). The letter from Thomas Jefferys identifying John Green as Bradock Mead is in the E. E. Ayer Collection of the Newberry Library (Ayer 7115, Ms. 447).

39. Bradock Mead, *The Construction of Maps and Globes* (London: Horne, Knapton, et al., 1717). From the dedicatory preface to Samuel Molyneux, Esq., secretary to his royal highness the prince.

CHAPTER 2

1. Joseph-Nicolas Delisle, "Mémoire sur la carte de Bretagne," Newberry Library, Chicago, ms. sc 1793, 1721.

2. Monique Pelletier, *Les cartes des Cassini: la science au service de l'État et des régions* (Paris: Comité des travaux historiques et scientifiques, 2002), 122. Forty thousand livres reflected Cassini de Thury's initial budgeting in 1748. By 1757, he had to revise this figure upward to 4,500 livres per sheet, with 1,000 livres held back until the work had been checked. He allowed 300 livres per sheet to the *verificateur.*

3. François de Dainville, S.J., "La levée d'une carte en Languedoc à l'entour de 1730," in *La cartographie reflet de l'histoire* (Geneve: Slatkine, 1986), 364.

4. Henri Berthaut, *Les ingénieurs-géographes militaires 1624–1831,* 2 vols. (Paris: Imprimerie du Service géographique, 1902).

5. Later in the century it was recognized that an understanding of local place names in Celtic could be an aid to topographic description of the area. J.-N. Buache de la Neuville, Bibliothèque de l'Institut, Paris, ms. 2311 (2), "Observations sur la géographie de la province de Bretagne."

6. De Dainville, "La levée d'une carte en Languedoc à l'entour de 1730," 363.

7. Sarah Tyacke, "Map-sellers and the London map trade c. 1650–1710," in *My head is a map: essays and memoirs in honour of R. V. Tooley,* ed. Helen Wallis and Sarah Tyacke (London: Francis Edwards and Carta Press, 1973), 70.

8. William Ravenhill, *A map of the County of Devon . . . by Benjamin Donn* (Exeter: Devon and Cornwall Record Society and the University of Exeter, 1965), 8.

9. J. B. Harley, "Origins of the ordnance survey," in *A history of the ordnance survey,* ed. W. A. Seymour (Folkestone: Dawson, 1980), 8. Harley cites Roy's proposal to the king, published in J. Fortescue, ed., *The correspondence of King George the Third from 1760 to December 1783*(London: Macmillan, 1927–28), 1:328–34.

10. A. H. W. Robinson, *Marine cartography in Britain* (Leicester: Leicester University Press, 1972), 77. The surveyor was Lewis Morris.

11. J. B. Harley, "The Society of Arts and the surveys of English counties 1759–1809," *Journal of the Royal Society of Arts* 112 (1964): 123.

12. Sarah Bendall, *Dictionary of land surveyors and local map-makers of Great Britain and Ireland 1530–1850,* 2nd ed. (London: British Library, 1997), 40.

13. Bernard Romans, *A concise natural history of east and west Florida* (New York, 1775), 193, 195–96.

14. R. A. Skelton, "The origins of the Ordnance Survey of Great Britain," *Geographical Journal* 128, no. 1962 (1962): 419.

15. AN, MC, XV (239), 22 February 1673, and MC, CXXII (500) 12 July 1700, cited in Mireille Pastoureau, *Les atlas français XVIe–XVIIe siècles* (Paris: Bibliothèque nationale, 1984), 231–33. Other prices for maps contracted by Jaillot from Sanson fall in the range of 60–100 livres per sheet.

16. Henri Cordier, "Du Halde et d'Anville (Cartes de la Chine)," in *Recueil de mémoires orientaux par les Professeurs de l'Ecole des Langues orientales* (Paris: E. Leroux, 1905).

17. Philippe Buache to M. Maries, 30 June 1756, in Philippe Buache, "Programmes, cahier, cartes pour les leçons géographiques des quatre fils du Dauphin, fils de Louis XV," BNF, Réserve Ge. DD. 2025, 1735–1760; Elizabeth Rodger, "An eighteenth century collection of maps connected with Philippe Buache," *Bodleian Library Record* 7, no. 2 (1963).

18. AN, Y 1903, 31 August 1772, d'Anville's evaluation of a *Carte des Postes de France,* to settle a dispute over pay between Mr Richard, the Contrôleur Générale des Postes, and the draftsman, le Sr. Scellier. "La journée d'un dessinateur dans un travail où il ne s'agit point des se signaler par de l'invention et du génie, sera bien payée sur le pied de 4#."

19. Giovanni Antonio Rizzi-Zannoni, "État du Travail relative à la Carte de la Mer Mediterranée," AN, MAR 1JJ/5/16.

20. Christian Sandler, *Die Reformation der Kartographie zum 1700* (Munich: R. Oldenbourg, 1905).

21. Guillaume Delisle moved to the quai de l'Horloge when he married in 1706; his shop was at the Couronne de Diamants. See Claude Delisle to Louis Renard, 11 December 1706, AN, MAR 2JJ/60/12.

22. This connection is elaborated by Nelson-Martin Dawson, *L'atelier Delisle: l'Amérique du Nord sur la table à dessin* (Sillery, Quebec: Editions du Septentrion, 2000), 103–8.

23. Dennis Reinhartz, *The cartographer and the literati: Herman Moll and his intellectual circle* (Lewiston: Edwin Mellen Press, 1997).

24. AN, MAR 1JJ/5/5 and 6/1 for 1773 and 1774. Rizzi-Zannoni, l'Huillier, Bonne, Méchain, and Le Moyne père each were budgeted at 4,000 livres per year.

25. "Qui est venue en france pour cet effet and pour m'offrir du Service en Angleterre." Cambernon de Bréville, "État de Services de S. Cambernon de Bréville," AN, MAR 1JJ/6/1, 16 March 1774. "J'ai refusé constamment ses propositions malgré les besoins d'argens où je me trouvais par Esprit de patriotisme et par un point d'honneur qui ne me fera jamais manquer à ceque je dois à mon Roy, à Ma patrie, et à moimême. M. Grante de Blairfraindy Ecossais demt. à Versailles Colonel à la Suite de la Legion Royale peut certifier s'il en est requis que je n'avance rien de trop car il a été témoin oculaire des propositions que l'on ma fait et de mon réfus."

26. "Une quantité considérable d'Estampes les plus indécentes" AN, Y 11585C, 7 and 8 December 1768. For the full story, Mary Sponberg Pedley, "Gentlemen abroad: Jefferys and Sayer in Paris," *Map Collector* 37 (1986).

27. Bibliotheca Apostolica Vaticana: Vat Lat 9812 (100–102), d'Anville to Passionei, 19 February 1752, 29 April 1752. Gabriel Marcel, "Correspondance de Michel Hennin et de d'Anville," *Bulletin de la Section de Géographie Historique et Descriptive* 3 (1907).

28. François de Dainville, S.J., "Cartes de Bourgogne du XVIIIe siècle," in *La cartographie reflet de l'histoire,* ed. Michel Mollat du Jourdin (Geneva: Editions Slatkine, 1986), 78.

29. Gilles and Didier Robert de Vaugondy, *Atlas universel* (Paris: Gilles and Didier Robert de Vaugondy and Antoine Boudet, 1757), 26a, 31b.

30. BL, Add. Ms. 9767 (53) Establishment & Accompts of the Committee for Trade and Plantations. The price here also includes sending Mr Randolph to Cornbury, perhaps to fetch it.

31. E. G. R. Taylor, "Robert Hooke and the cartographical projects of the late seventeenth century (1666–1696)," *Geographical Journal* 90 (1937): 539. Taylor does not give the documentary source of these figures.

32. BL, Egerton 3809 (153) and (154).

33. J. H. Andrews, *Plantation acres,* 85.

34. Ibid., 74. The chief protractor for the Trustees' Survey of 1700–3 was paid £200 per year. Andrews points out that only £50 in copying fees were gathered per year, making this method of dissemination an unsatisfactory way of paying the costs of the survey.

35. "Deux personnes peuvent copier cinquante plans ou desseins tandis que six ne feront point quatre cartes qui demandent beaucoup de travail, de discussions, de combinaisons et d'essais." Pierre-François Méchain to Joseph Bernard de Chabert, 23 March 1787, AN, MAR 1JJ/44/8, cited by Olivier Chapuis, *À la mer comme au ciel: Beautemps-Beaupré et la naissance de l'hydrographie moderne (1700–1850)* (Paris: Presses de l'Université de Paris-Sorbonne, 1999), 228, 953 n. 123.

36. This was the case for individual sheets of the *Carte de France,* for which many of its 174 sheets were only printed in runs of fewer than 200. Pelletier, *Les cartes des Cassini: la science,* 232–34.

37. Such presentations were also good advertising for the public distribution of maps. The French astronomer Joseph-Nicolas Delisle sent his map of the putative discoveries of Admiral de la Fuente to Thomas Birch, secretary of the Royal Society in London, in order to have it checked for veracity. He also sent copies to the geographer cum engraver Thomas Jefferys for speculative sale at c. 2 livres per map, with the hope this might stimulate demand for more. Delisle to Jefferis [*sic*], Paris: Bibliothèque de l'Observatoire, Manuscrits, B.1.7 (216), 6 October 1753.

38. "Lui a couté plusieurs années et a pas été entrepris par un motif pécuniaire, quoiqu'il ne fut pas raisonable d'exiger qu'il en fasse un pur don." Jean-Baptiste Bourguignon d'Anville, BNF, ms. fr. 22147 (3).

39. "Pour me recompenser des fraix de l'impression et de la gravure et me mettre en état de continuer à publier de pareils ouvrages." Delisle to Maire, Bibliothèque de l'Observatoire, Paris, B.I.7(84), 2 April 1753. Delisle was sending thirty copies of a small world map showing the visibility of the passage of Mercury across the sun to friends in Italy (ten to Florence, twenty to Turin).

40. Andrews, "French school."

41. De Dainville points out that tin was sometimes used for engraving, and the English engraver George Vertue claimed to have used pewter to engrave a map of London in order to make the map look like a woodcut map. Vertue's eight-sheet map, *Civitas Londinium Ano. Dmi. Circiter MDLX,* was based on "an ancient print in the possession of Sir Hans Sloane" attributed to Ralph Agas. James Howgego, *Printed maps of London circa 1553–1850,* 2nd ed. (London: Dawson, 1978), 11–12. This map was brought to my attention by Jonathan Potter. Item 151 in Jonathan Potter, *Antique maps from Jonathan Potter [catalogue 22]* (London: Jonathan Potter, 2003).

42. Coolie Verner, "Copperplate printing," in *Five centuries of map printing,* ed. David Woodward (Chicago: University of Chicago Press, 1975); David Woodward, *Maps as prints in the Italian Renaissance: makers, distributors, and consumers* (London: British Library, 1996). It is also well covered in François de Dainville, S.J., *Le langage des géographes: termes, signes, couleurs des cartes anciennes 1500–1800* (Paris: Editions A. et J. Picard, 1964).

43. Verner, "Copperplate printing," 52.

44. François Courboin, *L'estampe Française* (Brussels: Librairie d'Art et d'Histoire, G. Van Oest & Co., 1914), 33; François de Dainville, S.J., *Cartes anciennes de l'église de France* (Paris: Vrin, 1956), 254.

45. AN, MAR B3. 699: 24–25, cited in Jonathan Dull, *The French navy and American independence* (Princeton: Princeton University Press, 1975), 256–57.

46. "Lève, dessine, et grave avec goût." *L'année littéraire* 1 (1765), 143–44, cited in de Dainville, "Cartes de Bourgogne," 75.

47. Chapuis, *À la mer,* 534–36.

48. Laurence Worms, "Thomas Kitchin's 'journey of life': hydrographer to George III, mapmaker, and engraver," *Map Collector* 62 (1993). Worms points out the Nonconformist connection between Kitchin and his master and father-in-law, Emanuel Bowen, a Welsh Baptist.

49. In the lawsuit between the engravers Delahaye and Antoine Boudet, publisher of Robert de Vaugondy's *Atlas universel,* several engravers are mentioned: Baisiez and Laurent, who engraved italics; Guillaume Delahaye, who supervised and etched the mountains; and Mlle de la Rivière, who used a punch for the boundary divisions. Mary Sponberg Pedley, "New light on an old atlas: documents concerning the publication of the *Atlas universel* (1757)," *Imago Mundi* 36 (1984); BNF, Ge. FF. 13374, fols. 49, 53.

50. "L'ondulation de la mer et des rivières et la retouche au burin." François de Dainville, S.J., *La carte de la Guyenne par Belleyme 1761–1840* (Bordeaux: Delmas, 1957), 24.

51. A good summary of these techniques may be found in Antony Griffiths, *Prints and printmaking* (London: British Museum, 1980).

52. Lattré used the same technique for a double-hemisphere world map designed by the astronomer Joseph-Jérôme de La Lande (*Figure du Passage de Venus sur le Disque du Soleil qu'on observera le 3 juin 1769*). The geographical portion of the map is printed in black and the lines depicting the passage of Venus in red. My thanks to Ed Dahl for pointing out this map to me.

53. NL, Case Ms 5191.

54. Timothy Clayton, *The English print: 1688–1802* (New Haven: Yale University Press, 1997), 214.

55. Ibid., 176; J. B. Harley and Paul Laxton, introduction to *A survey of the County Palatine of Chester by P. P. Burdett (1777)* (Liverpool: Historical Society of Lancashire and Cheshire, 1974), 6. Burdett may have learned the technique in France, where he toured in 1771, though the prime developer of aquatint, Jean Baptiste Le Prince, was renowned for his reluctance to share knowledge of the technique.

56. Andrew Cook, "Alexander Dalrymple's *A Collection of Plans of Ports in the East Indies* (1774–1775): a preliminary examination," *Imago Mundi* 33 (1981): 51–58.

57. R. Campbell, *The London Tradesman: Being a compendious View of all the Trades, Professions, Arts, both Liberal and Mechanic, now practised in the Cities of London and Westminster: Calculated for the information of Parents, and instruction of youth in their Choice of Business* (London: T. Gardner, 1747), 113–14.

58. Pastoureau, *Atlas français,* 169.

59. Ibid., 231. One of the contracts between Jaillot and Cordier (AN, MC, XV [235], 7 March 1672) specified that Cordier's engraving would include the cartouches, the hachures for the sea, plans of fortified towns, and "générallement toutes les autres choses qui composent le plan d'une carte . . . comme aussy tous les grandz noms contenus en chascune desdites cartes. Et à l'esgard de la graveure des petites lettres, ledit Cordier s'oblige de les faire graver par les deux personnes qui seront par luy choisies (and generally all the other things which make up the plan of a map . . . as well as all the large names contained in each of the said maps . . . And with regard to the engraving of the small letters, the said Cordier is obliged to have them engraved by two people who will be chosen by him). Cordier had to promise to devote all his time to work from Jaillot, except for three months when he could work for Sanson. Jaillot, in turn, agreed to lodge the engraver "en un appartement sur le devant."

60. "Mon graveur, que je paie pour travailler toute l'année." Guillaume Delisle, draft of letter regarding a map of Provence, AN, MAR 2JJ/60/162.

61. Marianne Grivel, "Le cabinet du roi," *Révue de la Bibliothèque nationale,* no. 18 (1985): 43–45. The budget for the large plates that formed the reproduction of the royal collections allowed 500–600 livres per square foot of engraving with the burin, 200 livres for etching. Costs still varied depending on the size of the plate and the skill and celebrity of the engraver, ranging from as little as 120 livres for one of the plates of the *Médailles antiques* to 1,200 livres for *La sainte famille* of Raphael, engraved by Gérard Edelinck.

62. Corinne Le Bitouzé, "Le commerce de l'estampe à Paris dans la première moitié du XVIIIe siécle," unpublished thesis, Ecole nationale des Chartes, Paris, 1986, 1:20–21.

63. "Réduit à la condition de simple Manouvrier, puisqu'il n'est payé qu'au Cent, le Graveur en Lettres n'a d'autre interêt que d'aller vite en besogne. Cette préoccupation occasionne des fautes d'ortographe, des déplacements de noms et des Échappées de Burin qui dégradent plus ou moins, la pureté des contours et les teintes des ombres qui figurent sur le plan, les Routes, les eaux, les Montagnes, les Bois et les Terreins dans les divers États de Culture." Guillaume Delahaye, "Rapport sur les travaux du

Citoyen Delahaye, graveur en géographie et en topographie, 29 thermidor an II [16 August 1794]," Paris: Conservatoire nationale des arts et métiers, M-404.

64. "Lorsqu'on me donné un Dessin Topographique pour graver, je prends un papier huilé bien transparent que je pose sur le dessin, bien tenu, pour qu'il ne varie point, je tire le cadre interieure, dessine à l'Encre de la Chine le plan des Montagnes, Roches, Bois, et Rivieres, puis j'ecris tous les mots, donnant à tous la disposition la plus agréable, rectifians les mots mal placés, afin qu'il n'yait aucune lettre coupée par des Rivières ou cachée par des Bois.

"Cette operation faite avec beaucoup d'exactitude, je dispose sur la planche vernie qui doit recevoir le dessin, un papier serpente blanc, rougi du côté du vernis, puis je mis mon dessin huilé et papier blanc rougi sur la planche, en l'assûrant pour qu'il ne puisse varier, rabattant les bords du dessin par dessus la planche; puis avec une pointe d'acier arrondie, je dessine Seulement toute l'Ecriture indiquée dans le Dessin, et je tire le cadre intérieur très exact, Ensuite je lève de dessus la planche le papier huilé et celui rougi, et dessine sur le vernis, avec une autre pointe d'acier, entire le Cadre à la Règle et très juste; ce qui est essential, pour reposer les papiers huilé et rougi, après l'Ecriture gravée, que je grave avec burin.

"J'ai fait differens Elèves dans le plan et l'Ecriture, qui se sont distingués. Il y en a quatre dans le plan et deux dans l'Ecriture, mais ils ne réunissent pas toutes les parties de cet art. Cequi fait que très souvent ceux qui gravent le plan, ne sachent pas disposer l'emplacement de l'Ecriture, elle est très mal placée, communément, et à travers les Bois, Montagnes, etc. D'autres font dans le plan, des places Blanches pour mettre les mots lesquels sont mal remplis et de mauvais goût, n'ayant point de bonnes notions sur la disposition." Ibid.

65. Mary Sponberg Pedley, *Bel et utile: the work of the Robert de Vaugondy family of mapmakers* (Tring: Map Collector Publications, 1992), 54–60; Mary Sponberg Pedley, "New light on an old atlas: documents concerning the publication of the *Atlas universel* (1757)," *Imago Mundi* 36 (1984).

66. "La partie qui doit dominer de préférence dans une carte, ce sont les bois: après ce sont les villes, les rivierres, enfin les montagnes à proportion de leur elevation, tout le reste tel que les vignes, les près, les marais doit être traité légèrement." Pierre Patte, in a manuscript mémoire in the BNF, Cartes et Plans, Société de géographie, colis 33, 3884, pp. 43–65, quoted in Pelletier, *Les cartes des Cassini: la science,* 139–41.

67. Ibid., 157.

68. See especially those between Delisle and l'abbé Jean Bobé reproduced by Dawson, *Atelier Delisle,* 231–44.

69. Gilles Robert de Vaugondy, "Réponse analitique au mèmoire du Sr. Buache par le Sr Robert de Vaugondy Géog. ord. du Roi auteur du nouvel atlas du Sr Boudet Libraire imprimeur du Roi," BNF, Ge. EE. 4990, 1751, Section X. "O vere Punica/Buachica fides: . . . puisque ces mêmes cartes sont communiquées icy à des Savans et envoyées toutes enluminées dans les païs respectifs pour profiter des lumières de leurs savans et par ce moyen tendre à la plus grande perfections possible. Il faut attendre qu'un ouvrage soit achevé pour dire qu'on y a mis la dernière main."

70. Norbonne de Pelet, "Mémoire concernant le Dépôt de la Marine," 12 March 1771, AN, MAR G 234.

71. Mireille Pastoureau, "Confection et commerce des cartes à Paris aux XVIe et XVIIe siècles," in *La carte manuscrite et imprimée du XVIe au XIXe siècle: Journée d'étude sur l'histoire du livre et des documents graphiques,* ed. Frédéric Barbier (Munich: Saur, 1983); Pastoureau, *Atlas français,* 232.

72. Campbell, *The London Tradesman,* 114.

73. Louise Lippincott, "Arthur Pond's journal of receipts and expenses, 1734–1750," *Walpole Society* 54 (1988); Louise Lippincott, *Selling art in Georgian London: the rise of Arthur Pond* (New Haven: Yale University Press, 1983).

74. Lippincott, *Selling art,* 153.

75. Pond paid £1 2s for the lettering of titles for the plates of Anson's voyages. Clayton, *English print,* 21.

76. Monique Pelletier, *La carte de Cassini: l'extraordinaire aventure de la Carte de France* (Paris: Presses de l'École des Ponts et Chaussées, 1990), 109.

77. Gabriel Marcel, "À propos de la Carte des Chasses," *Revue de géographie* (1897): 12–15.

78. Berthaut, *Ingénieurs-géographes,* 2:60.

79. "Il va accélérer les planches moins avancées et profiter de la belle saison pour le travail de l'eau forte et réserver celui du burin pour l'hiver." Belleyme to the Intendant of Guyenne, 20 November 1788, Archives de Gironde, C 2417, 34, quoted in de Dainville, *Cartes anciennes de l'église,* 230 n. 54.

80. Henri Berthaut, *La Carte de France* (Paris, 1898), 1:225–26.

81. Johannes Dörflinger, "Time and cost of copperplate engraving: illustrated by early nineteenth century maps from the Viennese firm Artaria & Co.," *Imago Mundi* 35 (1983): 59, 63.

82. Woodward's estimate is for Forlani's workshop in 1566. Woodward, *Maps as prints,* 24. Schilder advances the notion of 154 square centimeters per day, based on contracts between engravers and H. Hondius and J. Jansson in 1630, but points out that engraving at that speed would be nearly impossible. Günter Schilder, *Dutch folio-sized single sheet maps with decorative borders, 1604–60* (Alphen aan den Rijn: Uitgeverij Canaletto/Repro-Holland, 2000), 26.

83. George Carhart, "The significance of craft practices for early modern map production," *Imago Mundi* 56 (2004): 194–97.

84. Kenneth Nebenzahl and Don Higginbotham, *The atlas of the American Revolution* (Chicago: Rand McNally, 1974), map 12, p. 87.

85. Clayton, *English print,* 149.

86. G. N. G. Clarke, "Taking possession: the cartouche as cultural text in eighteenth-century maps," *Word and Image* 4 (1988): 455–74.

87. "Les Ornemens ou cartouches pour servir aux titres, explications, notes nécessaires pour l'intelligence de chacune carte seront d'abord dessinés pour être donnés au graveur de signes. On distinguera dans ces cartouches le blazon de Messeigneurs les prelats et barons dans chaque diocèse, celui de la Province et celui de la Ville capitale du Diocèze. Ces armoiries seront soutenues de leurs propres supports au défaut

desquels elles seroient accompagnées de génies ou de figures allégoriques et attributs convenable à ce qui sera connu des productions de la terre, de l'histoire naturelle et du commerce des différents cantons, à quoi l'on ajoutera ce qui pourra se trouver de particulier par rapport aux monuments anciens et modernes." Archives de Hérault, C.4671, quoted by de Dainville, *Le langage des géographes,* 64.

88. "Des Sauvages surtout des testes plates, avec tous leurs beaux habits." AN, MAR 2JJ/56/X,26,K: Bobé to Delisle, 27 May 1718, quoted in Dawson, *Atelier Delisle,* 239.

89. Ibid., 170.

90. Mary Sponberg Pedley, ed., *The map trade in the late eighteenth century: letters to the London map sellers Jefferys and Faden* (Oxford: Voltaire Foundation, 2000), 28–29.

91. "Qui est parfaitement entré dans le goût des Peintures faites par les Chinois mêmes." Preface to du Halde, *Description geographique, historique, chronologique, politique de l'Empire de la Chine* (Paris: P. G. Lemercier, 1735), xlviii–xlix, quoted in Cordier, "Du Halde et d'Anville," 398.

92. De Dainville, "Cartes de Bourgogne du XVIIIe siècle," 73–74.

93. Hubert François Bourguignon Gravelot (1699–1773), like d'Anville himself, adopted a different surname from his patronymic, Bourguignon. Both men were the sons of Hubert Bourguignon, a master tailor, and Charlotte Vaugon. It is speculated that Gravelot's surname came from his godfather. *Dictionary of national biography,* 8:428.

94. British Museum, Prints and Drawings, French Roy XVIIIc, published in Pedley, *Bel et utile: the work of the Robert de Vaugondy family of mapmakers,* 63–64.

95. AN, O[1]. 1934, année 1752, no. 49, le mémoire de Cochin, cited in Christian Michel, *Charles-Nicolas Cochin et le livre illustré au XVIIIe siècle* (Geneve: Librairie Droz, 1987), 260–61. The same mémoire tells us the price for a *fleuron du titre* of two Ls intertwined, and for a *lettre grise,* an S *orné de laurier,* 18 livres each, design and engraving. The Tardieus were a large extended family of map and print engravers as well as *planeurs* (preparers or smoothers of copperplates). Alexandre Tardieu, "Notice sur les Tardieu, les Cochin et les Belle, graveurs et peintres," *Archives de l'art Français: receuil de documents inedits* 4 (1876).

96. Pastoureau, *Atlas français,* 229.

97. Ibid., 136. The inventory taken after Duval's death in 1683 includes "une press garnie de deux roulleaux, un bacquet, six ais, une pouelle, un gril, le tout tel quel servant pour imprimer . . . 12 livres." (AN, MC, XX (401), 8 October 1683). Pedley, *Bel et utile: the work of the Robert de Vaugondy family of mapmakers,* 23. Gilles Robert de Vaugondy had inherited a roller press from Pierre Moullart-Sanson, along with a stock of maps, globes, plates, and manuscripts.

98. "Je l'avois prévenu que c'étoit un des plus habiles, mais qu'il avoit besoin d'être veillé. En effet, par la négligence de Lévêque et des autres valets des Menus-Plaisirs, avec qui il buvoit, il fit faire l'ouvrage par un mauvais compagnon, au moyen de quoy les planches furent usées sans avoir donné à peine cent bonnes épreuves." Cochin, *Mémoires,* quoted in Courboin, *L'estampe Française,* 119.

99. "Par des hauteurs, des insultes, et mémes des violences." Paul Chauvet, *Les ouvriers du livre en France des origines à la Révolution* (Paris: Presses Universitaires de France,

1959), 332. Chauvet cites a case in 1772 in which Richomme le Jeune brought charges against two *compagnons,* Le Roy and Péguy, who spoiled between them three hundred and five hundred examples of a plate for the *Encyclopédie.* In one instance, they dampened the paper too much and applied ink to the plates so badly that one couldn't tell black from white on the printed sheet. In the other instance, they dampened the paper four days too soon, allowing it to become too dry to print on.

100. De Dainville, *Cartes anciennes de l'église,* 253–54. De Dainville's description of the engraving and printing process closely follows that of Louis Albert Ghislain Bacler-d'Albe, "Notice sur la gravure topographique et géographique," in *Mémorial du Dépôt de la Guerre* (Paris, 1803), 125–40.

101. Robinson, *Marine cartography,* 82.

102. "In 1740, soon after the war broke out with Spain, a map was published in our Magazine, of the West-Indies and the countries adjacent . . . That map, in which was exhibited every place of importance that could be attacked either by sea or land, was so well received that near 20,000 were sold in less than 12 months." Edward Cave, "Observations on the Map of America," *Gentleman's Magazine* 25 (1755), 296.

103. De Dainville, *Cartes anciennes de l'église,* 255, gives a chart showing the number of proofs that could be pulled by a worker in a day, set against paper size and preparation method of the plate. The largest copperplate, typical of folio maps (*grand aigle,* 97.5 × 66.5 cm) could produce 70–80 prints when prepared *au chiffon,* or 25–30 *à la main.* Berthaut, *Ingénieurs-géographes,* 1:62, cites a budget of 1807 allowing for thirty-two proofs per day to be pulled. Chapuis, *À la mer,* 297, cites the case of a budget proposed by Charles Coutadeur, maître imprimeur de la ville de Paris et du Dépôt de la Marine, to print maps for the comte de Fleurieu. Coutadeur figured that forty large sheets or eighty half sheets could be printed per day, at the rate of 36 livres per 100 *grand-aigle* or 18 livres per 100 of the half sheets. AN, MAR 1JJ/27/7, 1 September 1787.

104. Buache to the comte de Maurepas, BNF, Cartes et Plans, Ge. FF. 13732 (II), 10 February 1737. "Les cartes [de Mediterranée] ne peuvent être distribuées que six semaines après leur impression sans quoi Elles se maculeroient."

105. Philippe Buache, "Projet du recueil des cartes géographique des diocèses du Languedoc." Bibliothèque du Havre, ms. 338, 15 September–10 October 1748.

106. Harley and Laxton, introduction, 8.

107. Aaron Arrowsmith, "Map of England and Wales: the result of fifteen years labour/dedicated by permission to his Royal Highness the Prince Regent by H. R. Highness's dutiful servant and hydrographer, A. Arrowsmith" (London: A. Arrowsmith, 1815), [1:190,080] 1 inch to 3 English statute miles. My thanks to James Flatness of the Library of Congress for bringing this map to my attention.

108. Savary des Bruslons, *Dictionnaire universelle de commerce* (Paris, 1723), 1:878, cited in Clayton, *English print,* 21.

109. "Papier chapelet d'Auvergne que j'envoie à Nuremburg et à Augsbourg pour les imprimer; cequi fait qu'on les paye plus cher qu'en Allemagne." Roch-Joseph Julien, "Observations du Sieur Julien sur ce Catalogue," *Nouveau catalogue de cartes géographiques et topographiques . . . divisé en deux parties* (Paris: chez R.-J. Julien, 1763), 4.

110. D'Allensay to Pingré, AN, MAR 1JJ/5/7, March 1773.

111. "La Delicatesse qu'il mettoit tant dans le choix du Papier que dans l'impression et l'enluminure." Demanne to Faden, 10 June 1782, in Pedley, *Map trade,* letter I:59, p. 134.

112. For a detailed view of this world, see Robert Darnton, *The business of the Enlightenment: A publishing history of the Encyclopédie 1775–1800* (Cambridge: Harvard University Press, 1979), 185–96.

113. Delisle, "Mémoire Bretagne." AN, MC, XLIII (467), "inventaire après décès," 30 April 1772, of Marie Charlotte Loy, Desnos' wife.

114. BNF, ms. fr. 2219, pièce 17, "Mémoire pour les graveurs et marchands d'estampes," cited in Le Bitouzé, "Commerce de l'estampe à Paris," 1:76.

115. Ibid., 1:77. The local ink was "rude et graveleux."

116. "Il faudroit les assujetir à avoir toujours sous leurs coudes une peau de chamoy qui seroient [*sic*] etendüe sur la planche et ne laisseroit à découvert que l'endroit où ils travailleroient." Pierre Patte, quoted in Pelletier, *Les cartes des Cassini: la science,* 140.

117. Catherine Hofmann, "L'enlumineure des cartes et des atlas imprimés, XVIe–XVIIIe siècles," *Bulletin du comité français de cartographie* 159 (March 1999).

118. Pedley, *Bel et utile: the work of the Robert de Vaugondy family of mapmakers,* 96, map 15.

119. "Que dans trois ou quatre leçons on scait toute l'essence, les Dames et les Religieuses se peuvent occuper facilement à cette sorte d'exercise." Hubert Gautier, *Art du laver, ou Nouvelle manière de peindre sur le papier* (Lyon: T. Amaulry, 1687), 29–30, cited in Hofmann, "L'enlumineure des cartes," 37, 39.

120. Brion de la Tour uses the feminine noun in his critique of Lattré's *Atlas moderne,* 1766. AN, MAR 1JJ/3/11.

121. Arthur Pond, "Journal of Receipts and Expenses 1734–1750," BL, Add. Ms. 23724, 1734–1750, cited in Lippincott, "Arthur Pond's journal."

122. Clayton, *English print,* 130.

123. Ibid.

124. Buchotte, *Les regles du dessein et du lavis pour les plans particuliers des ouvrages et des bâtimens et pour leurs coupes, profils, elevations, & façades, tant de l'Architecture militaire que civile . . . ,* 2nd ed. (Paris: C. Jombert, 1755).

125. John Smith, *The art of Painting in Oyl, to which is added The whole art and mystery of coloring maps and other prints with Water colors* (London: Hawes, Clarke & Collis, 1769), 110.

126. Delisle to Renard, AN, MAR 2JJ/60/12, 11 December 1706.

127. Jacques-Nicolas Bellin, *Petit atlas maritime* (Paris, 1764), 1:3.

128. For example, "*A new and correct Map of the World* . . . six feet wide, 3 feet 2 inches in depth, 8s in sheets; on canvas, with rollers, and coloured, 16s; . . . *The World* in two large hemispheres . . . decorated with 20 small maps, 9 feet wide, 6 feet 4 inches deep, . . . 15s in sheets, uncoloured; on canvas with rollers and coloured, £1 15s. . . ; *England and Wales,* by De la Rochette, . . . engraved by Thomas Kitchen, £1 1s in sheets; coloured and pasted on Cloth with a roller and ledge, £1 10s." Robert Sayer and John Bennett, *Sayer and Bennett's enlarged Catalogue of new and valuable Prints, in sets, or single; also useful and correct Maps and Charts; likewise Books of architecture, views of antiquity, drawing and copy books, etc etc in great variety at No. 53, in Fleet-Street, London; . . .* (London: Sayer and Bennett, 1775), 25.

129. "La pluspart des Enlumineurs qui divisent les Cartes suivent les points qu'ils y trouvent & qui souvent y sont mis contre les regles de Geographie; & comme quelquefois ils n'y en trouvent pas, ils conduisent leurs pinceaux le long des plus grosses rivières, ou suivant leur caprice; & distribuent ainsi aux Souverains des Estats aussi grands & aussi petits qu'il leur plaist . . . Une exact Division est encore necessaire pour un bon usage des Cartes Geographiques. On la fait d'ordinaire avec des couleurs qui y marquent & renferment justement l'estenduë de chaque païs." P. Duval, *Traité de Géographie qui donne la connoissance et l'usage du globe et de la carte* (Paris, 1672), 58–59, cited in Hofmann, "L'enlumineure des cartes," 40.

130. "Qu'il faudra suivre dans ces Cartes . . . Elle fait distinguer du premier coup d'oeil les terres de la mer qui les enveloppe et cela sans cacher le détail des côtes comme font les enluminures des cartes ordinaires." Buache to the comte de Maurepas, 2 February 1737, BNF, Ge. FF. 13732 (II).

131. Josef Konvitz, *Cartography in France, 1660–1848: science, engineering, and statecraft* (Chicago: University of Chicago Press, 1987), 23.

CHAPTER 3

1. James Boswell, *Life of Johnson* (Oxford: Oxford University Press, 1980; originally published 1791), 522.

2. François de Dainville, S.J., "Cartes de Bourgogne du XVIIIe siècle," in *La cartographie reflet de l'histoire,* ed. Michel Mollat du Jourdin (Geneva: Editions Slatkine, 1986), 74.

3. Mary Sponberg Pedley, ed., *The map trade in the late eighteenth century: letters to the London map sellers Jefferys and Faden* (Oxford: Voltaire Foundation, 2000), letter I:8/9, p. 57.

4. Richard Gough, *British topography,* 2:477, regarding J. Dickinson's "New and correct map of the southern part of the County of York," 1750, quoted by J. B. Harley and J. C. Harvey, introduction to *A survey of the county of Yorkshire by Thomas Jefferys, 1775,* ed. J. B. Harley and J. C. Harvey (Lympne Castle: Harry Margary, 1973).

5. Delisle to Renard, AN, MAR 2JJ/60/12, 11 December 1706.

6. Guillaume Delisle, "Mémoire pour le Sr. Guillaume Delisle . . . contre le Sr. Louis Renard" AN, MAR 2JJ/60/13.

7. Renard to Delisle, AN, MAR 2JJ/60/20, 27 December 1706.

8. Pedley, *Map trade,* 15.

9. Henri Cordier, "Du Halde et d'Anville (Cartes de la Chine)," in *Recueil de mémoires orientaux par les Professeurs de l'Ecole des Langues orientales* (Paris: E. Leroux, 1905), 394–98. D'Anville's contract for the maps of China with le père Du Halde prevented him from selling his maps of China separately until Du Halde agreed to such a sale, with one-third of the profit going to Du Halde. D'Anville received 1,000 livres in their final contract for his design work and supervision of the engraving. The plates seem not to have belonged to d'Anville, since he requested ten copies of each of the maps, along with his monetary reimbursement. This supposition is strengthened by the fact that the *Nouvel Atlas de la Chine* was printed in Amsterdam and seen by d'Anville as a

counterfeit. According to Cordier (p. 398), the plates for the maps of China were in the possession of the Paris publisher Moutard.

10. "Aux étalages de ces gagnent-deniers, qui vendent sur le trottoirs des quays et le long des murs." Philippe Buache, "Procès entre Buache gendre de Delisle et les Heritiers de la Veuve Delisle . . . ," BNF, Cartes et Plans, Ge. FF. 13373, pt. 3, "Divers moyens pour prouver que les impressions sont indivisibles des planches" [1747?]. In Paris, the colporteurs were regulated as to where they could work and what they could sell. Twelve were installed at specific places, such as two at the end of the pont St Michel, two at the end of the pont du Marché neuf, and two in front of the Horloge du Palais. They could sell no book of more than eight pages. Thus they were very near the heart of the map trade and easily able to sell single-sheet maps in competition with the mapmakers. Etienne Martin Saint-Léon, *Histoire des Corporations de Métiers depuis leurs origines jusqu'à leur suppression en 1791* (Paris, 1922), 480.

11. Buache to de Maurepas, 10 February 1737. This material is also described in Josef Konvitz, *Cartography in France, 1660–1848: science, engineering, and statecraft* (Chicago: University of Chicago Press, 1987), 76–77. Konvitz suggests that the minister agreed to this plan, though Buache left the dépôt some time later in the year 1737.

12. AN, MAR G 234, Mélange (20), Conseil des Dépeches, à Votre Majesté. For the full story of reorganization of the dépôt and map sales after J.-N. Bellin's death in 1773 and the accession of Louis XVI in 1774, see Olivier Chapuis, *À la mer comme au ciel: Beautemps-Beaupré et la naissance de l'hydrographie moderne (1700–1850)* (Paris: Presses de l'Université de Paris-Sorbonne, 1999), 190–98.

13. Pastoureau, *Les atlas français XVIe–XVIIe siècles* (Paris: Bibliothèque nationale, 1984), 388, citing BNF, Imprimés Inv. F. 21258 (52) and ms. fr. 21733, fols. 57–58 in the following language: "voulant d'obliger les particuliers qui n'en ont besoin que de quelques unes, à acheter les livres entiers qu'ils ont formez de toutes leurs cartes, ce qui porte préjudice au public & particulièrement aux officiers de ses troupes, dont aucuns n'ayant pas le moyen d'acheter lesdits livres entiers ne peuvent avoir les cartes particulières dont ils ont besoin." De Dainville cites the case of Sanson, whose maps of the bishoprics of France were sold at 20 sous each separately, but 10 sous each when sold in a set. François de Dainville, S.J., *Cartes anciennes de l'église de France* (Paris: Vrin, 1956), 258.

14. Jean Lattré, "Catalogue du fonds du Sr Lattré Graveur rue St Jacques près St Severin, à la Ville de Bordeaux," in *Atlas moderne* (Paris: [1777]). Lattré offered his *Atlas moderne* "complet petit in-fol. de 78 feuilles, demi-reliure, 42 livres; relié en veau, 46 livres; papier fin, lavé à la Hollandoise, 60 livres." Coloring and paper quality rather than the binding raised the price.

15. "Elles trouveroient peu d'Amateurs en France au prix qu'on les vend en Angleterre." "Observations du Sieur Julien sur ce Catalogue," in Roch-Joseph Julien, *Nouveau catalogue de cartes géographiques et topographiques . . . divisé en deux parties* (Paris: chez R.-J. Julien, 1763), 4.

16. "Voyant que cette feuille restoit presque blanche et me rapellant ce que je dois au public qui depuis quarante ans acceuillit si favorablemt mes productions, j'ai cru devoir joindre ce cours de Rivière d'Hudson qui est un chef d'oeuvre de Sauthier, qui

se vend 3# 12s à Londres." Georges-Louis Le Rouge, *Atlas Amériquain septentrional* (Paris: Le Rouge, 1778), maps 21–22.

17. Johannes Dörflinger cites comparable statistics for map engraving in Vienna. Johannes Dörflinger, "Time and cost of copperplate engraving: illustrated by early nineteenth century maps from the Viennese firm Artaria & Co.," *Imago Mundi* 35 (1983): 66 n. 17.

18. Buache, "Procès entre Buache et les Delisles."

19. For Jaillot, AN, MC, XV (415), 15 November 1712, in Pastoureau, *Atlas français,* 233. Louis-Charles Desnos maintained an inventory of around 1,000 copperplates, 1,182 atlases, and about 70 reams of paper in 1778 (AN, MC, LXXV [745], 16 February 1778). The Verrier inventory was sold to his former partner, the engraver François Gabriel Perrier, for 7,010 livres. AN, MC, LXXXIX (793), 16 August 1784. The total 20,000 maps in the Verrier stock consisted of 400–500 examples of each type of map, according to the auditor of the inventory, Charles Louis Jabineau de Marolles. The lawyer remarked that most of the inventory consisted of four or five hundred copies of each title and that at auction or in retail only forty or fifty copies would be sold, leaving the rest to be reduced in price to next to nothing. "Les marchandises qui composent le fonds de Commerce consistant en quantité d'Exemplaires des mêmes Cartes de Geographie, que chaque espèce de Cartes est composée de quatre à cinq cent Examplaires, que si elles etoient venduës à l'encan et en détail, il en resulteroît que quelques Particuliers acheteroient peut être quarante à cinquante exemplaires de chaque espèce, et quant au surplus il ne se trouveroit point d'Encherisseurs et le tout se donneroit à vil prix." By buying the entire stock, Perrier paid just under 3 livres each for the maps, quite expensive by retail standards, but he was also acquiring many manuscript maps, some rare maps, and about 8,000 German, English, and Dutch maps with which to continue his business.

20. Pedley, *Map trade,* 24–25.

21. Jacques-Nicolas Bellin, "Observations faites à Mr le comte de Narbonne-Pelet," 18 December 1768, AN, MAR 1JJ/3/13. Some thought Bellin was taking unjust profits from his arrangements with the dépôt. See, e.g., AN, MAR 1JJ/25/7, cited by Bernard Le Guisquet, "Le Dépôt des cartes, plans et journaux de la Marine sous l'ancien régime (1720–1789)," *Annales Hydrographiques* 18, no. 765 (1992): 19 nn. 56, 57. Other evidence also suggests that his private sale of maps brought him little. His widow inherited nothing but debts from him, and he provided his daughter, an only child, with no dowry for her marriage to Etienne Taitbout, "Capitaine au régiment de la Marine." Contrat de mariage 1763, Arch. Ville de Paris 5-AZ 4002.

22. Merigot, "Etat des Comptes," AN, MAR 1JJ/5/15, 30 March 1773–12 November 1774. Examples of problems are: 18 August 1773, dépôt sent map 64 instead of map 63; 28 July 1773, only three maps of Marseilles sent instead of twelve; 3 June 1773, Merigot reminds Oudin, the clerk in the dépôt, to make sure that maps are not pierced or torn (of the last twelve sent, only three could be sold); 11 May 1773, M. Tanqueray in Nantes refused to buy charts for 28s, offering only 26s; 22 July 1773, Merigot requests the new map of the English Channel by Dicquemare, being sold for £4 10s.

23. Jean-Nicolas Buache de la Neuville, "Etat des cartes et ouvrages livrés par le Depot generale des Plans de la Marine au Sieur Buache Géographe du Roi chargé de la Vente des dites ouvrages pour le Depot pendant le 3 premiers mois de 1779," AN, MAR 1JJ/16/3, 1 April 1779. It appears that the retailers in the port cities would be making a greater profit than Buache de la Neuville. The archival material does not comment on this; perhaps the rationale for the wholesale prices was that Buache de la Neuville bought maps from the dépôt only when he had secure orders while the retailers bought maps on speculation from Buache de la Neuville, incurring a higher risk.

24. Jean Claude Dezauche, "Papiers Dezauche," BNF, Cartes et Plans, Ge. EE. 3119, 1788–1834, fols. 45–46. Buache de la Neuville sold his business in order to take up the post of garde-adjoint in the dépôt, anticipating his appointment as chief hydrographer in 1789.

25. François Courboin, *L'estampe Française* (Brussels: Librairie d'Art et d'Histoire, G. Van Oest & Co., 1914), 145–47.

26. Covens & Mortier to Faden, 18 September 1778 ("le Capitaine nous a dit qu'il vous plairat de faire retirer cette Roule par une de votre domestique [*sic*] pour prevenir de Declarer a la Douane"); Lattré to Faden, 2 April 1778 ("Envoyé cela je vous prie en rouleau pour eviter les frais"); Pedley, *Map trade,* 111, 95.

27. Konvitz, *Cartography,* 21–29, esp. 24. Louis XV's initial support of Cassini de Thury from 1747 included a subvention of 150,000 livres to be reimbursed by the provincial governments. The annual cost of mapmaking was estimated to be 80,000 livres, of which 56,000 were for salaries and 24,000 were for printing and supplies. Cassini de Thury estimated that 2,500 copies of each map would be printed for sale. If all could be sold, he estimated an income of 1,800,000 livres. The start of the Seven Years' War in 1756 caused the government to withdraw the subvention.

28. Ibid., 36.

29. Wilfrid Prest, *Albion ascendant: English history, 1660–1815* (Oxford: Oxford University Press, 1998), 157, has described Hanoverian government as decentralized to the point of being "polyarchic—with power widely and almost randomly distributed among a bewildering array of agencies and individuals, including Crown, ministry, both houses of Parliament, privy council, the Church of England, the lawcourts, the Bank of England, and the East India Company."

30. E. G. R. Taylor, "Robert Hooke and the cartographical projects of the late seventeenth century (1666–1696)," *Geographical Journal* 90 (1937): 533–35.

31. Peter Barber, "Maps and monarchs in Europe, 1550–1800," in *Royal and republican sovereignty in early modern Europe,* ed. Robert Oresko, G. C. Gibbs, and H. M. Scott (Cambridge: Cambridge University Press, 1996), 92–93, 95.

32. R. A. Skelton, "The origins of the Ordnance Survey of Great Britain," *Geographical Journal* 128, no. 1962 (1962): 422.

33. Barber, "Maps and monarchs," 90–98.

34. A. H. W. Robinson, *Marine cartography in Britain* (Leicester: Leicester University Press, 1972), 71–86. Robinson's particular example is Lewis Morris, antiquary, literary scholar, philologist, mineralogist, and customs officer, who also trained as a sur-

veyor. His brainchild, the survey of the Welsh coast, was supported during the 1730s and 40s only sporadically and reluctantly by the Admiralty and with grudging help from the Customs Board. Morris arranged for Emanuel Bowen in London to engrave his charts and harbor plans; they were published in 1748 as *Cambria's Coasting Pilot* and offered by subscription. Such was the *Pilot's* popularity that it was out of print by 1761.

35. Charles W. J. Withers, "The social nature of map making in the Scottish Enlightenment, c. 1682–c. 1832," *Imago Mundi* 54 (2002): 46–66.

36. J. B. Harley, "The Society of Arts and the surveys of English counties 1759–1809," *Journal of the Royal Society of Arts* 112 (1964).

37. John Green [Bradock Mead], *Remarks in support of the New Chart of North America in six Sheets* (London: Thomas Jefferys, 1753), 3.

38. De Dainville, *Cartes anciennes de l'église.* In a study of local mapping, Alice Stroup points out the ambiguities found in maps produced by the clergy. Alice Stroup, "Le comté Venaissin (1696) of Jean Bonfa, S.J.: a paradoxical map by an accidental cartographer," *Imago Mundi* 47 (1995).

39. Christine Petto, "Mapping the body politic in early modern France," Ph.D. thesis, Indiana University, 1966, studies the role of patronage in the work of Jaillot, Delisle, Bellin, and d'Anville. Nelson-Martin Dawson, *L'atelier Delisle: l'Amérique du Nord sur la table à dessin* (Sillery, Quebec: Editions du Septentrion, 2000), chap. 3 ("Le contexte clientéliste"), offers an even more nuanced study of Delisle and his patrons.

40. For the duke's scientific interests, see J. H. Shennan, *Philippe, duke of Orléans: regent of France 1715–1723* (London: Thames and Hudson, 1979), 14, 131. For a study of the duke as a student of Claude Delisle and his interests in geography, see Dawson, *Atelier Delisle,* 41–45.

41. Petto, "Mapping," 232–40.

42. "Je n'aurois encore osé mettre la main à un pareil ouvrage, si les bontés et la magnificence vraiment royale d'un Grande Prince ne me faisoient une loi de cette entreprise et n'en facilitoient l'exécution . . . [Le duc] se chargeoit des frais de l'ouvrage." Jean-Baptiste Bourguignon d'Anville, *Analyse géographique de l'Italie* (Paris: la Veuve Estienne et fils, 1744), viii–ix.

43. "Nouvelle carte entitulée première partie de l'Europe contenant la France, l'Alemagne, l'Italie, l'Espagne et les Isles Britanniques, publiée sous les auspices de Monseigneur Louis Philippe duc d'Orléans, premier Prince du Sang, par le sieur d'Anville." Announced in *Journal des Sçavans,* September 1758, 630. The advertisement continued: "Cette Carte est en même tems recommendable par la beauté de la gravure, pour laquelle la magnificence d'un grand Prince n'a épargné aucune dépense."

44. Mary Sponberg Pedley, *Bel et utile: the work of the Robert de Vaugondy family of mapmakers* (Tring: Map Collector Publications, 1992), 50.

45. R. A. Skelton, "The origins of the Ordnance Survey of Great Britain," *Geographical Journal* 128, no. 1962 (1962).

46. Adrian Johns, *The nature of the book: print and knowledge in the making* (Chicago: University of Chicago Press, 1998), 351. Johns makes this point about subscription lists and seventeenth-century books. Whether subscribers to eighteenth-century maps

and atlases thought these works were any good is worth further research. A subscriber may well have been pleased to have a map of that which had not been mapped, or mapped in such a way, before; he or she was not necessarily endorsing the value of the map.

47. Robert Plot, *The natural history of Oxfordshire,* 2nd ed. (Oxford and London: Leon Lichfield, Charles Brome, and John Nicholson, 1705), [ii].

48. William Ravenhill, "Joel Gascoyne, a pioneer of large scale county mapping," *Imago Mundi* 26 (1972): 67, fig. 4: "Proposals made for an Actual survey of the County of Devon . . . London, Printed in the year 1700."

49. Monique Pelletier, *Les cartes des Cassini: la science au service de l'État et des régions* (Paris: Comité des travaux historiques et scientifiques, 2002), 145–55. Konvitz points out that the company raised only enough capital to sustain its 80,000 livres annual budget for 10 or 12 years. Konvitz, *Cartography,* 24.

50. Pelletier, *Les cartes des Cassini: la science,* 153.

51. Mary Sponberg Pedley, "The subscription list of the 1757 *Atlas universel:* a study in cartographic dissemination," *Imago Mundi* 31 (1979).

52. F. J. G. Robinson and P. J. Wallis, *Book subscription lists: a revised guide* (Newcastle upon Tyne: Harold Hill & Son Ltd for the Project for Historical Bibliography, 1975). Robinson and Wallis list maps and atlases in their guide, providing a good starting point for research on this topic, which deserves much more detailed analysis. A recent example of such analysis is the paper of A. D. M. Phillips analyzing a subscription list for a map that was never published: "The local market for the late seventeenth century printed county map in England: the subscribers to Gregory King's map of Staffordshire," Monday, 16 June 2003, International Conference on the History of Cartography, Cambridge, Massachusetts.

53. John Andrews and Andrew Dury, *A topographical map of Wiltshire,* in 18 sheets (London, 1773). BL, Maps 5700.3 and K. Top.43.30.8; RGS, 1.C.108.

54. J. B. Harley, "Power and legitimation in the English geographical atlases of the eighteenth century," in *Images of the world: the atlas through history,* ed. John A. Walters and Ronald E. Grim (Washington: Library of Congress, 1997), 169–73. Within the caveat regarding misinterpretation of the percentage figures, one should note that Harley's statistics do not entirely support his claim that "the balance of power lay with patron rather than craftsman" in atlas production, when one-quarter to nearly one-half of the subscribers he studies were those very craftsmen.

55. Booksellers made up 3 percent of the subscribers to Senex's 1721 *New general atlas;* for Senex's 1728 *Atlas maritimus,* they number just 1 percent.

56. *The Connecticut Journal,* 14 January 1778. Romans offered the map to subscribers at 12s plain, 15s colored; those who bought twelve got one free. If shopkeepers bought fifty or more, they received one plain map free for every six they purchased. The price for nonsubscribers was 15s plain, 18s colored. Philip Lee Phillips, *Notes on the life and works of Bernard Romans,* (Deland: Florida State Historical Society, 1924), 89.

57. Monique Pelletier, *La carte de Cassini: l'extraordinaire aventure de la Carte de France* (Paris: Presses de l'École des Ponts et Chaussées, 1990), 120.

58. Taylor, "Robert Hooke," 531.

59. J. B. Harley, "John Strachey of Somerset: an antiquarian cartographer of the early eighteenth century," *Cartographic Journal* (June 1966): 5.

60. The addition of coats of arms, a detailed bit of engraving requiring great accuracy, added considerably to the cost of map or subscription. John Warburton charged 12s 6d for the addition of family shields on the proposed map of the Yorkshire in 1720. John Warburton, "Mr Warburton's list of subscribers to his map and survey of Yorkshire, with the several sums paid and remaining due," BL, Manuscripts, Landsdowne 916, [1720], 60r–v.

61. Taylor, "Robert Hooke," 531.

62. Edmund Thompson, *Maps of Connecticut before the year 1800* (Windham, Conn.: Hawthorn House, 1940), 37.

63. J. B. Harley and Donald Hodson, introduction to *The Royal English Atlas . . . by Emanuel Bowen and Thomas Kitchin* (Newton Abbot: David and Charles Reprints, 1971), 9.

64. Pastoureau, *Atlas français,* 230–31.

65. Gilles Robert de Vaugondy, "Réponse analitique," Ge. EE. 4990. My thanks to Catherine Hofmann of the BNF for allowing me to study this manuscript while it was being catalogued by the Département des Cartes et Plans.

66. Mary Sponberg Pedley, " 'Commode, complet, uniforme, et suivi': problems in atlas editing in Enlightenment France," in *Editing early and historical atlases,* ed. Joan Winearls (Toronto: University of Toronto Press, 1995), 102–4.

67. J. B. Harley, "The bankruptcy of Thomas Jefferys: an episode in the economic history of eighteenth-century map-making," *Imago Mundi* 20 (1966); Pedley, *Bel et utile: the work of the Robert de Vaugondy family of mapmakers,* 113–23.

68. Laurence Worms, "Thomas Kitchin's 'journey of life': hydrographer to George III, mapmaker, and engraver," *Map Collector* 62 (1993); Laurence Worms, "William Faden," in *Dictionary of national biography: missing persons* (Oxford: Oxford University Press, 1993). Faden left an estate valued at about £30,000. National Archives, London, PROB 11/1860.232, 259 ff. My thanks to Laurence Worms for sharing this information.

69. Guillaume Delisle, draft of letter to unknown party, AN, MAR 2JJ/60/162, [early 1720s?]. Delisle asks for advice regarding details on the roads and boundaries of a one-sheet map of Provence he is preparing. He hopes to have these details soon, since, as he comments, he doesn't have any work to give to his engraver at present, "que je paie pour travailler toute l'année."

70. Henri Berthaut, *La Carte de France* (Paris, 1898), 2:226.

71. Fleurieu, memoranda, AN, MAR 1JJ/17/14, 8 May 1780; 24 May 1780. Claro's death was all the more public since he had fallen into the pond of the Thuileries, "conduit par quelques verres de vin qu'il aura but de trop, avant de boire l'eau des Thuileries."

72. Jacques-Nicolas Bellin to Chabert, AN, MAR 1JJ/3/1, 24 February 1759.

73. La veuve Bellin to the minister of the marine, AN, MAR 1JJ/5/7, 31 January 1773; 20 March 1773. A similar complaint is found in Bellin's personnel file. Jacques-

Nicolas Bellin to the duc de Praslin, AN, MAR C7 24, dossier Bellin, 19 September 1766.

74. Gabriel Marcel, "À propos de la Carte des Chasses," *Revue de géographie* (1897): 11, 14. Of total costs of 41,551 livres, Delahaye had been paid 27,400 livres but was still owed 14,151 livres, about 34 percent of the total cost.

75. Courboin, *L'estampe Française,* 159. See also Christian Michel, *Charles-Nicolas Cochin et l'art des Lumières* (Rome: École Française de Rome, 1993), 179 n 3. In a letter to Descamps of 6 June 1780, Cochin complained that the Trésor owed him 22,000 livres. As directeur des bâtiments du roi and garde des dessins du roi, Cochin should have received annual pensions totaling 3,400 livres, "s'ils avaient été versés régulièrement, mais ce n'était guère le cas et Cochin évoque constamment dans sa correspondance les retards des payements des Bâtiments." Michel, *Charles-Nicolas Cochin et l'art des Lumières,* 165.

76. J. B. Berthier, "Tableau de conviction, 16 Octobre 1771," in *Observations de M. Berthier, gouverneur de l'Hotel de la Guerre, en réponse à celles de M. DuPeron, directeur de l'Imprimerie Royale* ([Paris], [1771]).

77. D'Anville to [DeSaint and Saillant and Durand], 24 November 1759, BNF, ms. fr. 22147, (2).

78. Pelletier, *La Carte de Cassini: l'extraordinaire aventure,* 112.

79. Bauche [bookseller], "[letters concerning the maps prepared by Brion de la Tour for the *Topographie de l'Univers* de l'abbé Expilly]," BNF, ms. fr. 22068 (24,25,26,39), 18 May 1758; 29 May 1758. Bauche expressed surprise at the geographer's attitude about payment: "il ny a point d'exemples d'un pareil procedé. Tous les graveurs laissent leurs planches aux Libraires et fort souvent n'en sont payés qu'à bout de 3 mois et meme beaucoup davantage. Le Sr Brion qui a voulu malgré le Sr. Bauche lui laisser ses planches et en le Lendemain exige son Payment avec tant d'Empressement que si il y avant une année."

80. J. B. Harley and Paul Laxton, introduction to *A survey of the County Palatine of Chester by P. P. Burdett (1777)* (Liverpool: Historical Society of Lancashire and Cheshire, 1974), 4–7.

81. Louis DeVorsey, ed., *De Brahm's report of the general survey in the southern district of North America* (Columbia: University of South Carolina Press, 1971), 54–55.

82. G. N. D. Evans, *Uncommon obdurate: the several public careers of J. F. W. Des Barres* (Salem: Peabody Museum of Salem; Toronto: University of Toronto Press, 1969), 59–78; Alex Krieger, David Cobb, and Amy Turner, eds., *Mapping Boston* (Boston: Leventhal Foundation, 1999), 106.

83. *Dublin Journal,* 14 March 1761; Andrews, "French school," 283.

CHAPTER 4

1. "Cette modicité de prix étant d'ailleur le moyen le plus sûr d'en empecher la contrefaçon chez l'étranger." Jacques-Nicolas Bellin, *Petit atlas maritime* (Paris, 1764), 1:3.

2. "Le marchand d'estampes qui a dépensé huit à neuf mille francs et qui est obligé

de vendre son estampe dix ou douze francs, sait très bien que pour douze ou quinze cents francs, on peut faire copier son estampe par un jeune homme qui commence à avoir du talent, et qu'ainsi ce contrefacteur peut donner son estampe pour cinquante sols, tandis que lui ne peut la vendre moins de dix francs, et cette copie est suffisante pour le grand nombre qui n'est pas connaisseur." Cochin to J. B. Descamps, 20 January 1778, 3–4; Rouen, Bibliothèque municipale, quoted by Christian Michel, *Charles-Nicolas Cochin et l'art des Lumières* (Rome: École Française de Rome, 1993), 445, n.56. The letter is also cited by Peter Fuhring, "The print privilege in eighteenth century France, pt. 1," *Print Quarterly* 2 (1985): 178 n. 21. Cochin specifically used the example of a painting of a French harbor by the well-known artist Joseph Vernet as prepared for engraving. Cochin would charge 1500 livres for such a design. The print publisher would then hire a distinguished and skilled engraver to work the copper (e.g., Jacques-Philippe Le Bas) for about 5–6000 livres. With printing costs, the publisher makes a total investment of 8–9000 livres. To recoup his investment, he would have to charge 10–12 livres per print and sell at least 750 or 800.

3. These problems of course are still with us, as witnessed by the lawsuit brought in 2001 against the Automobile Association of Great Britain by the Ordnance Survey, the British government mapping agency. The automobile association was required to pay £20,000,000 in damages to the Ordnance Survey for using the OS base map without seeking permission. Joint Statement by Centrica and Ordnance Survey, Monday 5 March 2001, OS 16/01: Centrica and Ordnance Survey settle AA copyright case.

4. Maurice Daumas, *Les instruments scientifiques au XVIIe et XVIIIe siècles* (Paris: Presses Universitaires de France, 1953), 339–85.

5. Mary Sponberg Pedley, "The map trade in Paris, 1650–1825," *Imago Mundi* 33 (1981).

6. AN, MAR G 234, Mélange (20), Conseil des Dépeches, à Votre Majesté, [c. 1772].

7. Daumas, *Les instruments scientifiques au XVIIe et XVIIIe siècles,* 299–323.

8. Laurence Worms, "Location in the London map trade: a talk given to members of IMCOS on 3rd June 2000," *International Map Collectors' Society Journal* 82 (autumn 2000). Adrian Johns discusses importance of location of the bookseller in seventeenth-century London. Adrian Johns, *The nature of the book: print and knowledge in the making* (Chicago: University of Chicago Press, 1998), 108–26.

9. The translator's preface to the *Compendium of Ancient Geography by Mons. D'Anville* (London: R. Faulder, 1791), xii.

10. "J'ai essuyé plus d'une fois des reproches sur ce que je n'avois point formé un elève, pour me donner (comme on vouloit bien s'exprimer) un successeur. . . . Mais, si je n'ai point formé d'Elève, j'ai fait bien des Copistes . . . L'estime que l'Angleterre veut bien faire des Cartes que j'ai publiées, en a fait paroître à Londres une copie universelle, mais d'une exécution grossière, comme on m'en a parleé, et très-éloignée de l'élégance, qui dans les Cartes originales ajoute un prix à ce qu'elles pouvoient en avoir par leur composition." Jean-Baptiste Bourguignon d'Anville, *Considérations générales sur l'étude et les Connoissances que demande la composition des Ouvrages de géographie* (Paris: Lambert, 1777), 110.

11. The main source for the following section is David Hunter, "Copyright pro-

tection for engravings and maps in eighteenth century Britain," *Library* 9, no. 2 (1987): 128–47.

12. For extended discussion of the privilege and letters patent in seventeenth-century London, see Johns, *Nature of the book,* 248–62.

13. R. A. Skelton, *County atlases of the British Isles, 1579–1850: a bibliography* (Folkestone: Dawson, 1978), 101.

14. Ibid., 164, 69.

15. Timothy Clayton, *The English print: 1688–1802* (New Haven: Yale University Press, 1997), 82, n.28.

16. Hunter, "Copyright protection," 145.

17. What follows is largely taken from Fuhring, "The print privilege in eighteenth century France, pt. 1."

18. Ibid., 177.

19. Ibid., 178.

20. Ibid., 190.

21. Renard to Delisle, AN, MAR 2JJ/60/27, October 1707.

22. "Mortier ne rougit pas pour cela: car il est riche et sans honneur: il dit par devant toute la Compagnie qu'il vendoit ses cartes au gré des achepteurs et qu'il en avoit toujours en provision avec le nom de tous les autheurs les plus fameux, accommodant ainsi avec les mêmes cartes ceux qui vouloient—Delisle, Samson, Defer, Jaillot Baillieu etc. Il dit qu'il avoit plus de 50 cartes sous le nom de Mr deLisle. . . . Et voila sur quel fondement il abuse la credulité du Public. Il negotie ses cartes sous vôtre nom en Allemagne, sous celui de Sanson en Angleterre; sous celui de De Fer en Pologne et en Italie, et partout ainsi." Renard to Delisle, AN, MAR 2JJ/60/28, 22 November 1707.

23. David Garrioch, *Neighbourhood and community in Paris: 1740–1790* (Cambridge: Cambridge University Press, 1986), 8.

24. "Twists in the plot cost AA map cheats £20m," *The Times,* London, Tuesday, 6 March 2001. In a letter to *Printing World,* 21 May 2001, Ordnance Survey senior press officer Philip Round reiterated that the OS's "fingerprinting" techniques—particular styles and designs that are unique to OS maps—did not constitute deliberate errors.

25. Quoted in Miles Harvey, *The island of lost maps* (New York: Random House, 2000), 139–40.

26. Numa Broc, "Une affaire de plagiat cartographique sous Louis XIV: le procès Delisle-Nolin," *Revue d'Histoire des Sciences et de leurs applications* 23 (1970). Also in Nelson-Martin Dawson, *L'atelier Delisle: l'Amérique du Nord sur la table à dessin* (Sillery, Quebec: Editions du Septentrion, 2000), 30–37.

27. "Depuis que les cartes ont commencé à se multiplier et sont entrées dans un commerce pécuniaire, on a vu plusieurs ignorants qui, par avidité du gain, se sont mêlés de faire ce qu'ils n'entendaient pas, et des vendeurs d'images qui ont voulu faire de la géographie une dépendance de la taille douce. . . . La profession de géographie est devenue un véritable brigandage . . . les plaigaires se gardent bien de copier une carte trait pour trait . . . [c'est delicat de] discerner lequel des deux est le voleur et lequel le volé." Broc, "Plagiat cartographique," 144, citing BNF, ms. fr. 21733, 82–95.

28. For example, he had reduced the breadth of the Mediterranean, Asia, and

Asia Minor; given new shapes to South America and New Holland (Australia); and, by reducing the breadth of the Atlantic and Indian Oceans, made the Pacific Ocean much larger than on previous maps. Ibid., 144.

29. "Était une des mes découvertes, mais comme il n'est pas toujours à propos de publier ce qu'on sait ou que l'on croit savoir, je n'ai pas fait graver cette mer sur les ouvrages que j'ai rendus public, ne voulant pas que les étrangers profitassent de cette découverte." Ibid., 145.

30. '"D'une manière nette et précise (sur) l'usage que l'on doit faire des mémoires pour la construction des cartes." Ibid., 148.

31. "Parce qu'on peut avoir communication des ouvrages d'un auteur sitôt qu'ils sont entre les mains des graveurs & imprimeurs, ou que leurs originaux ont été communiquez à quelques-unes." Dawson, *Atelier Delisle,* 37.

32. Buache's pursuit of Dheulland and Vallet is found among the papers in Philippe Buache, "Procès entre Buache gendre de Delisle et les Heritiers de la Veuve Delisle. . . ." BNF, Cartes et Plans, Ge. FF. 13373, [1747?].

33. Philippe Buache, "Transaction entre Philippe Buache, quay de l'Horloge, et Guillaume d'Heulland, rue Serpente, and Jerome Vallet, place Maubert, graveurs," AN, MC, LVII (383), 18 August 1747.

34. R. A. Skelton, "Copyright and piracy in eighteenth century chart publication," *Mariners Mirror* 46 (1960): 207–12. Skelton gives summaries of both the case brought by Sayer and Bennett and a later case brought by David Steel in 1789 The details of both cases were published by Moore himself in later editions of his works.

35. "Voilà donc une nouvelle preuve que la connaissance des Sciences et des Beaux Arts n'est point un privilège exclusif qui n'en permettre l'accès qu'à une sexe. Les femmes se distingueront aussi dans cette noble carrière, lorsqu'elles voudront s'occuper utilement, ou plutôt lorsqu'on aura senti l'abus de ne leur donner qu'une education frivole." *Journal de Trévoux,* June 1762, 1,534.

36. BNF, ms. fr. 22120 (62), 9 February 1764, in which Lattré complains of several instances of infringements on his privilege by Desnos and Rizzi-Zannoni.

37. "Desnos se plaint de ceque par la voie de la Gazette d'Utrecht on a publié contre lui une calumnie . . . l'Eclipse du premier Avril prochaine dont ce Géographe a donné une carte qui a paru à peu près dans le meme tems que celle du sieur Lattré. On veut que le S. Desnos ait été plagiaire et condamné comme tel par arrêt du Parlement . . . nous pensons très sincérement que M. Rizzi Zannoni auteur de cette carte n'a pas besoin des aîles d'autrui pour s'elever à la region des Astres. Nous croyons aussi devoir publier que Desnos continuer toujours à débiter cette carte de l'Eclipse a 1# 4s." *Journal de Trévoux,* February 1764, 147–48. My thanks to Catherine Hofmann and Mireille Pastoureau for alerting me to this notice.

38. AN, MAR C7 24, dossier Bellin, cited in Bernard Le Guisquet, "Le Dépôt des cartes, plans et journaux de la Marine sous l'ancien régime (1720–1789)," *Annales Hydrographiques* 18, no. 765 (1992): 17.

39. See note 1 of this chapter.

40. William Faden, *The Province of New Jersey Divided into East and West, commonly called The Jerseys* (London: Faden, 1777).

41. Mary Sponberg Pedley, ed., *The map trade in the late eighteenth century: letters to the London map sellers Jefferys and Faden* (Oxford: Voltaire Foundation, 2000), 29.

CHAPTER 5

1. David Bosse, "The Boston map trade of the eighteenth century," in *Mapping Boston,* ed. Alex Krieger, David Cobb, and Amy Turner (Boston: Muriel G. and Norman B. Leventhal Family Foundation, 1999), esp. 38–40.

2. Elaine Forman Crane, *A dependent people: Newport, Rhode Island, in the revolutionary era* (New York: Fordham University Press, 1985).

3. D. W. Meinig, *The shaping of America: a geographical perspective on 500 years of history,* vol. 1: *Atlantic America, 1492–1800* (New Haven: Yale University Press, 1986), 106.

4. The only history of the cartography of Rhode Island or of Narragansett Bay remains Howard Millar Chapin, *Cartography of Rhode Island* (Providence, 1915). At seven pages with two illustrations, it provides only the bare bones of the history of printed maps of Rhode Island. Charles Blaskowitz is mentioned in passing as having done his survey after that of "Des Barras" (a reference to Joseph Frederick Wallet Des Barres, who will be discussed here shortly). I thank Susan Danforth and David Bosse for alerting me to Chapin's work. A bibliography of works relating to Newport, Rhode Island, lists only the manuscript map of Newport harbor, surveyed by Peter Harrison, in 1745. Charles E. Hammett, *A contribution to the Bibliography and Literature of Newport, R.I. comprising a list of books published or printed in Newport, with notes and additions* (Newport, R.I.: Charles E. Hammett, Jr., 1887).

5. Carl Bridenbaugh, *Peter Harrison: first American architect* (Chapel Hill: University of North Carolina Press, 1949). Harrison's manuscript map is in the National Archives, London, C.O. 700, Rhode Island No. 5.

6. Bridenbaugh, *Peter Harrison,* 97. An earlier survey of 1741 is recorded in 174 lines of verse printed as a broadside. W[illiam] C[handler], "A journal of the survey of Narragansett Bay and parts adjacent Taken in the month of May and June A.D. 1741 By order of the Hon. Court of Commissioners, appointed by King George the Second. Poetically Described by one of the Surveyors." Published by Early American Imprints, Micro-F 2112–40232 from a copy in the Rhode Island Historical Society.

7. Willis Chipman, "The life and times of Major Samuel Holland, surveyor-general, 1764–1801," *Ontario History* 21 (1924), 22.

8. Louis DeVorsey, "William Gerard De Brahm," in *Geographers biobibliographical studies,* ed. T. W. Freeman (London: Mansell, 1986); Louis DeVorsey, ed., *De Brahm's report of the general survey in the southern district of North America* (Columbia: University of South Carolina Press, 1971).

9. G. N. D. Evans, *Uncommon obdurate: the several public careers of J. F. W. Des Barres* (Salem: Peabody Museum of Salem; Toronto: University of Toronto Press, 1969).

10. WLCL, Henry Young, Loyalist Index, Blaskowitz, W.O.42/59/B.13; Ind. 5603 (1783).

11. Douglas W. Marshall and Howard H. Peckham, *Campaigns of the American Revolution: an atlas of manuscript maps* (Ann Arbor: University of Michigan Press, 1976), 132.

12. Peter J. Guthorn, *British maps of the American Revolution* (Monmouth Beach, N.J.: Freneau, 1972); D. C. Harvey, *Holland's description of Cape Breton Island and other documents* (Halifax: Public Archives of Nova Scotia, 1935), 54–55. National Archives, London, AO 3/140, document IX, "Account of Expences for the Survey of the Northern District of North America, incurred from 24th Decr. 1763 to 24th Decr. 1764," refers to "Charles Blaskowitz, Ensign Goldfrap, and James Grant as Volunteers at 1 s. per day each from 24th March to 24th Decr." Many thanks to Ed Dahl for his assiduous digging for sources on Blaskowitz.

13. Nathaniel Shipton, "General James Murray's map of the St Lawrence," *Cartographer* 4, no. 2 (1967). There are five manuscript copies of the Murray map; of the two in the Public Archives of Canada, one has sheets signed by D. Hamilton, Charles M'Donnell, and Charles Blaskowitz.

14. *A survey of Lake Champlain including Crown Point and St John's . . . by John Collins, depy. Surv. Genl., May 21th 1765. Charles Blaskowitz, draughtsman.* Manuscript map in Library of Congress, G3802 C45 1765 .C6 Vault.

15. F. J. Thorpe, "Holland, Samuel Johannes," in *Dictionary of Canadian biography* (Toronto: University of Toronto Press, 1983). Thorpe says Holland organized this unit in March 1777.

16. Murtie June Clark, *Loyalists in the southern campaign in the Revolutionary War* (Baltimore: Genealogical Publishing, 1981), 2:291, 292, 293, 295.

17. Peter J. Guthorn, *British maps of the American Revolution* (Monmouth Beach, N.J.: Freneau, 1972), 12. Guthorn follows the attribution of the report to Robert Melville. See note 18 below.

18. Anonymous, "A British navy yard contemplated in Newport, R.I. in 1764," *Rhode Island Historical Magazine* 6, no. 1 (1885): 42–47. This article is without author and the letter/report lacks a provenance. It is dated "J—— 16, A.D. 176——." The letter begins, "I arrived here after a passage of sixty days from the Lands' End, and from that time to the present, a period of two months, I have been constantly engaged in obtaining the surveys and drafts of this harbour . . . and Narragansett Bay . . . in conformity to your Lordship's directions, and in furtherance of His Majesty's views, very explicitly noted in my instructions by desire of the Board of Admiralty." The writer of the *Rhode Island Historical Magazine* article suggests that the author of the report was Robert Melville, who was appointed governor of the Ceded Islands in the West Indies in October 1763; he arrived in Grenada in December 1764. However, the dates mentioned in the report do not coincide with the dates when Melville left Portsmouth (25 August 1764), not Lands End; he arrived first in Barbados (23 October 1764), then Tobago, and finally Grenada (13 December 1764). These dates do not allow for a sixty-day passage and a further sixty days in Rhode Island. For his departure, see Melville to Charles Jenkinson, joint secretary to the Treasury under George Grenville, 18 August 1764, BL, Add. Ms. 38203 (93). For his arrival in Grenada, see Melville to the Lords Commissioner of the Board of Trade and Plantations, 3 January 1765, National Archives, London, CO 101/1 (181–83). There is no record of any report similar to the *Rhode Island Historical Magazine* account in the Index and Digest of Admiralty in Letters (ADM 12). It is true, however, that George Grenville, as first

lord of the Treasury, would have been interested in Rhode Island because of his deep concerns about clandestine trade and smuggling in the colonies and its attendant loss of revenue for the Treasury. Rhode Island was particularly notorious for such activities and in 1763–64 the Treasury had requested that the Admiralty station ships off Rhode Island to prevent illicit trade there. John L. Bullion, *A great and necessary measure: George Grenville and the genesis of the Stamp Act 1763–1765* (Columbia: University of Missouri Press, 1982), 76 n. 46. As to the "young" "Mr Blaskerwich," he would have been about twenty-two in 1764 if we follow the earlier year for his birth. R. A. Skelton, perhaps relying on the War Office notice of Blaskowitz's death, stated that Blaskowitz "was only twelve years old when, as Ordnance Cadet, he surveyed Narragansett Harbour in 1764." R. A. Skelton, "The origins of the Ordnance Survey of Great Britain," *Geographical Journal* 128, no. 1962 (1962): 418. For the War Office death notice, see WLCL, Henry Young, Loyalist Index, Blaskowitz, W.O.42/59/B.13.

19. The list of churches under the plan of Newport has "G. Fourth Baptist Meeting House [founded] 1770."

20. Anonymous, "A British navy yard," 44.

21. Ibid., 44–46.

22. Crane, *A dependent people,* 111.

23. Ibid., 112, 113.

24. Carl Bridenbaugh, in ibid., 146 n. 27.

25. Ibid., 114.

26. Stiles specified in his diary that the cannons included six twenty-four-pounders, eighteen eighteen-pounders, fourteen six-pounders, and six four-pounders. Ezra Stiles, *The literary diary of Ezra Stiles,* ed. Frank Dexter (New York: Scribner's Sons, 1901), 1:499–500, 9 December 1774. Ezra Stiles (1727–1795) was a graduate and later president (1778) of Yale College. After practicing law briefly, he returned to the ministry, becoming pastor of the Second Congregational Church of Newport in 1756 until the exigencies of war and the British occupation of Newport in 1777 moved him to New Hampshire prior to his translation to Yale.

27. *Massachusetts Gazette and the Boston Weekly Newsletter,* 5 May 1774.

28. *Newport Mercury,* 3 October 1774: "Captain Holland's company of surveyors is now taking a survey of the coasts of this colony."

29. Stiles, *Literary diary of Ezra Stiles,* 1:466, 25 October 1774. It is surprising that Stiles didn't comment further on the survey. He himself had made a map of Newport in 1758 (dated 9 August 1758, 15 × 12.5 inches with many notes, noted by Charles Hammett; see note 4) and records in his diary on 1, 2, 4, 7, 9,and 11 March 1769 various maps he compiled for a history of New England or North America that he was composing. Stiles, *Literary diary,* 1:4–5.

30. Grace S. Machemer, "Headquartered at Piscataqua: Samuel Holland's coastal and inland surveys, 1770–1774," *Historical New Hampshire* 57, nos. 1 & 2 (2003), 12 nn. 36, 37. Machemer also says that deputy surveyors each had a servant, probably a young boy, building credit for sea time.

31. Chipman, "Samuel Holland," 23. The estimated expenses for one year's work on the Northern District survey were submitted to George III on 4 February 1764 by

Lord Hillsborough and his fellow commissioners for trade and plantations, the ministry responsible for colonial affairs. The estimates were: allowance for a deputy surveyor, £100 per annum; assistant surveyors, 7s per day; draftsman, 5s per day; private men to assist in the survey as camp, color, and chainmen, to make signals, 6d per day; extraordinary expenses for horses and guides, £100 per annum; annual total, £700. The first year's total added £416 15s for two sets of "proper instruments" for the surveyors.

32. A wide variety of soundings may be noted on the various manuscript maps of Narragansett Bay that followed from the initial Blaskowitz survey. Few of the soundings from one map to another match.

33. For comparison, John Montresor records various durations of time for drafting maps. G. D. Scull, ed., *The Montresor journals* (New York: New York Historical Society, 1881): p. 372 (June 1766, four days for drawing a fair draft of Crown Point and environs); 386–87 (September 1766, nineteen days for a plan of the harbor at Philadelphia with soundings and islands); pp. 342–44 (December 1765 to January 1766, for his survey and fair draft of New York, from 16 December 1765, until 8 February 1766, when he presented the results to General Thomas Gage).

34. Des Barres requested reimbursement for the costs of his instruments, which included a sextant, telescope, astronomical quadrant at £60, quadrant with a reflecting telescope at £40, a pair of seventeen-inch globes at 6 guineas, and a pair of beam compasses. His total house and fuel costs in 1767 had been £50, or a little over £4 per month. Evans, *Uncommon obdurate,* 63 n. 11, 64.

35. Machemer, "Holland's surveys," 15, quoting Governor John Wentworth's address to the assembly in January 1772. The survey and manuscript map were not published until 1784, by William Faden of London.

36. The Faden collection was purchased by Congress in 1864 for $1,000 from Edward Everett Hale, who had inherited them from his father, Nathan Hale, who had acquired them from the Rev. Mr Converse, who purchased the collection at a sale. William Dawson Johnston, *History of the Library of Congress,* vol. 1, 1800–1864 (Washington: Government Printing Office, 1904), 340. Many of the maps in the collection are original manuscripts drawn by officers in the British army during the Seven Years' War and the American Revolution.

37. Thorpe, "Holland, Samuel Johannes," 427. Holland sent copies of the surveys of Boston Harbor, Martha's Vineyard, Nantucket, the Elizabeth Islands, and Narragansett Bay to Lord Dartmouth in May 1775. Holland to Lord Dartmouth, 27 May 1775, National Archives, London, C.O. 5/76 (122–23).

38. Marshall and Peckham, *Campaigns of the American Revolution,* 30.

39. For example, *An original plan of Brenton's Neck and all the ground to the southward of the Town of Newport,* by Edward Fage, Captain, Royal Artillery, 1779, WLCL, 3-J-18, matches the Blaskowitz survey for the names of the proprietors and the topography of the Neck. *Plan of Rhode Island surveyed and Drawn by Edwd Fage . . . in the years 1777–1778–1779,* WLCL, 3-J-14, Clinton Ms. 62, is also based on Blaskowitz, showing houses of principal farmers used as barracks or military headquarters during the occupation of Rhode Island.

40. Evans, *Uncommon obdurate,* esp. 59–78.

41. Ibid., 68.

42. This contrasts with the French ships that arrived in Narragansett Bay during the summer of 1778 under the command of le comte d'Estaing, whose ship was provided with a printing press.

43. Shipton, "General James Murray's map," 96. Montresor had erased Holland's name from the copy about to be shipped to London. Holland suspected something and demanded an inspection before it left, revealing Montresor's act.

44. Chipman, "Samuel Holland," 18–19. Chipman quotes Holland's account, dated 11 January 1792, in extenso.

45. Scull, *The Montresor Journals,* 392, quoted in J. B. Harley, Barbara Petchenik, and Lawrence Towner, *Mapping the American Revolutionary War* (Chicago: University of Chicago Press, 1978), 85.

46. Samuel Holland to General Haldimand, governor of Canada, 16 March 1773, Public Archives of Nova Scotia, vol. 367½, doc. 3, transcripts from the Haldimand collection, published in Harvey, *Holland's Description of Cape Breton Island,* 131–32.

47. The presumed Blaskowitz manuscript in the Faden collection of the Library of Congress has an annotation at the bottom: "latitude 41°27′14″ north variation 6°15′ west." Library of Congress, G3772 N3 1777 .B49 Faden 87.

48. Compare the topographic detail to plans of ports that Alexander Dalrymple, hydrographer to the East India Company, was having engraved at the same time as Des Barres was working on the *Atlantic Neptune.* The engravers in both workshops seem to have been experimenting with similar techniques. See Andrew Cook, "Alexander Dalrymple's *A Collection of Plans of Ports in the East Indies* (1774–1775): a preliminary examination," *Imago Mundi* 33 (1981): 46–64.

49. "Lord Howe having signified in his letter of 21 March that he is informed that an accurate chart of Rhode Island with all the sounding was presented to the Board of Ordnance by Harrison formerly collector at Boston, he therefore requested Major Holland may be furnished with that draft; ordered that if any such papers in the drawing room, that a copy be made for Ld Howe." WLCL, W.O. 47/87, 22 March 1776, Map Division card catalogue, annotation by D. W. Marshall, map curator. See above for Harrison's map.

50. National Archives, London, C.O. 700 Rhode Island no. 5. See note 5 above.

51. This is the amount cited by Des Barres as the allowance made by the government for the 257 plates of charts and views contained in the *Atlantic Neptune.* Joseph Frederick Wallet Des Barres, *A statement submitted by Colonel Desbarres for Consideration* ([London]: [1796]), 5.

52. The cost per plate, equivalent to approximately 735 livres, is the cost of engraving each chart of the *Neptune Americo-septentrional* by the Dépôt de la Marine. See appendix 1.

53. Evans, *Uncommon obdurate,* 68. Evans says the maps were "so costly private buyers were hard to find." Yet at 1s per sheet, the maps from the *Atlantic Neptune* were at par or slightly below the market average. The later Dutch version of the *Atlantic Neptune* maintained a similar price structure at 1 guilder per sheet, roughly equivalent

to 1s. Compare John J. McCusker, *Money and exchange in Europe and America, 1600–1775* (Chapel Hill: University of North Carolina Press, 1978), 60.

54. Evans, *Uncommon obdurate,* 68.

55. WLCL, 3-J-15. The Gage map is from the Gage papers; the Clinton map is Clinton 104.

56. WLCL, Atlas E-6/7-B. Map no. 29 in *The Atlantic Neptune,* volume 3, adds soundings and shoals. Map number 14 in *Charts of Coasts and Harbours of New England* shows the naval positions of French and British forces in 1778.

57. Anonymous, "A British navy yard," 45: "The climate is the most salubrious of any part of his Majesty's possessions in America. Rhode Island, which is truly styled the Garden of America, is so celebrated for the healthiness of its climate, and the beauties of its scenery, that it is made the resort every summer of numerous wealthy inhabitants of the Southern Colonies, and the West Indies, seeking health and pleasure."

58. The thirty-three "principal farmers" were John Banister, Thomas Banister, Barker, Bowler, Benjamin Brenton, Jahleel Brenton, James Brenton, Church, John Collins, Dudley, Dyre, Jonathan Easton, Nicolas Easton, Walter Easton, Elam, Gould, Harrison, Honyman, Jepson, Isaac Lawton, Robert Lawton, Lopez, Malbone, Overing, Pease, James Potter, Abraham Redwood, William Redwood, Rome, Scott, Tillinghast, Wanton, and Charles Wickham.

59. Such material may lie in papers of several of the families listed here now in the Rhode Island and Newport historical societies.

60. Many Newporters profited particularly from the manufacture and trade in spermaceti, the waxlike substance found in the head of a sperm whale, used for making candles.

61. Crane, *A dependent people,* table 1, 25–29. "The basis for tax assessment was 'An Act for taking a just Estimate of the Rateable Estates in this Colony in order that the rates and taxes may be equally assessed upon the Inhabitants,' June 1767, Proceedings of the Rhode Island Assembly. The following were ratable assets: real estate, including land, dwelling houses, distill houses, sugar houses, ropewalks, warehouses, wharves, mills, spermaceti works, lime kilns, tan yards, iron works, pot ash and pearl ash works, slaves, trading stock, money, wrought plate, and livestock." Ibid., 29.

62. Peter Wilson Coldham, *American loyalist claims . . . abstracted from the PRO, Audit Office series 13, bundles 1–35 & 37* (Washington: National Geneological Society, 1980), passim. They are John Bannister, Samuel Dyre, Thomas Gould, Peter Harrison—of the survey of 1758—Isaac Lawton, Godfrey Malbone, Henry Overing, Simon Pease, George Rome, and Joseph Wanton.

63. Stiles, *Literary diary of Ezra Stiles,* 2:131–34. Stiles lists the inhabitants left in Newport when it was taken by the British on 8 December 1776.

64. Jane Clark, "Metcalf Bowler as a British spy," *Rhode Island Historical Magazine* 23, no. 4 (1930).

65. Gov. Thomas Hutchinson to the earl of Hillsborough, Boston, 12 June 1772, in *Documents of the American Revolution,* ed. K. C. Davies, Colonial Office Series (Dublin, 1972), 5:119, quoted in Crane, *A dependent people,* 84.

66. Ibid., 88 – 89.

67. Chief Justice Frederick Smyth of New Jersey to Lord Dartmouth, 8 February 1773, Chalmers Papers, p. 75, New York Public Library, quoted ibid., 85 n. 144.

68. William Faden, *A Catalogue of Maps, Charts, and Plans* (London: William Faden, 1778), 10 – 11. It is one of the most expensive maps in the catalogue, the average price being 1s or 2s. Only the "Plan of the City of New York," surveyed in 1765 – 66 by Bernard Ratzer, also an engineer in the Royal American Regiment, printed on one sheet of Grand Eagle paper, is similarly advertised by Faden at 5s. Ibid., 13.

69. Des Barres' paper used on WLCL, Clinton 104, Maps 3-J-15, bears the same watermark, a crowned fleur-de-lys over LVG, as maps published by Sayer in his *Coasting Pilot* and the *West Indian Atlas* (1780), suggesting that this was a common size from a common source in London. It matches Heawood number 1861. E. A. Heawood, *Watermarks* (Hilversum, Holland: Paper Publication Society, 1950). No watermark was visible on WLCL, Map 3-J-11 (Faden), or on WLCL, Map 4-J-29 (Dépôt de la Marine).

70. Mary Sponberg Pedley, ed., *The map trade in the late eighteenth century: letters to the London map sellers Jefferys and Faden* (Oxford: Voltaire Foundation, 2000), 10.

71. Faden, *A Catalogue of Maps, Charts, and Plans*, e.g., *Plan of the Operations of the King's Army . . . wherein is particularly distinguished the Engagement on the White Plains*, 13; and *A Topographical map of the northern part of New York Island, exhibiting the plan of Fort Washington, survey'd by order of Lord Percy, by J. C. Sauthier and published . . . for the Benefit of the Soldiers Fund*, 13 – 14. Several more Sauthier maps printed by Faden are listed in Guthorn, *British maps of the American Revolution*, 41 – 42.

72. Guthorn, *British maps of the American Revolution*, 41; *A Plan of the Town of NewPort with its Environs Survey'd by order of . . . Earl Percy, . . . March 1777*, a manuscript by Sauthier, is in the Duke of Northumberland Collection in Alnwick Castle. See William P. Cumming, *British maps of colonial America* (Chicago: University of Chicago Press, 1974), appendix B, 79 – 84.

73. Library of Congress, G3772 N3 1777 .B49 Faden 87.

74. Ira D. Gruber, *The Howe brothers and the American Revolution* (New York: Atheneum, 1972), 192 – 93. Percy wrote to his superior, General Sir Henry Clinton, now on leave in London, that he was so hurt by Sir William's charges "that nothing on earth shall make me stay here, to subject myself to such another indignity. Nay, I had rather quit the Service intirely than remain here any longer." Percy to Clinton, 20 February 1777, WLCL, Clinton Papers, vol. 20, 32.

75. London *Morning Post*, 24 and 27 May and 7 and 11 June 1777, and the *Public Advertiser* (London), 7 June 1777, cited ibid., 212. "The Howes are not in fashion. Lord Percy is come home disgusted by the younger." Horace Walpole to Sir Horace Mann, 18 June 1777, quoted in Charles Knowles Bolton, ed., *Letters of Hugh Earl Percy from Boston and New York, 1774 – 1776* (Boston: Charles E. Goodspeed, 1902), 80.

76. The specific location of Jefferys and later Faden's shop is found in R. A. Skelton, *James Cook: surveyor of Newfoundland* (San Francisco: David Magee, 1965), 28. Skelton asserts that Jefferys's shop was 5, Charing Cross, located on the north side of the Strand, adjacent to the Northumberland Coffee House, facing Northumberland

House. He reproduces the Richard Horwood map of London (1799), with the house numbers shown and also a mezzotint by T. Malton (1787) of a view of Northumberland House and the shops opposite.

77. Faden received his gold medal from the Royal Society of the Arts in 1796 for his publication of a map of Sussex, surveyed by Thomas Yeakell and William Gardner. See J. B. Harley, "The Society of Arts and the Surveys of English Counties 1759–1809," *Journal of the Royal Society of Arts* 112 (1964): 273.

78. J. B. Harley, "The bankruptcy of Thomas Jefferys: an episode in the economic history of eighteenth-century map-making," *Imago Mundi* 20 (1966). Harley discusses the role of such surveys in the bankruptcy of Jefferys.

79. AN, MAR 1JJ/14/3.

80. Ibid.

81. Ellen Cohn, "Benjamin Franklin, Georges-Louis Le Rouge, and the Franklin/Folger chart of the Gulf Stream," *Imago Mundi* 52 (2000). Le Rouge was a heartfelt advocate of mastering foreign languages. In the introduction to the text of his *Parfait Aide de Camp où l'on traite de ce que doit sçavoir tout jeune Militaire qui se propose de faire son chemin à la Guerre . . .* (Paris, 1760), he recommends learning one's own language well, then English and German "so necessary in all wars"; the energy required for study would leave the young aide-de-camp with little for the seductive charms of "les amourettes," who brought nothing but destruction.

82. *Le Théatre de la Guerre entre les Turcs et les Russes en 2 feuilles* and *La Russie meridionale, 3 flles.* Georges-Louis Le Rouge, *Catalogue des cartes et plans de Le Rouge, Ingénieur-géographe du Roi, ancien Lieutenant au Regiment de Saxe* (Paris: Le Rouge, 1773). BNF, Estampes Yd. 904 (45). Also relevant for comparison purposes is *Le Plan de Lisbonne . . . 2 grandes feuilles,* offered for 3 livres in the *Journal des Sçavans,* March 1756.

83. AN, MAR 1JJ/18/4: "Fourni par Buache en 1777, 1778, 1779 au Dépôt Gen. De C. Pl. J."

84. AN, MAR 2JJ/39, ms. de Chabert: Compte rendu du Dépôt général des Cartes, Plans, et Journaux de la Marine, et du Dépôt des Plans des Colonies, pp. 10, 17: Notice des Services et des Titres de tous les Employés au Dépôt et proposition d'une nouvelle organization consequemment à la demande de réduction et de suppression possible [c. 1791]. "*M. Petit, graveur:* 7–8 ans chargé de graver des cartes marines, ~~sur cequ'il passoit pour habile dans ce genre~~ mais comme on a reconnue qu'il ne gravoit aussi bien que les Echelles, et que peu capable de faire le trait et nullement les relevés, il se servoit pour le suppler d'ouvriers inconnus . . . il en résultoit que la gravure des cartes, entreprise et non executée par lui étoit chere et sujette à beaucoup d'inconveniens, on a pris la partie il y a 5 ans [i.e., 1786] de ne pas se servir de lui et d'employer des graveurs veritablements ouvriers exactes, et dociles quoique plus habiles, qui [sont] à meilleur marché."

85. AN, MAR 1JJ/17/10, "Neptune Americo-Septentrional . . . dressés au Dépôt générale des Cartes, Plans, et Journaux de la Marine," 15 December 1780. "Quoique ce Neptune ne soit pas terminé on a pensé que, pour répondre à l'empressement du Public et des Navigateurs, il convenoit de publier la partie de cet ouvrage qui est déjà éxécutée et que les circonstances de la guerre actuelle rendent intéressante."

86. 26 March 1780: "Etat des Cartes nouvelles de Cotes de l'Amerique septentrionale délivrés par ordre de Monseigneur pour un objet particulier de Service." 26 March 1780, AN, MAR 1JJ/17/10. The maps were sent to M. le Comte Hector.

87. "Envoyé deux Receuils de l'Amerique septentrionale en papier doré à Mr le Mis Chabert Capt. Vaisseau du Roi à Brest, le 11e janvier 1781." Ibid.

88. Fleurieu [to Le Moyne, garde du Dépôt], 28 November 1780, AN, MAR 1JJ/17/10. Fleurieu asked for the nineteen sheets of the Neptune, including the frontispiece "Comme jai deja été obligés de donner pour le Roi les feuilles d'epreuves que j'avais à Versailles . . . il voudra bien fair faire . . . six receuils brochés comme je les ai demandés en faisant travailler le relieur la nuit . . . il faut absolumment que j'emporte . . . trois receuils jeudi matin avec moi à Versailles."

89. Pedley, *Map trade,* 34.

90. Ibid.

91. Ibid., 157–60: from Famitte, II: 11 and II: 12.

92. Ibid., 40.

CHAPTER 6

1. *Journal des Sçavans,* January 1700, 187–88.

2. Quoted in *Examen Impartial de la critique, sans nom d'auteur, des cartes de la mer Baltique et du golfe de Finlande, présentées à M. le Maréchal de Castries, ministre et secrétaire d'Etat de la marine, par Le Clerc, écuyer, chevalier de l'Ordre du Roi et membre de plusieurs Académies* (Paris: l'Imprimerie de Clousier, imprimeur de Roi, 1786), 64.

3. David C. Jolly, *Maps in British periodicals, pt. 1: Major monthlies before 1800* (Brookline, Mass.: David C. Jolly, 1990). Jolly reproduces some of the critiques levied by Cave, the *Gentleman's Magazine* editor, against maps published by the competitor *Universal Magazine,* 36–41.

4. The Rev. Patrick Murdoch, "On the best form of geographical maps," *Philosophical Transactions of the Royal Society* 50, pt. 2 (1758): 553–62; William Mountaine, "A short dissertation on maps and charts: in a letter to Rev. Thomas Birch, DD and Secretary," *Philosophical Transactions of the Royal Society* 50, pt. 2 (1758): 563–68.

5. Table of contents of *Journal Book,* 1741–1748, ms. 213. In the seven-year period represented in ms. 213, there are seventeen mentions of maps or globes in the *Journal Book* minutes, most of them new maps presented to the Society. A further survey of the Index of Papers, ms. 704, 1716–1738, found twenty-three mentions of maps and globes presented to the Society. These counts do not include the discussions of longitude or the proposals for various surveys and new instruments for measuring, nor do they include the presentation of astronomical observations that would have an effect on longitude measurement. At the Académie Royale des Sciences, elections for the adjunct position of geographer took place in 1751, 1753, 1754, 1772, 1773, 1774, 1781, and 1782. In those years, many geographers presented maps and work to the Académie for consideration. See the *procès-verbaux* for those years in the Académie's archives.

6. "Le nombre prodigieux de Cartes, dont l'Europe est inondée, met les Amateurs

& même les premiers Géographes dans l'impossibilité de connoître celles qui méritent la préférence. Le meilleur moyen d'acquerir cette connoissance, étoit sans doute d'en faire l'objet d'un commerce; mais comme une pareille entreprise demande des avances très considerables & onéreuses pendant les premieres années, les Géographes se sont contentés de vendre leurs cartes & celles qu'ils étoient à portée d'avoir d'un moment à l'autre." Roch-Joseph Julien, *Nouveau catalogue de cartes géographiques et topographiques . . . divisé en deux parties* (Paris: chez R.-J. Julien, 1763), Observations, 3.

7. "Un bon Atlas doit être composé selon l'usage qu'on veut en faire, la dépense qu'on veut y mettre, etc. Le meilleur moyen est de se servir de mon Catalogue, pour choisir soi-même les Cartes, où me consulter ensuite sur le choix des Auteurs: le mien doit être d'autant moins suspect, que les planches des Cartes (qui composent ma collection) ne m'appartenant pas, il m'est indifférent de fournir l'une ou l'autre." Ibid.

8. "Les Souverains et les parfaits Politiques ne peuvent sans elle bien gouverner leurs Estats, & parfaitement demesler les Interests de leurs Voisins; ny leurs Genéraux et Officiers qui ont quelque commandement dans les Armées faire la Guerre avec Succez. Les Gens d'Eglise sçavent assez de quelle utilité elle leur est pour le Gouvernement Politique des choses Ecclesiastiques. Les Magistrats connoissent par elle l'Estendue de leurs Jurisdictions. Les Gens de Finance ne s'en peuvent passer pour les Impositions et les Receptes des Deniers. Les Negocians ne feroient leur Commerce qu'avec desavantage, si elle ne les instruisoit des Routes qu'il leur faut tenir. Les Voyageurs qui n'entreprennent leurs Voyages que par curiosité ne pourroient reussir dans leur dessein, s'ils n'estoient conduits par elle. Et ceux-la meme qui n'ont aucune teinture des belles Lettres, ni aucun Employ dans la Vie Civile, croiroient ne pouvoir passer agréablement une Partie de leur grand loisir, s'ils ne l'employoient quelquefois à lire ou les Histoires, ou les Voyages, ce qu'ils ne peuvent faire sans le secours de la Geographie. Poets, philosophes, et historiens ne peuvent negliger la Géographie sans tomber dans des béveuës qui ne sont nullement excusables." Guillaume Sanson, *Introduction à la Géographie* (Paris: [1681]), n.p.

9. "Au plus grand nombre des personnes studieuses, de celles qui lisent l'Histoire, comme de celles qui ne sont attentives qu'aux évenemens modernes." Preface to Didier and Gilles Robert de Vaugondy, *Atlas universel* (Paris: Gilles and Didier Robert de Vaugondy and Antoine Boudet, 1757), 35.

10. Mary Sponberg Pedley, "The subscription list of the 1757 *Atlas universel:* a study in cartographic dissemination," *Imago Mundi* 31 (1979).

11. Introduction to S. Bolton, *A complete system of geography* (London, 1747).

12. "Apprendre d'une maniere aisée avec ce livre à connoistre des distances des lieux, à lever un plan géographique, à mesurer sur mer le chemin & la route d'un vaisseau, à observer les longitudes & les latitudes." Advertisement in the *Journal des Sçavans* (1715}: 620–623 for N. Chamereau, *Géographie pratique contenant les instructions suffisantes pour rendre une personne assez habile pour dresser lui-même des Cartes,* 4 vols., 440 pp. (Amsterdam, 1715).

13. For example, A.-H. Jaillot, AN, O[1] 30, 247, 20 July 1686, cited in Mireille Pastoureau, *Les atlas français XVIe–XVIIe siècles* (Paris: Bibliothèque nationale, 1984), 232.

Buache received an additional 600 livres as *premier géographe.* AN, O[1] 116, f° 1140, 24 November 1770.

14. Robert Darnton, *The literary underground of the old regime* (Cambridge: Harvard University Press, 1982), 136.

15. Guy Chaussinand-Nogaret, *La noblesse au XVIIIème siècle: de la féodalité aux lumières* (1976; reprint, Brussels: Editions Complexe, 1984), 77–81.

16. Ibid., 101.

17. Ibid., 106–7. Similar figures are cited in Michel Marion, *Bibliothèques privées à Paris au milieu du XVIIIe siècle* (Paris: Bibliothèque nationale, 1978), 138. However, maps themselves often slip out of these statistics if they do not constitute an atlas. Marion's research for the decade 1750–59 studies the inventories taken after death; separately listed in many inventories are the portfolios of prints and maps, which are not counted among the books in a library. Similarly, many works that were very popular in eighteenth-century libraries, such as Charles Rollin's *Histoire Romaine,* contained many maps. Suffice it to say, much research is yet to be done on map collecting in France in the eighteenth century.

18. Michael Sonenscher, *Work and wages: natural law, politics, and the eighteenth century French trades* (Cambridge: Cambridge University Press, 1989), 204, citing BNF, ms. Joly de Fleury 557, fol 27.

19. Darnton, *The literary underground of the old regime,* 136.

20. Jean Lattré, "Catalogue du fonds du Sr Lattré Graveur rue St Jacques près St Severin, à la Ville de Bordeaux," in *Atlas moderne* (Paris: [1777]).

21. Marie-Pierre Dion, *Emmanuel de Croÿ (1718–1784): Itinéraire intellectuel et réussite nobiliaire au siècle des Lumières* (Brussels: Editions de l'Université de Bruxelles, 1987), 131.

22. Daniel Roche, *Les républicains des lettres: gens de culture et lumières au XVIIIe siècle* (Paris: Librairie Arthème Fayard, 1988), 54.

23. Philip Lee Phillips, *Notes on the life and works of Bernard Romans* (Deland: Florida State Historical Society, 1924), 48.

24. R. Campbell, *The London Tradesman: Being a compendious View of all the Trades, Professions, Arts, both Liberal and Mechanic, now practised in the Cities of London and Westminster: Calculated for the information of Parents, and instruction of youth in their Choice of Business* (London: T. Gardner, 1747), 114, 275.

25. Court and City Register, 1776, p. 170, cited in Edward E. Curtis, *The organization of the British army in the American Revolution* (New Haven: Yale University Press, 1926), 158.

26. David Hancock, *Citizens of the world: London merchants and the integration of the British Atlantic community, 1735–1785* (Cambridge: Cambridge University Press, 1995), 383–84.

27. Stella Tillyard, *The aristocrats: Caroline, Emily, Louisa, and Sarah Lennox 1740–1832* (New York: Farrar, Straus and Giroux, 1994), 50–51.

28. John Brooke, "The library of King George III," *Yale University Library Gazette* 52, no. 1 (1977): 38–40. I am indebted to Peter Barber for this reference. The maps and prints in the king's topographic collection show prices of between roughly 2d and 5s, within the range of map prices we have used.

29. Frederick Augusta Barnard, *Bibliothecae Regis Catalogus* (London, 1820), 1:v–vi.

30. Wilfrid Prest quotes the Frenchman Jean-Paul Grosley, who observed that in 1765 London artisans earned twice what their counterparts were paid in France. Wilfrid Prest, *Albion ascendant: English history, 1660–1815* (Oxford: Oxford University Press, 1998), 168 n. 14, citing *A Frenchman's year in Suffolk,* ed. N. Scarfe, 1988, 78.

31. Timothy Clayton, *The English print: 1688–1802* (New Haven: Yale University Press, 1997), 22–23, citing the *Norwich Gazette* and Corporation of London Records Office, Court of Orphans records, nos. 3265, 3306. These prices reflect the first half of the century.

32. Louise Lippincott, "Arthur Pond's journal of receipts and expenses, 1734–1750," *Walpole Society* 54 (1988): 264.

33. James Raven, *Judging new wealth: popular publishing and responses to commerce in England, 1750–1800* (Oxford: Clarendon Press, 1992), 51.

34. Prices from *The Ambulator* (London, 1782), cited in Louis-Sébastien Mercier, *Parallèle de Paris et de Londres: un inédit de Louis-Sébastien Mercier* ([1789?]; reprint, Paris: Didier Erudition, 1982), 195 n. 56.

35. Robert Sayer and John Bennett, *Sayer and Bennett's enlarged Catalogue of new and valuable Prints, in sets, or single; also useful and correct Maps and Charts; likewise Books of architecture, views of antiquity, drawing and copy books, etc etc in great variety at No. 53, in Fleet-Street, London; . . .* (London: Sayer and Bennett, 1775).

36. "Cette grand facilité à s'instruire de la Géographie a fait naître la pensée à quantité de gens de s'eriger en Géographes, ce qui leur a semblé d'autant plus aisé, qu'en changeant ou ajoûtant quelque chose, ils ont creu que l'on ne s'appercevroit point de leur larcin. Mais leur insuffisance leur a fait faire d'étranges méprises. Car non seulement ils ont erigé Royaumes de leur propre autorité, ils ont confondu differentes sortes de Gouvernement de Milice, de Justice, et de Finances; les Jurisdictions ecclésiastiques et temporelles; les provinces, les uns avec les autres, donné de fausses limites, inventé des capitales dans des Pais où sont plusieurs Etats Independans." Guillaume Sanson, preface to Sanson, *Introduction à la Géographie,* n.p.

37. "Ce n'est jamais sur la foi des Cartes qu'il faut prononcer quand elles ne sont pas acompagnées d'instructions et de raisonemens." Claude Delisle, "Seconde lettre de M. de Lisle à M. Cassini pour justifier quelques endroits de ses Globes et ses Cartes," *Journal des Sçavans* 28 (1700): 384. Delisle continued this thinking the following month: "que les Cartes, quand elles ne sont pas accompagnées d'instructions, ne doivent servir tout au plus qu'à nous donner quelque scrupule, si elles ne sont pas conformés à nos idées . . . qu'il faut plus que des Cartes pour établir une verité Géographique." Claude Delisle, "Troisième lettre du S. de Lisle à M. Cassini sur la question que l'on peut faire si la Japon est une Isle," *Journal des Sçavans* 28 (1700): 413–14.

38. "Dresser des cartes: Se déstiner à ce travail, est une condition bien plus difficile à remplir . . . par l'obligation de prendre un parti, au risque de pouvoir se tromper, dans un détail infini, sur lequel il sera facile de glisser en écrivant, ou même ne pas sentir les difficultés faute de les apercevoir par ce qu'il peut suffire d'avoir de connoissances pour écrire. En supposant que les matériaux sont rassemblés en quantité et qu'on n'en sauroit trop avoir, quel chaos à débrouiller? Quel fond d'intelligence n'a-t-on pas dû devoir acquérir pour les juger?" Jean-Baptiste Bourguignon d'Anville,

Considérations générales sur l'étude et les Connoissances que demande la composition des Ouvrages de géographie (Paris: Lambert, 1777), 53.

39. "Mais il est que bon de juger de la valeur de cette Carte, & de savoir si elle est faite sur de bons & de fidèles mémoires." Delisle, "Seconde lettre," 381.

40. "J'ai pris la pris la précaution de representer sur mes Globes et sur mes Cartes, la Côte coupée & interompue dans cet endroit, tant du côté du Cap Mendocin, que du côté de la Mer Vermeille. J'ai laissé dans ces deux endroits comme des pierres d'atente *pendent opera interrupta,* et je n'ai pas cru devoir me déterminer sur une chose qui est encore si incertaine: ainsi je n'ai fait de la Californie ni une Isle ni une parte du Continent. Le Sieur Nolin qui m'a copié trait pour trait dans cet endroit comme en plusieurs autres, ne sachant pas ce qu'il faisoit ni pourquoi il le faisoit." Ibid., 385.

41. Delisle claimed he interviewed two men who had been on La Salle's voyage to the Mississippi, M. de Beaujeu and La Salle's brother, M. Cavalier, and had seen their manuscript maps. He also based his placement of the mouth of the Mississippi on the manuscript map and letters of M. d'Iberville, who traveled to the Gulf of Mexico in 1698, as well as letters and a map from others on d'Iberville's vessels. Delisle, "Lettre de M. Delisle à M. Cassini, sur l'embouchure de la rivière de Mississippi," *Journal des Sçavans* 28 (1700), 368, 372–73.

42. "Par raisonement, par conjecture, par estime, & par raport aux pays voisins." Ibid., 372.

43. Christine Petto, "Mapping the body politic in early modern France," Ph.D. thesis, Indiana University, 1966), 154–202.

44. "Par-tout la Boussole à la main, il a relevé les principaux gisemens de pointe en pointe; toutes les fois, que le tems lui a permis, il a observé la hauteur du Pole, il a estimé avec le plus de précision, qu'il étoit possible, les distances d'un lieu à un autre; enfin il n'a rien négligé de tout ce qui pouvoit servir à la connoissance de ce Pays." Jacques-Nicolas Bellin, "Remarques de M. Bellin . . . sur les Cartes et les Plans, qu'il a été chargé de dresser . . . ," in *Journal d'un voyage fait par ordre du Roi dans l'Amérique Septentrionale,* ed. le révérend père de Charlevoix (Paris: chez Didot, libraire, 1744), 3:xii.

45. "Je ne suis point content de ce que j'en ai donné: mais il ne m'a pas été possible de faire mieux." Ibid., 3:xiii.

46. "Les Mémoires particuliers, qui sont au Dépôt, m'ont donné les moyens de la représenter un peu plus fidelement, qu'on ne l'a vû jusqu'à présent. Cependant je crois, qu'il faut attendre encore d'autres éclaircissemens, car toutes les Parties ne m'en paroissent pas également constatées." Ibid., 3:xiv. As to those curious islands in Lake Superior: "Il est inutile de remarquer que les François sont les seuls qui puissent donner des connaissances fidèles de ces Lacs; les noms des Isles, qui y sont répanduës, & des Rivieres, qui s'y déchargent, . . . ils font voir que ce n'est qu'à nos Voyageurs, & sur-tout aux Missionnaires, qu'on est redevable de leurs découvertes." The fact that several of the larger islands in Lake Superior are named for government ministers connected with the marine does not draw comment from Bellin. See Robert Karrow, "Lake Superior's mythic isles: a cautionary tale for users of old maps," *Michigan History* 69, no. 1 (1985). Karrow points out the connection between these islands and the reports made by the French naval captain Louis Denys de la Ronde, who explored parts

of Lake Superior in search of copper deposits. Two manuscript maps exist in the archives of the marine, one of which is by the French-Canadian engineer Chaussegros de Lery fils, showing Lake Superior with the islands named for the government ministers. Karrow suggests very plausibly that these maps and the de la Ronde reports provided Bellin with the geography of the lake.

47. "On sera peut-être surpris de ne pas trouver des sondes sur mes plans; c'est-à-dire, la quantité de brasses, ou de pieds d'eau; je sçais que ces détails sont extrêmement utiles, & il m'auroit été facile de les remplir avec exactitude: mais des raisons particulières, qui n'ont rien de commun avec la Géographie, m'en ont empêché. A l'égard des plans des Ports, qui n'appartiennent pas à la France, j'y ai mis des sondes." Bellin, "Remarques," 3:vi.

48. "Ces sortes de travaux littéraires trouvent trop peu de lecteurs." Preface to Robert de Vaugondy, *Atlas universel,* 19b.

49. Even the dour d'Anville joked about his own style when embarking on a discussion of projections: "il seroit blâmable de vouloir, par une délicatesse déplacée, éviter la sécheresse attachée à une discussion très importante." At the end of ten pages he closed: "J'ai annoncé de la sécheresse dans cette discussion sur les Projections et il paroîtra bien que je tiens parole." D'Anville, *Considérations générales,* 20, 29.

50. Mary Sponberg Pedley, ed., *The map trade in the late eighteenth century: letters to the London map sellers Jefferys and Faden* (Oxford: Voltaire Foundation, 2000), I:18, p. 74.

51. "On auroit lieu d'espérer qu'elles seroient plus exactes & moins souvent répétées; le Public se trouveroit plus en état de juger & de rendre à l'Auteur la justice qu'il merite." Anonymous reviewer, *Carte des nouvelles découvertes au nord de la mer du sud . . .* par M. (J.N.) de L'Isle," *Journal des Sçavans,* December 1752, 806–11.

52. From Herman Moll, *The World Described: or, a New and Correct Sett of Maps . . .* (Dublin: George Grierson, 1732), plates 2, 18; and Herman Moll, *[Atlas of Europe]* ([London]: [1715?]), plate 17.

53. For the criticism of Mitchell's map by Bradock Mead (under the pseudonym of John Green) and a detailed study of Mitchell's response, see Matthew Edney, *John Mitchell's map: an irony of empire* (Web site), Osher Map Library, University of Southern Maine, 1997; http://www.usm.maine.edu/~maps/mitchell/.

54. D'Anville, *Analyse géographique,* x–xi, xiii, xxviii, xl.

55. D'Anville, *Considérations générales,* 21–27.

56. Ibid., 68–80.

57. "Elle ne donne point une idée de la variété que la Nature a mise dans son ouvrage." Ibid., 66.

58. "Modeles des Formes et Hauteurs des Caracteres pour servir à l'Exécution de La Carte de la Guyenne," illustrated in François de Dainville, S.J., *Le langage des géographes: termes, signes, couleurs des cartes anciennes 1500–1800* (Paris: Editions A. et J. Picard, 1964), 76.

59. On Moses Pitt and his ill-fated enterprise, see Michael Harris, "Moses Pitt and insolvency in the London booktrade in the late seventeenth century," in *Economics of the British booktrade 1605–1939,* ed. Michael Harris and Robin Myers (Cambridge: Chadwick-Healy, 1985); E. G. R. Taylor, "The *English Atlas* of Moses Pitt 1680–

1683," *Geographical Journal* 95 (1940); and Leona Rostenberg, "Moses Pitt, Robert Hooke, and the *English Atlas*," *Map Collector* 12 (1980).

60. Robert Hooke, "[A proposal for] "A true Representation of the Universe and the severall parts thereof. . . ." BL, Manuscripts, Sloane 1039.

61. Mary Sponberg Pedley, *Bel et utile: the work of the Robert de Vaugondy family of mapmakers* (Tring: Map Collector Publications, 1992), 51. See also the preface to Robert de Vaugondy, *Atlas universel*, 19b.

62. M. de Guignes, review of *Atlas universel pour l'étude de la Géographie* . . . , in *Journal des Sçavans*, December 1787, 2568–72.

63. Josef Konvitz, *Cartography in France, 1660–1848: science, engineering, and statecraft* (Chicago: University of Chicago Press, 1987), 109, citing "Mémoire sur les travaux géographiques de la ville; mémoire concernant le receuil géographique du cours de la Seine, exécuté en 1767 par Phil. Buache," BNF, Ge. DD. 2334.

64. "Toutes les fautes sont couvertes d'un beau voile; c'est élégant du burin dans certaines parties & qui fait honneur aux artistes que le Sr *** a employés, de lá vient que les Cartes plaisent à l'oeil et font pitié à l'esprit." Brion de la Tour, *Errata de l'Atlas Moderne ou Appel au Public de l'accusation de Plagiat intentée par le Sieur *** [Bonne] contre M. Brion, ingénieur géographe du roi* ([Paris], 1766). Brion's atlas had been previously titled *Atlas moderne dressé pour l'étude de la Géographie*, but the accusations had caused him to change the title to *Atlas général civil ecclésiastique et militaire*. Both atlases, along with many others in Paris in the 1760s, seemed to have been created to accompany a new edition (1762) of the abbé Nicolle de la Croix's *Géographie moderne*.

65. "Cette carte est séduisante par la manière dont les Montagnes y sont traitées, quoique sans proportion; mais elle est rébutante par la finesse excessive de ses caractéres." Ibid., 29.

66. "Est-il permis à des géographes d'ignorer les changemens politique qui arrivent dans les Etats? Un géographe italien, qui plus est, est-il excusable de ne pas bien connoitre son pays? Les negligences en géographie font de conséquence et un Auteur ne scauroit trop veiller à ce que les enlumineuses suivent exactement les modèles qu'il doit donner." Ibid., 31–33. Brion points out many errors on the map of Italy: for example, Rizzi-Zannoni annexed the duchy of Guastella to the state of Modena when it was in fact ceded to the duke of Parma; he gave Bresci, Bergamo, and Cremona to Milan instead of to the Venetian Republic. Brion notes that Rizzi-Zannoni has copied the two-sheet map of Italy by d'Anville, but "on peut appeller cela un plagiat mal-adroit." Note that Brion uses the feminine form to describe the map colorers: *les enlumineuses*.

67. "Beauté exquise ensorte que l'exterieur portât la marque de ce qu'elles sont pour leur exactitude. Vous savey que les cartes sont un fond sur lequel tout le monde a droit: et de bons ouvrages comme les vôtres devoient se faire rechercher des ignorants par les beauté du dehors, comme les scavants parcequ'elles sont en elles memes." Louis Renard to Delisle, AN, MAR 2JJ/60/21, c. 1706. Delisle's correspondent in Madrid, l'abbé Jouin, also encouraged Delisle to make a big map of North America, as it was intended for a Spanish market, which loved big images, though they knew nothing about them. AN, MAR 2JJ/60 (139), l'abbé Jouin to Delisle, 24 June 1724. "Les Espagnols estant amys de grandes images quoyquils ny connoissant rien."

68. "Il n'est pas vray que jaye toujours vendües mes cartes à Paris en beau et grand papier. Au contraire elles sont été toujours vendües sur du papier tel que je vous lay envoyé, et il n'y a que depuis quelques tems que quelques particuliers ayant souhaité d'en avoir plus grandes marges, j'en ay fait tirer exprès pour eux. Je leur ay fait payer 2 s davantage. Pour cequi est des conseils que vous mavez donné, il est bien inutile de me recommander la beauté de cartouches lorsque les Cartes etoient gravées, et quand elles ne l'auroient pas été, je ne les aurois pas faites autrement que je les ay faits. Un géographe qui fait son metier doit savoir cequi doit dominer dans son ouvrage." Guillaume Delisle, "[list of items for the procès against Renard]," AN, MAR 2JJ/60/16. Letter of 27 January [1708?]

69. Pedley, *Map trade,* 24.

70. J. B. Harley, introduction to *The County Maps from William Camden's Britannia by Robert Morden (1695): a facsimile* ([London]: David and Charles Reprints, 1972), viii, citing Oxford University, Bodleian Library, Wood 658, fol. 806.

71. "Nos cartes, soit générales, soit particulieres, sont toutes sur des échelles ou des mesures differentes. Ce qui cause une espèce de confusion et d'embarras dans l'esprit. Est-on obligé d'examiner de suite cinq ou six Cartes du même Auteur sur cinq ou six Provinces voisines; il faut, en changeant de Carte, changer de mesure, et perdre l'idée de celle dont on vient de se servir, & à laquelle cependant il faudra revenir l'instant d'après." L'abbé Nicolas Lenglet Du Fresnoy, *Méthode pour étudier la géographie . . . et un Catalogue des Cartes géographiques, Relations, Voyages et Descriptions nécessaires pour la Géographie,* 3rd ed. (Paris: chez Rollin, fils, 1741), 1:292.

72. Ibid., catalogue, 1: 96, 108. "Carte curieuse et très bien faite" ("*Carte general de l'Empire des Goths pour l'Histoire d'Espagne* de Don Juan de Ferreras, Robert de Vaugondy," [1742]); "bonne Carte, mais de peu d'usage" (*Delphinatus,* Guillaume de Lisle, 1715); "carte originale, très bonne" (*Carte de l'Isle de Corse,* Jaillot, 1740); "mauvaise carte" (*La Russie blanche ou Moscovie,* Jaillot and Sanson, 1700).

73. "Bonne carte faite aux dernieres observations." Ibid., Catalogue, 1:66–67.

74. "En neuf feuilles. Carte excellent, qui a été élevée sur les lieux mêmes. Elle n'a contre elle que d'être mal orientée." Ibid., catalogue, 1:53. The map is oriented to the south. Nevertheless, it is now considered one of the most accurate maps of the Pyrenees of its time. Numa Broc, *Les montagnes vues par le géographes et les naturalistes de langue française au XVIIIe siècle* (Paris: Bibliothèque nationale, 1969), 24. Lenglet Du Fresnoy's criticism reflects the extent to which an expectation of a north orientation had permeated map readership by mid eighteenth century.

75. "C'est la Carte la plus particulière qui ait encore paru: ses principales Villes ont leurs noms en plusieurs langues modernes et même les noms anciens, avec les dates des prises et reprises des Places par les Chrétiens et par les Infideles. Il y a une petite carte qui fait voir l'étenduë de l'ancien Royame d'Hongrie, avant que les Turcs y eussent fait leurs conquêtes." Lenglet Du Fresnoy, *Méthode . . . Catalogue,* catalogue, 1:173.

76. "Ces cartes si belles et si détaillées, sont très nécessaires pour les Campemens et marche des Armées. Elles sont toutes sur une même échelle, et peuvent s'assembler

en une même Carte. Ce sont les Cartes de Generaux des Armées du Roi, qu'un Ingenieur François remit à Bruxelles pour les faire graver; on les a contrefaites en Hollande; mais elles ne sont pas si exactes, ni si justes." Ibid., catalogue, 1:54–55.

77. "Cette carte seroit encore beaucoup meilleure si elle étoit en plus grand Volume. M. de Lisle l'a faite deux fois; mais l'Edition que je marque est la meilleure. La seconde Edition de 1718 est très mal gravée." Ibid., catalogue, 1:48.

78. "Carte belle & bien gravé" (*Mappemonde,* Nicolas and Guillaume Sanson, 1674); "fort mal gravée" (*Globe terrestre, gravée pour l'usage du roi,* Guillaume de Lisle, 1720). Lenglet Du Fresnoy, *Méthode . . . Catalogue,* catalogue, 1: 4, 5.

79. "Cette carte est belle & bien gravée; mais que de découvertes n'a-t'on pas faites dans la Géographie depuis près de 120 ans qu'elle a paru?" Ibid.

80. "Mais quoique bonne, elle est durement et pesamment gravée: ce qui ne laisse pas d'y répandre de la confusion." Ibid., catalogue, 3:492.

81. "Par une heureuse disposition de faire des oppositions, le blanc fait ressortir le noir . . . la multiplicité des objets et surtout des métayries semées çà et là, le détail de la nomenclature jettent d'abord en certains endroits une espèce de confusion, que la petitesse de l'échelle rend inévitable. Le pays est d'ailleurs très coupé de bois, de montagnes, de vallons et de landes qui, par leur détail, obscurcissent le fond et ne donnent point lieu à des oppositions et à un effet piquant pour l'oeil." Archives d'Ille-et-Vilaine, C.4924, cited by François de Dainville, S.J., *Cartes anciennes de l'église de France* (Paris: Vrin, 1956), 224–25 n. 31. Cassini IV found the officials of Brittany to be obtuse and difficult in keeping their part of the bargain of surveying the province. "Rien de pis que de traiter d'affaire avec des gens qui n'y entendent rien," he commented. ("There is nothing worse than doing business with people who understand nothing.") Monique Pelletier, *Les cartes des Cassini: la science au service de l'état et des régions* (Paris: Comité des travaux historiques et scientifiques, 2002), 191.

82. "Il ne suffit pas que les cartes soit curieuses et nouvelles ou pour le moins interessantes et utiles au progrez de la géographie, comme je me propose de les faire; mais il est encore avantageux qu'elles plaisent aux yeux étant gravées avec gout et netteté." Joseph-Nicolas Delisle to Bellin, 24 October 1752, Bibliothèque de l'Observatoire, B.1.7 (XII), no. 3.

83. Paul Laxton, introduction to *Two hundred and fifty years of map-making in the County of Hampshire* (Lympne Castle, Kent: Harry Margary, 1976), citing Richard Gough, *British Topography* (1780), 407, and John Hutchins to Gough, in Oxford University, Bodleian Library, Gough Ms., General Topography, 363, fol. 348v–349r.

84. Susan Danforth, "The first official maps of Maine and Massachusetts," *Imago Mundi* 35 (1983): 40–42, citing Massachusetts. Resolves. 1798, chap. 71 (29 June 1798).

85. Taylor, "The 'English Atlas' of Moses Pitt 1680–1683," 298.

86. The seminal article for this view is J. B. Harley, "Silences and secrecy: the hidden agenda of cartography in early modern Europe," *Imago Mundi* 40 (1988), reprinted in J. B. Harley, *The new nature of maps: essays in the history of cartography,* ed. Paul Laxton (Baltimore: John Hopkins University Press, 2001), 84–107. Harley's work has encouraged a cottage industry of "silences" articles, too numerous to list here. Dennis

Reinhartz, "An exploration of the 'silences' on various late eighteenth-century maps of northern New Spain," *Terrae Incognitae* 35 (2003), studies the aesthetic and design questions in the silences problem.

87. "Pour ne pas trop mulitiplier les objets et éviter la confusion quand le pays est trop chargé." Cassini IV to the procureurs du pays de Provence, 2 August 1785, Archives, Bouche de Rhone, C 1064, cited by Jean Reynaud, "Les Cassini et la carte de Provence (1776–1790)," *Bulletin de la Section de Géographie* 37, no. 1922 (1922): 121–22.

88. "Comment remplir les vides occasionnés par les contours sinueux des limites? Pour rendre le vide moins choquant on indiquera les villes et bourgs remarquables et on prolongera les routes, mais on supprimera les détails (montagnes, bois, près)." François de Dainville, S.J., "Cartes de Bourgogne du XVIIIe siècle," in *La cartographie reflet de l'histoire,* ed. Michel Mollat du Jourdin (Geneva: Editions Slatkine, 1986), 73, citing Archives de Côte d'Or, C 3529.

89. Preface to Robert de Vaugondy, *Atlas universel,* 32.

90. A problem still encountered among modern readers of maps. A recent study of 187 fifteen-year-olds in France found that while 74 percent knew what the legend of a topographical map was for, 63 percent could not understand the abbreviations. Only 37 percent of the students found reading the map easy; others were baffled by the orientation, scale, and representation of relief (63 percent of the sample thought the contour lines were roads). Michèle Véchambre, "Lire la carte topographique à 15 ans," in *L'oeil du cartographe et la représentation géographique du Moyen Age à nos jours,* ed. Catherine Bousquet-Bressolier (Paris: Comité des travaux historiques et scientifiques, 1995).

91. De Dainville, *Le langage des géographes,* xiii.

92. Catherine Delano-Smith and Roger J. P. Kain, *English maps: a history* (London: British Library, 1999), 51, 73, 95, 97, 106, 93. The authors point out the instances of innovation, particularly on the county surveys of William Yates and Isaac Taylor, as well as the representation of towns in plan by Robert Morden (1701, 1704).

93. "Il a pris un soin extraordinaire de contenter la curiosité la plus delicate et pour cet effet il a inventé des caracteres qui . . . à distinguer les villes, les bourgs, les paroisses, les villages, les châteaux, fermes, les abbayes, les prieurez, les moulins à eau & à vent, les croix & les arbres. Il n'a pas omis comme les autres Geographes, les parcs, les chaussées, . . . les routes & les chemins, & a meme ajouté plusieurs noms de ruisseaux; des contrées & de pays, avec la division de banlieue de Paris & de l'Election." Anonymous, "*Environs de Paris,* dediez à Monseigneur le Dauphin par son très-humble serviteur N. de Fer, son Géographe, . . . 1690," *Journal des Sçavans* (1690): 279–80. Such detail in signs, continued the critic, made it all the more surprising that de Fer had neglected to include longitude and latitude on his map, an error of omission soon forgiven because of the abundance of other information.

94. The influence of Vauban's engineering work on English mapmaking is explored in A. Stuart Mason and Peter Barber, " 'Captain Thomas, the French engineer': and the teachings of Vauban to the English," *Proceedings of the Huguenot Society* 25, no. 3 (1991).

95. "Cette manière commence à s'introduire. Elle est fort estimée." Renard to Delisle, 27 December 1706, AN, MAR 2JJ/60/20.

96. Ibid.

97. "Comme les travaux de ces Sçavans [les Homann] ne doivent pas être renfermés dans l'enceinte de l'empire, ils jouïroient d'une plus grande réputation, s'ils étoient publiés en latin; d'ailleurs, quel avantage ne retireroient point de leurs lumieres beaucoup de sçavans qui sont étrangers dans la langue allemande!" Preface to Robert de Vaugondy, *Atlas universel,* 14a. Markus Heinz points out that many Homann maps were printed with Latin titles and Latin notes. Markus Heinz, "A research paper on the copperplates of the maps of J. B. Homann's First World Atlas (1707) and a method for identifying different copperplates of identical looking maps," *Imago Mundi* 45 (1993).

98. "Ces cartes étant faites plutôt pour les Européens que pour les Asiatiques, auroient dû représenter les noms qu'on les connoît ici." Preface to Robert de Vaugondy, *Atlas universel,* 31a.

99. "Une enluminure riche et des plus fines à quoy un grand nombre de curieuses employent de grosses sommes." Renard to Delisle, 25 July 1707, AN, MAR 2JJ/60/25. On Dutch taste in color, see Lisa Davis-Allen, "The national palette: painting and map coloring in the seventeenth-century Dutch republic," *Portolan,* spring 1999, 23–36.

100. Edmund Gibson on his edition of William Camden's *Britannia* (1695), quoted in Harley, introduction to *County Maps,* viii.

101. "Une planchette de bois d'environ 8 pouces ⅔ de diamètre et garnie d'un limbe de cuivre divisé en 360 degrés qui ont tout au plus chacun ⅚ de ligne." De Bonnecamps to Delisle, 23 October 1755–30 January 1756, Bibliothèque de l'Observatoire, Paris, Bigourdan, Resumé de la Correspondence de J.-N. Delisle, XIII (154), Ms. 1029a.

102. Thomas Pownall, *A topographical description of the dominions of the United States of America [being a revised and enlarged edition of] a Topographical description of such parts of North America as are contained in the (annexed) map of the Middle British Colonies etc. in North America,* ed. Lois Mulkearn (1775; reprint, Pittsburgh: University of Pittsburgh Press, 1949), 6.

103. *An inventory of the contents of the Governor's Palace taken after the death of Lord Botetourt: an inventory of the personal estate of his Excellency, Lord Botetourt, Royal Governor of Virginia, 1768–1770,* (Williamsburg, VA: Colonial Williamsburg Foundation, 1981), 5–7, 11. The governor hung the Fry and Jefferson map of Virginia in the front parlor along with Bowen and Mitchell's map of North America. He had five maps in the "little middle room" and two in the pantry, one of which may have been a map of New England. Benjamin Donn's map of Bristol hung "in the Garrett Room over his lordship's bed chamber." My thanks to David Bosse for alerting me to this inventory.

104. Robert Plot, *The Natural History of Oxfordshire,* 2nd ed. (Oxford and London: Leon Lichfield, Charles Brome, and John Nicholson, 1705), iii.

105. Ibid.

106. AN, MAR 2JJ/60.

107. AN, MAR 2JJ/60/2 (69–77).

108. Preface to Robert de Vaugondy, *Atlas universel,* 25b.

109. Quoted from Gibson's preface in Harley, introduction to *County Maps,* viii.

110. Ralph Hyde, "The making of John Rocque's map," in *A-Z of Georgian London* (London: London Topographical Society, 1982).

111. E.g., Peter Perez Burdett's survey of Derbyshire. Not only did members of the Royal Society question Burdett closely about his techniques, but one of the members "had also made judicious inquiries in the county." J. B. Harley, "The Society of Arts and the surveys of English counties 1759–1809," *Journal of the Royal Society of Arts* 112 (1964): 120–21.

112. De Dainville, *Cartes anciennes de l'église,* 78–79, reproduces the questionnaire for the diocese of Narbonne, found in BNF, mss., coll. de Languedoc (Bénedictins), t.2, folios 158 ff., and t.21, folios 25 ff. The collections also include the responses from two dioceses.

113. A contribution thoroughly analyzed ibid.

114. Ludovic Drayperon, "Enquête à instituer sur l'éxécution de la grande carte topographique de France de Cassini de Thury," *Revue de géographie* 38 (1896): 11. "J'aurais désiré plus de précautions de la part des ingénieurs, plus de secours de la part des seigneurs et des curés, qui ont refusé quelquefois aux ingénieurs l'entrée des clochers, et les éclaircissements qu'ils demandaient." The Jesuit priests–cum–surveyors Christopher Maire and Ruggiero Boscovich encountered similar resistence from the clergy in their triangulation of the Papal States in 1750. See Mary Sponberg Pedley, "'I due valentuomini indefessi': Christopher Maire and Roger Boscovich and the mapping of the Papal States (1750–1755)," *Imago Mundi* 45 (1993): 66.

115. John Andrews, Andrew Dury, and William Herbert, *A Topographical Map of the County of Kent* (London, 1769), BL, Maps 1 TAB.21.

116. "Il est si facile de pécher quand on est dans un pays dont la langue et la pronunciation ne sont pas familières." C. F. Cassini to the Procureurs du Pays de Provence, 2 August 1785, Archives des Bouches du Rhone, C 1064, cited in Reynaud, "Les Cassini."

117. Pelletier, *Les cartes des Cassini: la science,* 157. The *vérificateurs* could be fined for missing any errors or for not turning in the "certificates of conformity" signed by local gentry or clergy.

118. Coolie Verner, "Mr. Jefferson makes a map," *Imago Mundi* 14 (1959): 99, 102.

119. Ibid., 104.

120. Mary Sponberg Pedley, "Map wars: the role of maps in the Nova Scotia/Acadia boundary disputes of 1750," *Imago Mundi* 50 (1998): 98, quoting the *Memorials of the English and French Commissaires concerning the limits of Nova Scotia or Acadia* (London, 1755), 1:71–73.

121. "L'autorité [des géographes] ne doit point être décisive. Ils sont plus occupés de donner un air de système et de verité à leurs cartes, ainsi qu'une apparence de science and de recherche, qu'à fixer les droits de Princes and les véritables limites des pays." Ibid.

122. Preface to Robert de Vaugondy, *Atlas universel,* 31a. He writes with regard to his map of Egypt and the the sources of the Nile: "Il vaut mieux suivre un sentiment reçu que d'en forger de chimérique pour donner du nouveau en matière géographique."

123. "Il avoit cru devoir respecter le préjugé, & ne le choquer que sur les points, où la force de ses preuves alloit jusqu'à l'espece de demonstration, qui a lieu dans la Géo-

graphie." Jean-Pierre Niceron, *Mémoires pour servir à l'histoire des Hommes illustres dans la république des lettres . . .* (Paris: Briasson, 1727), 225.

124. "Mais j'espère bien que les vacances une fois passées et les additions et corrections que vous serés a portées d'y faire le renouvelleront et piqueront la curiosité des amateurs." Dion, *Emmanuel de Croÿ,* 228, citing Robert de Vaugondy to Croÿ, 1 September 1774, Dülmen, Archives de duc de Croÿ, D.K24. A second edition of the map was published in 1777. Pedley, *Bel et utile: the work of the Robert de Vaugondy family of mapmakers,* 79, 84–86.

125. Pastoureau, *Atlas français,* 231.

126. "Un assortiment complet de Cartes originales Angloises sur tout ce qui est relatif à la guerre présente dans l'Amérique Septentrionale." *Journal de Paris,* 15 May 1777, no. 135. I am grateful to Catherine Hofmann and Mireille Pastoureau for making these advertisements available to me.

127. "Ne m'envoyé point de batailles. On fait icy peu de Cas." Pedley, *Map trade,* 104, 24, 25, I:37, Lattré to Faden, 20 July 1778. Similar examples in the correspondence with Faden are found in I:52, Perrier and Verrier to Faden, 15 December 1780 (asking him to send "toutes les Cartes qu'il publiées sur la guerre d'Amerique, en Exceptant les plans de batailles"); and I:53, Lattré to Faden, 2 March 1781 ("vous trouverès y Joint deux actions de Bunkers Hill et 3 de la Bataille de Brandywine que je ne vous ai pas Demandé et que je ne pourois pas vendre ici").

128. "On s'y accoutumeroit . . . le temoignage des savants serviroit peu." Renard to Delisle, 27 December 1706.

129. Preface to the 2nd ed. (1775), Pownall, *Topographical Description,* 10.

130. Ibid.

131. Thomas Jefferys to *The London Evening Post,* 15 February 1776. My thanks to Donald Hodson and Laurence Worms for alerting me to this letter.

132. Pownall, *Topographical Description,* 11.

133. "Je m'en rapportes en cela à votre gout et à votre discernment à connaitre au coup d'oeil une carte geometrique et La distinguer de celles qui ne le sont pas et dont je me soucis encor moins." Pedley, *Map trade,* I:8/9, p. 58.

134. Bradock Mead, *The Construction of Maps and Globes* (London: Horne, Knapton, et al., 1717), 148–49. Mead's appendix considered "the present state of Geography . . . being a seasonable Enquiry into Maps, Books of Geography and Travel, intermix'd with some necessary Cautions, helps, and directions for future map-makers, geographers, and travellers."

135. Ibid., 149.

136. David Bosse, " 'To promote useful knowledge': An *Accurate Map of the Four New England States* by John Norman and John Coles," *Imago Mundi* 52 (2000): 153.

137. "Le renversement de la géographie contre ceux qui commerce d'imagerie." Buache to le comte de St Florentin, 21 February 1763, Bibliothèque municipale du Havre, ms. 1886, (210, 126–130), quoted in Ferdinand Vuacheux, "Un mémoire et une lettre du géographe Philippe Buache (1763)," *Bulletin de la Section de Géographie* (1901), 271–73.

138. "Un assemblage confus et grossier de pièces rapportées sans choix et sans or-

dre, aucun des principes qu'exige la science de la géographie n'y étant observée . . . l'avidité du gain de quelques artistes de la main." Ibid., 272.

139. "Si l'on faisoit subir un examen exact et rigoureux sur toutes ces parties à ceux qui veulent prendre le titre de Géographe, avant qu'ils puissant donner aucun ouvrage au Public, l'on n'en verroit pas tant de ce nom, ni de cartes qui sortent journellement de la main des Plagiaires et des Copistes. De là ce même Public ne seroit pas trompé, comme il l'est continuellement; la nation en recevroit plus d'honneur ches les Etrangers, où ces mauvais ouvrages passent aussi, les sçavans et bons géographes ne seroient pas confondus avec les ignorans et se trouveroient mieux récompensés du fruit de leurs études, de leurs travaux, et des grandes dépenses qu'ils sont obligés de faire pour enrichir le Public." Gilles Robert de Vaugondy, *Introduction à la géographie des Srs. Sanson, Géographes du roi, Quatrième édition, revûe, corrigée, et augmentée* (Paris: Durand, 1743), xiii.

140. Preface to Robert de Vaugondy, *Atlas universel,* 14a: "De semblables sociétés, établies dans les pays où l'on cultive les sciences et surtout la géographie, seroient un préservatif contre les mauvais ouvrages en ce genre qu'on produit sans honte et qui répandant beaucoup d'erreurs, ne servent qu'à déshonorer le pays d'ou on les voit sortir. La Suède ressent à présent l'utilité du bureau géographique établie à Stockholm." Robert de Vaugondy was quoted by Buache in his criticism of the *Atlas universel,* in Gilles Robert de Vaugondy, "Réponse analitique au mémoire du Sr. Buache par le Sr Robert de Vaugondy Géog. ord. du Roi auteur du nouvel atlas du Sr Boudet Libraire imprimeur du Roi," BNF, Ge. EE. 4990, 1751.

141. Robert de Vaugondy, "Réponse analitique." Robert de Vaugondy could not resist suggesting that the title and pension of *premier géographe du roi* be given to someone "whose diligence and work earned them," a slighting reference to Buache, the current *premier géographe,* criticized by Robert de Vaugondy for his lack of published work.

142. Dedication to Samuel Molyneaux, Mead, *Construction of Maps and Globes,* n.p.

143. Emanuel Bowen, "A new and correct map of Ireland" (c. 1745), RGS, E.H.13.

144. Preface to Robert de Vaugondy, *Atlas universel,* e.g., 29a (map of Prussia), 32a (map of North America).

145. "Savant bien connu dont le travail peut etre utile et avec mon agrément." Joseph-Bernard, le marquis de Chabert to Bompar, AN, MAR 1JJ/2/6, 22 October 1758.

146. "Toutes les cartes marines, portulans et instructions nécessaires pour la conduite des vaisseaux, tant de guerre que de commerce du royaume, soient exclusivement composés dressés et publiés au Dépôt de Sa Majesté par des personnes capables de s'en bien acquitter et que ces ouvrages soient toujours accompagnés d'analyses imprimées et indicatives des autorités dont on se sera appuyé, non seulement afin d'inspirer aux navigateurs une juste confiance en leur exposant au vrai le degré d'exactitude ou de doute que comportent ces cartes dans chacune de leurs parties, mais encore afin de les garantir de l'incertitude dangereuse où les jetterait un amas de cartes que pourraient publier sans cela des particuliers qui, quoique dénués de matériaux suffisants pour les construire, les annoncent cependant sous des titres

fastueux et exagérés pour en activer la vente." Arrêt du conseil du roi, 5 October 1773, quoted in Olivier Chapuis, *À la mer comme au ciel: Beautemps-Beaupré et la naissance de l'hydrographie moderne (1700–1850)* (Paris: Presses de l'Université de Paris-Sorbonne, 1999), 190.

147. Konvitz, *Cartography*, 235–38.

148. "Tous géographes, graveurs et autres personnes quelconques qui désireront faire graver, publier, et débiter des cartes géographiques quelles qu'elles soient, ou même des plans des villes, ports, havres, bayes, côtes, frontières ou autres, seront tenus d'en obtenir la permission de Mr. Le chancelier ou du garde des Sceaux. Et pour y parvenir, d'en remettre le dessin manuscrit ou gravé en épreuves des dits cartes ou plans, avec leurs fondemens ou preuves à l'appui, afin qu'il en soit fait, avant d'accorder ladite permission, l'examen par celui des départemens respectifs dont les dites cartes interesseront plus particulièrement l'administration . . . sous peine, contre les contrevenants, de six cents livres d'amende, et de la saisie et confiscation des cartes, épreuves et planches gravées au mépris des présentes dispositions." BNF, ms. fr. 22121, Arrêt du Conseil d'Etat du Roi qui ordonne la communication aux départements des cartes géographiques avant de les publier, 10 June 1786, quoted in Chapuis, *À la mer*, 202–3.

149. Ibid., 207–11; *Examen impartial*.

150. "Le fruit of vos préoccupations pueriles sera donc de tenir dans l'ignorance tous les hommes qui se serviroient honnetement des connoissances qu'on leur interdit, lorsque vous n'en sauriez priver aucun de ceux qui peuvent les employer à vous nuire. La définition moderne du mot patrie justifie, dit-on, cette exclusion et bien d'autres, mais quant l'inutilité d'un monopole est démontrée par une long expérience, on feroit aussi bien d'y renoncer." [Julien?], "Discours Préliminaire," Thomas Jefferys, *Atlas des Indes Occidentales* (London: Robert Sayer and John Bennett; Paris: Roch-Joseph Julien, 1777), 6.

151. Harley, "Society of Arts," 273.

152. J. B. Harley and Yolande O'Donoghue, introduction to *Old Series Ordnance Survey* (Lympne Castle: Harry Margary, 1975), xxxii.

153. Mary Sponberg Pedley, " 'Commode, complet, uniforme, et suivi': problems in atlas editing in Enlightenment France," in *Editing early and historical atlases*, ed. Joan Winearls (Toronto: University of Toronto Press, 1995).

154. For a full discussion of the intellectual transition of geography from a subsection of history and astronomy to its own discipline, see Anne Godlewska, *Geography unbound: French geographic science from Cassini to Humboldt* (Chicago: University of Chicago Press, 1999).

CONCLUSION

1. Norman J. W. Thrower, "Edmond Halley and thematic geo-cartography," in *The compleat plattmaker: essays on chart, map, and globe making in England in the seventeenth and eighteenth centuries*, ed. Norman J. W. Thrower (Berkeley: University of California Press, 1978).

2. The maps were published as part of a presentation to the Académie Royale des

Sciences entitled *Considérations géographiques* (1753). Numa Broc, "Un géographe dans son siècle: Philippe Buache (1700–1773)," *Dix-huitième siècle* 3 (1971); George Kish, "Early thematic mapping: the work of Philippe Buache," *Imago Mundi* 28 (1976). For more on Buache using cartography to work out scientific problems, see E. Clouzot, "Une enquête séismologique au XVIIIe siècle," *La Géographie* 24 (1914).

3. Ellen Cohn, "Benjamin Franklin, Georges-Louis Le Rouge, and the Franklin/Folger chart of the Gulf Stream," *Imago Mundi* 52 (2000).

4. Josef Konvitz, *Cartography in France, 1660–1848: science, engineering, and statecraft* (Chicago: University of Chicago Press, 1987), esp. chap. 6.

5. Mark Monmonier, *How to lie with maps* (Chicago: University of Chicago Press, 1991), 1.

APPENDIX I

1. Joseph-Nicolas Delisle, "Mémoire sur la carte de Bretagne," Newberry Library, Chicago, ms. Sc. 1793, 1721. This figure did not include the one-time purchase of instruments or engraving and printing costs for ten plates per year.

2. François de Dainville, S.J., "La levée d'une carte en Languedoc à l'entour de 1730," in *La cartographie reflet de l'histoire* (Geneva: Editions Slatkine, 1986), 370.

3. Jacques Cassini, "État de la Depense faite pour le Voyage de la Meridienne de Paris depuis le 4 May de l'année 1739 jusqu'au 31 Mars 1740," Smithsonian Institution Libraries, Washington, Dibner Collection, ms. 311A, 1740. This sum included 1,933 livres 7 sous for horses and equipment; 1,607 livres for instruments; 2,420 livres for mules to carry instruments, at 300 livres per month plus eight days; 23,684 livres 12 sous for domestic servants, food, and other miscellaneous expenses; and 5,000 for the salaries of the astronomers, which ranged from 600 to 1,200 livres.

4. Jacques Cassini, "État de la Depense faite pour le Voyage de la Perpendiculaire," Smithsonian Institution Libraries, Washington, Dibner Collection, ms. 311A, 1742. This sum included 1,746 livres for horses and accoutrements, 221 for instruments, 9,731 for domestic servants, food, and miscellaneous expenses, and 2,400 for four draftsmen at 600 livres each. The survey team hoped to recoup 435 livres by selling the horses.

5. Henri Berthaut, *La Carte de France* (Paris, 1898), 1:48. Cassini proposed paying his surveyors 4,000 livres a plate, but this amount would vary according to the difficulty of the terrain. Monique Pelletier, *La carte de Cassini: l'extraordinaire aventure de la Carte de France* (Paris: Presses de l'École des Ponts et Chaussées, 1990), 94.

6. Berthaut, *Carte de France,* 1:50.

7. Cassini to unnamed recipient, 22 May 1776, in BL, Add. Ms. 24210 (43), letters and papers relating to literature and art in France (1595–1823).

8. Delisle, "Mémoire Bretagne."

9. Ibid.

10. De Dainville, "La levée d'une carte en Languedoc à l'entour de 1730," 364.

11. Ibid., 370.

12. Cassini, "Depense meridienne."

13. Roger Desreumaux, "Relations entre les arpenteurs et leurs employeurs au XVIIIe siècle dans la région lilloise," *Revue du Nord, histoire et archéologie: Nord de la France, Belgique et Pays Bas,* no. 66 (April–September 1984): 534. This figure included "le papier-terrier des terres et seigneuries de Cobrieux et dépendances, ainsi que les cartes et plans figuratifs."

14. Desreumaux, "Relations entre les arpenteurs et leurs employeurs," 538.

15. Henri Berthaut, *Les ingénieurs-géographes militaires 1624–1831* (Paris: Imprimerie du Service géographique, 1902), 1:74, 86.

16. Josef Konvitz, *Cartography in France, 1660–1848: science, engineering, and statecraft* (Chicago: University of Chicago Press, 1987), 42.

17. Desreumaux, "Relations," 539. Ronald Edward Zupko, *French weights and measures before the Revolution* (Bloomington: Indiana University Press, 1978), 106.

18. Mireille Pastoureau, *Les atlas français XVIe–XVIIe siècles* (Paris: Bibliothèque nationale, 1984), vii.

19. Pastoureau, *Atlas français,* 230, citing the contract between A. H. Jaillot and Guillaume Sanson in AN, MC, XV (229), 11 December 1670.

20. Pastoureau, *Atlas français,* 231. Contract in AN, MC, XV (235), 6 May 1672.

21. Ibid.

22. Ibid.

23. Ibid. AN, MC, XV (239), 22 February 1673.

24. Nelson-Martin Dawson, *L'atelier Delisle: l'Amérique du Nord sur la table à dessin* (Sillery, Quebec: Editions du Septentrion, 2000), 83 n. 44, citing AN, 315 AP-2, 16 November 1717.

25. Henri Cordier, "Du Halde et d'Anville (Cartes de la Chine)," in *Recueil de mémoires orientaux par les Professeurs de l'Ecole des Langues orientales* (Paris, 1905), 394.

26. Ibid. By 1734, d'Anville's price had risen to 1,000 livres. Ibid., 396.

27. AN, MAR 1JJ/2/3, "État des Dépenses . . . par le Sr. Bellin sous les ordres de M. de la Galissonière . . . pendant le cours de trois années . . . depuis le mois d'aoust 1750 . . . jusques à la fin de juin 1753": manuscript maps purchased for the Dépôt de la Marine.

28. AN, MAR 1JJ/2/3, "État des Dépenses," 1750–53.

29. BNF, Ge. DD. 2025, Buache to d'Argenson, 30 June 1756, regarding thirty maps for an atlas for the education of the Dauphin at about 40 livres per design.

30. AN, MAR 1JJ/2/6, "État des Dépenses . . . juin 1757 . . . octobre 1758."

31. BNF, ms. fr. 22147 (1–10), proposal and contract between d'Anville and publishers Desaint and Saillant regarding the expenses for preparation of map for *Notice de la Gaule,* 24 November 1759.

32. BNF, ms. fr. 22120 (57), summary of lawsuit brought by Rizzi-Zannoni against de Lancelles, charging plagiarism of a map of Germany, 1763. Rizzi-Zannoni had contracted with Jean Lattré, engraver and print seller, to produce a map of Germany in six or eight sheets, at 200 livres per sheet.

33. BNF, ms. fr. 22120 (57).

34. Ibid. (62). Paid by Jean Lattré.

35. Vladimiro Valerio, *Società uomini e istituzioni cartografiche nel mezzogiorno d'Italia* (Florence: Istituto Geografico Militare, 1993), 92 nn. 94, 95.

36. AN, Y 1903: 31 July 1772, Jaillot's estimate of the value of the draftsman's work on the six-foot square manuscript map of the *Tableau du Service des Postes;* 31 August 1772, d'Anville's estimate for the *Carte des Postes de France* (i.e., *Tableau du service des Postes*). In d'Anville's estimation, a good draftsman would take two days to design a square foot; but in the case of this postal map, with the additional information required, a square foot might take three days.

37. AN, MAR 1JJ/5/16, "État des dépenses," 1 January–15 February 1773.

38. Ibid., copy of a letter from Rizzi-Zannoni to Sartine, [1774].

39. Jean Reynaud, "Les Cassini et la carte de Provence (1776–1790)," *Bulletin de la Section de Géographie* 37, no. 1922 (1922): 115, 22–23. This very high figure would seem to include surveying, compilation, and engraving; costs of paper and printing were figured at 700 livres extra.

40. BNF, Ge. DD. 3399 (1). Barbié du Bocage, for designs for fourteen maps in quarto for *Voyage de Pausanias* to be published in Paris by Claude Marin Saugrain and Pierre Nicolas Firmin Didot fils. The payment, however, was made in *assignats,* worth roughly ¼–⅓ the value of a livre.

41. BNF, Ge. DD. 3399 (77). To be published by Alexis Eymery, libraire, rue Mazarine. The copies were 5 in vellum in octavo; 20 on ordinary paper in octavo, 50 on ordinary paper in duodecimo.

42. BNF, Ge. DD. 3399 (67), Barbié du Bocage and Pierre Plassau, Etienne Regent, Joseph Jean Bastide Bernard, and Jean Grégoire, for designs of maps for the *Voyage de Chardin,* for the literary work relating to geography, and for overseeing the work.

43. Berthaut, *Carte de France,* 1:225.

44. Pastoureau, *Atlas français,* vii.

45. François de Dainville, S.J., *Cartes anciennes de l'église de France* (Paris: Vrin, 1956), 254. François Courboin, *L'estampe Française* (Brussels: Librairie d'Art et d'Histoire, G. Van Oest & Co., 1914), 33.

46. AN, MAR C7 24, dossier Bellin, 5 November 1765, "État des Dépenses que J'ay faites pour les cinq volumes du Petit Atlas Maritime sur les ordres qui m'ont été donnés par Monseigneur le duc de Choiseul." This was an average price for each of the 610 "planches de cuivre polies et brunies, de différentes grandeurs."

47. Rizzi-Zannoni, *Carta geografica del Regno di Napoli;* Valerio, *Società uomini e istituzioni cartografiche nel mezzogiorno d'Italia,* 92.

48. J. B. Berthier, "Tableau de conviction, 16 Octobre 1771," in *Observations de M. Berthier, gouverneur de l'Hotel de la Guerre, en réponse à celles de M. DuPeron, directeur de l'Imprimerie Royale* (1771).

49. AN, MAR 1JJ/15/7, "Mémoires des cuivres fournis par le Sr. Tardieu, Planeur au Dépôt de France de la Marine pendant le cours de l'Année 1778." The rate of 2 livres 17 sous per pound of copper is consistent throughout the year. The plates weighed from 23 to 25 pounds, with the exception of one *demi-cuivre* at 11½ pounds, costing 32 livres 15 sous.

50. Gabriel Marcel, "À propos de la Carte des Chasses," *Revue de géographie* (1897): 13.

51. BNF, Ge. DD. 3399, Contrats d'éditions, traités, récépissés, projets concernant J.-D. Barbier du Bocage, les libraires Saugrain, Panckoucke, et autres, An II-1813, p. 62. Mémoires des cuivres fournis au citoyen Saugrain par Tardieu, planeur, 2 pluviôse l'an 4. These detailed contracts and agreements between Barbié du Bocage and various publishers are rich in details for his map compositions. I have not drawn on them extensively here as they date from after the Revolution and reflect the inflation and different currencies (francs, assignats) used in that period. This figure shows the great increase in the price of copper in the late eighteenth century. The largest plate purchased by the publisher Saugrain weighed 10½ pounds, costing 670 livres 6 sous, nearly twenty times the same plate fifteen years earlier. (See note 49.)

52. AN, MC, CVXII (374), 17 July 1730, inventory after the death of Pierre Moullart-Sanson. Copper at 15 sous per pound from the list of 78 plates in quarto weighing 119 pounds, 78 plates in folio weighing 466 pounds. The average weight of a quarto plate was 1.5 pounds, that of a folio plate 6 pounds.

53. AN, MC, LVII (375), 20 November 1744, inventory after death of Marie Darbisse, widow of Guillaume Delisle.

54. BNF, Ge. EE. 3119, Papiers Dezauche.

55. Ibid., 20 August 1780, payment to J. L. Barbeau de la Bruyere from Dezauche for the plates of the *Mappemonde historique* along with the printed *Explication historique*, in octavo, plus all proof sheets.

56. Ibid., 15 November 1783, payment to De Mauviel de Bouïlon from Dezauche for four plates of the diocese of Coutances by Mariette.

57. Ibid., 30 June 1791, payment to Chauchard from Dezauche "en espece ayant-cours" for two maps: Germany in nine sheets plus a general map and a supplement, "en grandeur Jésus," and a four-sheet map of Italy in grand-aigle, plus all the proofs.

58. De Dainville, *Cartes anciennes de l'église*, 259, contract between Tavernier and Danckaerts.

59. Delisle, "Mémoire Bretagne."

60. AN, MAR 2JJ/60 (139) l'abbé Jouin, Madrid, to Guillaume Delisle, 24 June 1724. The map was for Don Andres de Barcias, a member of the royal Council of War of the king of Spain, who wanted the map for his book on the history of the Indies.

61. BNF, Ge. FF. 13732 (II), Philippe Buache to the comte de Maurepas, 2 February 1737. Buache was at this time working in the Dépôt de la Marine.

62. By Philippe Buache. De Dainville, *Cartes anciennes de l'église*, 79.

63. To the engraver Guillaume Delahaye. BNF, Ge. FF. 13374, *Procès de géographie entre Boudet Libraire et Delahaye graveur de cartes*, 1752. Concerning the engraving of the maps for the *Atlas universel* of Gilles and Didier Robert de Vaugondy. Discussed in Mary Sponberg Pedley, *Bel et utile: the work of the Robert de Vaugondy family of mapmakers* (Tring: Map Collector Publications, 1992), 51–68. Also Mary Sponberg Pedley, "New light on an old atlas: documents concerning the publication of the *Atlas universel* (1757)," *Imago Mundi* 36 (1984).

64. AN, MAR 1JJ/2/3, "État des Dépenses," 1750–53. Plates for the *Hydrographie Française*.

65. AN, MAR 1JJ/2/3, "État des Dépenses," 1750–53.

66. AN, MAR 1JJ/2/4, "État des Dépenses," 1753–56.

67. Berthaut, *Carte de France,* 1:48.

68. AN, MAR 1JJ/2/4, "État des Dépenses," 1753–56.

69. In the agreement between d'Anville and the publishers Desaint and Saillant for maps to accompany the *Notice de la Gaule,* 24 November 1759. BNF, ms. fr. 22147 (1–10).

70. AN, MAR 1JJ/3/11, 21 December 1766, costs of engraving specific maps from the *Hydrographie Française:* viz., *Islande pour pêche de la Baleine,* 550 livres; *Isle Madagascar,* 650 livres; *Partie Septentrionale de la Mer du Sud avec découvertes Russes,* 772 livres; *Ocean Occidental,* 4th ed, 1,200 livres; and *Presqu'île de l'Inde,* 1,320 livres.

71. AN, MAR 1JJ/3/12, 1767. Engraving various maps: *Grand canal de Mozambique,* 660 livres; *Mers d'Écosse et Norwège pour la pêche,* 850 livres; and grande carte, *Rivière de Bordeaux,* 1,050 livres.

72. De Dainville, *Cartes anciennes de l'église,* 80. This very high price for engraving may be accounted for by the elaborate decorative engraving on it, the inclusion of a plan of Norbonne, an alphabetical table of principal places in the diocese, and notes.

73. Marcel, "Carte des Chasses," 13.

74. AN, MAR 1JJ/4/1, 1770.

75. BL, Add. Ms. 74237, 15 August 1777, Cassini de Thury to Dupain—Triel.

76. AN, MAR 1JJ/14/3.

77. Berthaut, *Ingénieurs-géographes,* 2:45.

78. Berthaut, *Carte de France,* 1:225.

79. Académie Royale des Sciences, Pochette de Séance, 1754, note submitted by J. Guettard for engraving of the mineralogical maps prepared by Philippe Buache for the *Mémoires* of the Académie, 1746.

80. BNF, Ge. DD. 2025, Buache to d'Argenson, 30 June 1756, regarding thirty maps for an atlas used to tutor the Dauphin in geography.

81. AN, MAR C7/24 (14), dossier Bellin, 5 November 1765. Bellin's accounts for the preparation of the 585 maps of the five-volume *Petit atlas maritime.*

82. Bibliothèque de l'Institut, Paris, ms. 2324, papiers Delisle/Buache. Buache's expenses for a map of the coast of Guinea (c. 1763).

83. AN, MAR C7/24, dossier Bellin.

84. AN, MAR 1JJ/3/14, 1769. Engraving 14 plates of maps, plans, and views of Iceland and Norway for the *Voyages of Kerguelen.*

85. Marcel, "Carte des Chasses," 13.

86. Pastoureau, *Atlas français,* 231. Contract between map publisher A. H. Jaillot and engraver Louis Cordier. AN, MC, VI (542), 3 April 1672; (543) 1 May 1672. Cordier was to do the engraving himself, though he was allowed to choose two other engravers to do the *petits noms* (i.e., the smaller place names).

87. Valerio, *Società uomini e istituzioni cartografiche nel mezzogiorno d'Italia,* 92 nn. 94, 95. Single sheets of Rizzi-Zannoni's *Carte geografica del Regno di Napoli.*

88. François de Dainville, S.J., *La carte de la Guyenne par Belleyme 1761–1840* (Bordeaux: Delmas, 1957), 24. The engravers of the plan were Radu, Glot, Chalmandrier, Joseph

Perrier, Dupuis, Barrière, Vicq, Rousseau, and Houdeau. The engravers of the lettering were Bourgoin, Beaublé, Bertin, Bellanger, and Macquet.

89. Marcel, "Carte des Chasses," 13

90. De Dainville, *Guyenne,* 38. For each of four plates for the *Carte réduite de la Généralité d'Aquitaine.* De Dainville does not list the cost of engraving the plan of plate one. The engravers of plan and lettering were Dupuis, Dubuisson, and Beaublé.

91. Mireille Pastoureau, "Confection et commerce des cartes à Paris aux XVIe et XVIIe siècles," in *La carte manuscrite et imprimée du XVIe au XIXe siècle: Journée d'étude sur l'histoire du livre et des documents graphiques,* ed. Frédéric Barbier (Munich: Saur, 1983), 15. Charges of the engraver Jean Liébaux.

92. Bibliothèque de l'Institut, Paris, ms. 2324, papiers Delisle/Buache.

93. De Dainville, *Cartes anciennes de l'église,* 324, archives de Gironde C 2411.

94. De Dainville, *Guyenne,* 24.

95. Berthier, "Tableau."

96. AN, MAR 1JJ/2/5, "État des Dépenses," 1756–57.

97. AN, MAR 1JJ/2/4, "État des Dépenses," 1753–56. Rhumb lines were extended on the plates with charts of the Channel and the coasts of Ireland.

98. AN, MAR 1JJ/2/4, "État des Dépenses," 1753–56. Retouching added new observations to the *Carte de la Mer du Sud* and added more detail to the *Carte générale de la Côte de Guinée.* The charts of the Channel and the coasts of Ireland were reengraved.

99. AN, MAR 1JJ/2/6, "État des Dépenses," 1757–58.

100. Marcel, "Carte des Chasses," 13. The wages included food.

101. Ibid., 14. Delahaye supervised.

102. Coolie Verner, "Mr. Jefferson makes a map," *Imago Mundi* 14 (1959): 103.

103. Berthaut, *Ingénieurs-géographes,* 1:225.

104. Berthaut, *Carte de France,* 1:62.

105. Pastoureau, *Atlas français,* 231. AN, MC, XV (235), 7 March 1672.

106. Claire Lemoine-Isabeau, "Note sur la gravure de la carte de Ferraris," *Gazette des Beaux-Arts* 80 (1972): 372. The comte de Ferraris hired Jean-Baptiste-Pierre Tardieu to lead a team of engravers to work on the *Carte chorographique des Pays-Bas autrichiens* (1777), a map in twenty plates. This figure uses McCusker's rate of exchange for the florin or guilder. John J. McCusker, *Money and exchange in Europe and America, 1600–1775* (Chapel Hill: University of North Carolina Press, 1978), 44, 60.

107. Marcel, "Carte des Chasses," 13.

108. Berthaut, *Carte de France,* 1:225.

109. BNF, Ge. FF. 13374.

110. AN, MAR 1JJ/2/3, "État des Dépenses", 1750–53.

111. AN, MAR 1JJ/2/4, "État des Dépenses," 1753–56.

112. Ibid. Effacing the plate in order to add corrections.

113. Ibid.

114. Ibid.

115. Bibliothèque de l'Institut, Paris, ms. 2324, papiers Delisle/Buache, receipt for engraving, *Carte générale de la Côte de Guinée,* n.d.

116. AN, MAR 1JJ/3/12, 1767. For changes and corrections, *Islande et Mers du Nord,* 150 livres; on four plates of *Terre Neuve* with details, 300 livres.

117. Valerio, *Società uomini e istituzioni cartografiche nel mezzogiorno d'Italia,* 92, Rizzi-Zannoni, *Carta geografica del Regno di Napoli.*

118. AN, MAR 1JJ/15/7, 26 March 1778. To Petit, engraver, to retouch plates of plate for "quartiers de réduction." These sheets engraved with grids overlaid with quarter circles were devised as worksheets for graphically resolving problems of setting course with longitude and latitude. A moveable thread passed through the center of the quarter circle, allowing one to find the sine, cosine, tangent, etc., of the arc. For illustration and explanation, see Olivier Chapuis, *À la mer comme au ciel: Beautemps-Beaupré et la naissance de l'hydrographie moderne (1700–1850)* (Paris: Presses de l'Université de Paris-Sorbonne, 1999), 72.

119. AN, MAR 1JJ/16/13, 1780. To Petit, engraver, for retouching chart of *l'Océan occidental.*

120. Marcel, "Carte des Chasses," 13.

121. AN, MAR 1JJ/2/4, "État des Dépenses," 1753–56. The higher price for the plate of the flags may reflect the fact that the *Tableau* was made up of two sheets and that the hachuring work to achieve the effect of the various designs on the flags required more skill than the figured work of the frontispiece.

122. AN, MAR 1JJ/2/5, "État des Dépenses," 1756–57: *Découvertes géographiques des Antilles Angloises.*

123. AN, MAR 1JJ/2/6, "État des Dépenses," 1757–58.

124. Bibliothèque de l'Institut, Paris, ms. 2324, papiers Delisle/Buache. Receipt for engraving, n.d., *Carte Générale de la côte de Guinée.* One engraved stamp for the Dépôt de la Marine consists of a circle with an anchor and three fleurs-de-lys, surrounded by the legend "Depot General de la Marine."

125. AN, MAR C7/24, dossier Bellin. "État des Depenses . . . pour le *Petit atlas maritime."*

126. AN, MAR 1JJ/6/7, 1774. For "vues des Terres" to accompany the *Voyage de la Flore,* by Pingré, Borda, and Verdun de la Crenne. Paid to M. Goas.

127. AN, MAR 1JJ/6/7, 1774. Views for the *Voyage de la Flore,* by Pingré, Borda, and Verdun de la Crenne. Paid to P. Martini.

128. De Dainville, *Guyenne,* 38. Engraved by Charles-Nicolas Varin.

129. Paris, Académie Royale des Sciences: dossier d'Anville, Regarding maps of ancient Gallia and Italia.

130. AN, MAR 1JJ/15/7, September–October 1778. To Tardieu, *planeur.* Probably Pierre-Joseph Tardieu who was a *maître planeur,* and the father of four *graveurs géographes.*

131. AN, MAR 2JJ/60 (12). Claude Delisle in Paris to Louis Renard, Amsterdam, 11 December 1706.

132. Delisle, "Mémoire Bretagne."

133. AN, MAR 2JJ/60 (146), Delisle to l'abbé Jouin, 3 November 1724. For printing a map of America for Don Andreas de Barcias.

134. BNF, ms. fr. 22147 (1–10). Proposal and contract, 24 November 1759, be-

tween d'Anville and publishers Desaint and Saillant for maps for the *Notice de la Gaule* (1759).

135. Berthaut, *Carte de France,* 1:62.

136. De Dainville, "La levée d'une carte en Languedoc à l'entour de 1730," 363.

137. Buchotte, *Les regles du dessein et du lavis pour plans particuliers des ouvrages et des batiments et pour leurs coups, profils, elevations, & facades, tant de l'Architecture militaire que civile . . .* 2nd ed. (Paris: C. Jombert, 1743), 195–96. "Battu" meant with a surface so smooth the grain of the paper could not be perceived. "Serpente" was tracing paper. The other names refer to paper sizes; see de Dainville, *Cartes anciennes de l'église,* 255.

138. AN, MAR 2JJ/60 (146).

139. AN, MAR 1JJ/2/5, "État des Dépenses," 1756–57, for the *Essai Géographique sur les Isles Britanniques* in sixty sheets.

140. BNF, ms. fr. 22147 (1–10). D'Anville's maps for Desaint and Saillant, *Notice de la Gaule.*

141. AN, MAR 1JJ/3/14, 1769. Six reams of paper purchased for the *Voyages de Kerguelen.*

142. AN, MAR 1JJ/5/7, 19 May 1773. D'Allensay on the difficulties of procuring *grand aigle* (24 inches × 36 inches) from Angoulême; local manufacture would require a doubling of workers at the paper mill. He points out that the paper from the Angoulême region is good for printing because of its solidity. A ream of such paper weighed 125 to 130 pounds and cost one livre per pound.

143. AN, MAR 1JJ/5/15, June 1774. J. B. Berthier's summary accounts for paper and printing hydrographic charts at Versailles.

144. Berthier, "Tableau."

145. AN, MC, LXXV (745), 16 February 1778, inventory after the death of Marie Charlotte Loy, wife of Louis-Charles Desnos.

146. AN, MAR 1JJ/15/7, September–December 1778. From Dubois.

147. Verner, "Mr. Jefferson," 103.

148. Berthaut, *Carte de France,* 1:62.

149. De Dainville, "La levée d'une carte en Languedoc à l'entour de 1730," 363.

150. Delisle, "Mémoire Bretagne."

151. AN, MAR 2JJ/60 (146). Delisle to l'abbé Jouin, 3 November 1724, regarding map of America for Don Andreas de Barcias.

152. AN, MAR 1JJ/2/5, "État des Dépenses," 1756–57.

153. BNF, ms. fr. 22147 (1–10).

154. Valerio, *Società uomini e istituzioni cartografiche nel mezzogiorno d'Italia,* 91. For Rizzi-Zannoni's map, *Carta geografica del regno di Napoli.*

155. AN, MAR 1JJ/3/14, 1769. Printing costs for the maps, plans, and views for the *Voyages de Kerguelen:* 787 livres 10 sous for 3,150 maps.

156. AN, MAR 1JJ/15/7, September 1778. État des Cartes tirées par le S. Aubert, imprimeur en taille-douce.

157. AN, MAR 1JJ/16/4, December 1779. To Aubert, imprimeur en taille-douce: 240# 6 sous to print 1602 maps at 15# per 100.

158. De Dainville, *Cartes anciennes de l'église,* 255.

159. Verner, "Mr. Jefferson," 103.

160. Chapuis, *À la mer,* 297. Fleurieu printed hydrographical charts with Charles Coutadeur, the master printer for the city of Paris and for the Dépôt de la Marine.

161. BNF, Ge. FF. 13732 (II), Buache to the comte de Maurepas, 2 February 1737, for maps for the Dépôt de la Marine.

162. AN, MAR 1JJ/2/4, "État des Dépenses," 1753–56.

163. BNF, ms. fr. 22147 (1–10). D'Anville's contract with Desaint and Saillant for maps of Gaul.

164. Reynaud, "Les Cassini," 115.

165. Marcel, "Carte des Chasses," 14.

166. Verner, "Mr. Jefferson," 103.

167. Berthaut, *Carte de France,* 1:62.

168. AN, MAR 1JJ/2/3, "État des Dépenses," 1750–53. Separate plates engraved only with rhumb lines were printed in red, green, or black over a sheet already printed with the specific chart. See plate 1 for an illustration of printed red rhumb lines without a chart overprinted.

169. AN, MAR 1JJ/2/5, "État des Dépenses," for 100 examples of a six-sheet map of England with the green rhumb lines printed separately.

170. AN, MAR 1JJ/2/1–6 passim. Paid to a *garçon du bureau.*

171. AN, MAR 1JJ/2/4, "État des Dépenses," 1753–56. Quantity unknown.

172. Ibid. Quantity unknown.

173. AN, MAR 1JJ/2/5, "État des Dépenses," 1756–57. Quantity unknown.

174. AN, Y 1903, 31 August 1772. Evaluation by d'Anville, Seguin, and Jaillot of the *Tableau du Service des Postes.* Quantity of maps unknown.

175. AN, MAR 1JJ/2/3, "État des Dépenses," 1750–53. Cleaning the plates of the *Neptune François,* which were clogged with old ink and full of verdigris.

176. AN, MAR C7/24, dossier Bellin. Cleaning the plates of the *Petit atlas maritime,* including polishing and burnishing the plates.

177. AN, MAR 1JJ/16/4, December 1779: 30 livres for the workers, 15 livres for the powder, 3 livres 10 sous for water and rags: total 48 livres 10 sous.

178. BNF, ms. fr. 22121 (10).

179. AN, MAR 2JJ/60 (13), [1708?]. "Mémoire pour Le Sr. Guillaume Delisle contre le Sr. Louis Renard."

180. Verner, "Mr. Jefferson," 103.

181. Smithsonian Institution Libraries, Washington, ms. 313A, CF Cassini to Huquier, engraver: 210 livres 19 sous for 220 pounds.

182. AN, MAR 1JJ/2/5, "État des Dépenses."

183. Pastoureau, *Atlas français,* 136. At Duval's house: "une presse garnie de deux roulleaux un bacquet, six ais, une pouelle, un gril le tout tel quel servant pour imprimer."

184. AN, MC, CXVII (374), 17 July 1730, inventory after the death of Pierre Moullard-Sanson.

185. Total 866 livres. AN, MAR 1JJ/5/5, 1773. "État des frais et de la dépense faitte pour la construction de deux presses de taille douce et une troisième pour presser les ouvrages, commandés par M. Aubert, imprimeur du Bureau de la Marine."

186. Total 3,220 livres. De Dainville, "La levée d'une carte en Languedoc à l'entour de 1730," 363–64. Arch. Hérault C. Etats (138).

187. AN, MAR 1JJ/2/4, État des depenses ordinaire, 1755. "30 grands portefeuilles de fort carton pour de mettre les cartes venant de l'acquisition du cabinet de M. Delisle."

188. AN, MAR 1JJ/5/13, 19 November 1773. Requested by Messier, astronomer, for the Observatoire de la Marine. It was to be made to Messier's specifications and be similar to one Dolland made for the king of England. Messier points out that the telescope costs more than his salary, which is only 1,700 livres per year.

189. AN, MAR 1JJ/5/11, 9 January 1778. Chabert requested authorization from Sartine, minister of the marine, to have this quadrant constructed, with all its "précisions et améliorations."

APPENDIX 2

1. Sarah Tyacke, "Map-sellers and the London map trade c. 1650–1710," in *My head is a map: essays and memoirs in honour of R. V. Tooley*, ed. Helen Wallis and Sarah Tyacke (London: Francis Edwards and Carta Press, 1973), 70; E. G. R. Taylor, "Robert Hooke and the cartographical projects of the late seventeenth century (1666–1696)," *Geographical Journal* 90 (1937), from a document untraced.

2. Tyacke, "Map-sellers," 70.

3. William Ravenhill, "Joel Gascoyne's Stepney: his last years in pastures old yet new," *Guildhall studies in London history* 2, no. 4 (1977): 206. The length of plates ranged from c. 470 to 615 mm, to width from around 247 to 510 mm.

4. Sarah Bendall, *Dictionary of land surveyors and local map-makers of Great Britain and Ireland 1530–1850*, 2nd ed. (London: British Library, 1997), 40.

5. A. H. W. Robinson, *Marine cartography in Britain* (Leicester: Leicester University Press, 1972), 77, citing National Library of Wales, Aberystwyth, Add. Ms. 607A.

6. A. Stuart Mason and Peter Barber, " 'Captain Thomas, the French engineer': and the teachings of Vauban to the English," *Proceedings of the Huguenot Society* 25, no. 3 (1991): 286.

7. BL, Add. Ms. 74237 EE, n.d. "Henry Moore Esq., Lieutenant Governor of Jamaica, for trouble he hath been at in making a survey of the said island."

8. Willis Chipman, "The life and times of Major Samuel Holland, surveyor-general, 1764–1801," *Ontario History* 21 (1924), 23.

9. William Ravenhill, *A map of the County of Devon . . . by Benjamin Donn* (Exeter: Devon and Cornwall Record Society and the University of Exeter, 1965), 8, citing Donn to the Society of Arts, 26 August 1765, in Royal Society of Arts, London, Guard Book X, 31.

10. W. A. Seymour, *A history of the Ordnance Survey* (Folkestone: Dawson, 1980), 8;

Peter Barber, "Maps and monarchs in Europe, 1550 – 1800," in *Royal and republican sovereignty in early modern Europe,* ed. Robert Oresko, G. C. Gibbs, and H. M. Scott (Cambridge: Cambridge University Press, 1996), 88.

11. J. B. Harley, "The Society of Arts and the surveys of English counties 1759 – 1809," *Journal of the Royal Society of Arts* 112 (1964): 271.

12. G. N. D. Evans, *Uncommon obdurate: the several public careers of J. F. W. Des Barres* (Salem: Peabody Museum of Salem; Toronto: University of Toronto Press, 1969), 62 – 64, 67.

13. Elizabeth Rodger, *Large scale county maps of the British Isles 1596 – 1850: A union list* (Oxford: Bodleian Library, 1972), viii, citing a note in the annotated copy of Richard Gough, *British Topography,* in Oxford University, Bodleian Library, Gough Ms., 363 – 65; T. R. Holland, "Yeakell & Gardner maps of Sussex 1778 – 83," *Sussex Archaeological Collections* 95 (1957): 94 – 104.

14. Harley, "Society of Arts," 541.

15. R. A. Skelton, "The origins of the Ordnance Survey of Great Britain," *Geographical Journal* 128 (1962): 419.

16. Sarah Bendall, "Estate maps of an English county," in *Rural images: the estate plan in the Old and New worlds,* ed. David Buisseret (Chicago: University of Chicago Press, 1996), 76. The charge was £6 6s [6 guineas] for three days work plus £2 17s expenses.

17. BL, Add. Ms. 9767 (53). "Establishment and Accompts of the Committee for Trade and Plantations."

18. Taylor, "Robert Hooke," 539. Paid to John Seller et al.

19. BL, Egerton 3805 (153), invoice from Gabrill de la Hay to Prince George of Denmark, cited in Barber, "Maps and monarchs," 112.

20. BL, Egerton 3805 (154), accounts of Prince George of Denmark, cited in Barber, "Maps and monarchs," 112.

21. Stuart Mason and Barber, "Capt. Thomas," 286.

22. Bendall, "Estate maps," 76. Paid to Joseph Freeman.

23. Tyacke, "Map-sellers," 66. Figures derived from assessment of Thomas Jenner's stock by his successor, John Garrett, and Garrett's brother-in-law, John Overton.

24. Ibid., 74.

25. Taylor, "Robert Hooke," 539.

26. Ravenhill, "Joel Gascoyne's Stepney," 206, 209.

27. Arthur Pond, "Journal of Receipts and Expenses 1734 – 1750," BL, Add. Ms. 23724, 1734 – 1750. The period 1745 – 48 includes accounts of buying copper and paying engravers for the plates to accompany Lord Anson's *Voyages.* The engravers mentioned are Canneau [Pierre-Charles Canot], Fougeran/Foujeran [Ignace Fougeron], Grignion [Charles Grignon], Mason [James Mason], Müller [Johann Sebastian Müller], Seale [Richard W. Seale], Trucher [Marcus Tuscher], and Wood [John Wood]. They are identified in Louise Lippincott, "Arthur Pond's journal of receipts and expenses, 1734 – 1750," *Walpole Society* 54 (1988): 220 – 333. Folios 107, 108, 116, 117, 118, 119, 120, 121, 122, 123, 124, 125, 126, 130, 132, 133, 135, 136, 137, 139, 141, 144, 145, 146, 150, 159.

28. Harley, "Society of Arts," 271.

29. Edmund Thompson, *Maps of Connecticut before the year 1800* (Windham, Conn.: Hawthorn House, 1940), 37. Captain Abner Parker to Abel Buell, October, 1774. £6 5s for two large copperplates.

30. Philip Lee Phillips, *Notes on the life and works of Bernard Romans* (Deland: Florida State Historical Society, 1924), 25. Paul Revere to Bernard Romans, 4 May, 9 July, and 21 October 1774. The receipts do not specify how many plates were required to produce the map, which Romans planned to measure 12 by 7 feet.

31. Tyacke, *London map-sellers,* 118.

32. Tyacke, "Map-sellers," 73, citing BL, Stowe Ms. 746 (26).

33. Ibid., 122.

34. Robinson, *Marine cartography,* 81.

35. Sarah Tyacke, *London map-sellers 1660–1720* (Tring: Map Collector Publications, 1978), 122.

36. Ibid.

37. Evans, *Uncommon obdurate,* 66.

38. Tyacke, "Map-sellers," 66. Figures derived from assessment of Thomas Jenner's stock by his successor, John Garrett, and Garrett's brother-in-law, John Overton.

39. Harley, "Society of Arts," 270. Andrew Armstrong offered the plates to the Society of Arts, Manufactures, and Commerce.

40. Oxford University, Bodleian Library, Gough Ms., General topography 365, fol. 381, cited by A. D. M. Phillips, introduction to *A map of the county of Stafford from an actual survey begun in the year 1769 and finished in 1775 [by William Yates], Collections for a History of Staffordshire* (Lympne Castle: Harry Margary, 1984), xxiv. The plates were probably bought by William Faden who brought out a second edition of Yates's 1775 map of Staffordshire in 1795.

41. Tyacke, "Map-sellers," 74.

42. Taylor, "Robert Hooke," 539.

43. E. G. R. Taylor, "Notes on John Adams and contemporary map makers," *Geographical Journal* 97 (1941): 183; Tyacke, "Map-sellers," 74. Gregory King engraved the plates in three months.

44. William Ravenhill, "Joel Gascoyne, a pioneer of large scale county mapping," *Imago Mundi* 26 (1972): 67, fig. 4.

45. Robinson, *Marine cartography,* 81. All prices were given to Lewis Morris by Bowen.

46. Pond, "Journal of Receipts and Expenses," fols. 120, 121, 123, 144, 145.

47. Charles Welsh, *A bookseller of the last century: being some account of the life of John Newbery and of the books he published* (London, 1885), 361.

48. Harley, "Society of Arts," 271.

49. Phillips, *Notes on Bernard Romans,* 25. Totals from three separate billings (May, July, and October 1774).

50. Thompson, *Maps of Connecticut,* 37. Capt. Abner Parker, the surveyor, to Abel Buell, engraver.

51. J. B. Harley, Barbara Petchenik, and Lawrence Towner, *Mapping the American Revolutionary War* (Chicago: University of Chicago Press, 1978), 88, 163–64 n. 54. Harley, Petchenik, and Towner note that Des Barres was living in Soho from 1774 to 1779, an area in London with many engravers of French Huguenot extraction. The complexity of the plates of the *Atlantic Neptune,* which combine several different engraving and etching techniques, may account for the higher charge for engraving.

52. Reported by Francis Place to George Vertue, Vertue Notebooks, vol 2, 58, quoted in Tyacke, "Map-sellers," 74.

53. Reported by Francis Place to George Vertue, Vertue Notebooks, vol 1, 34–5, quoted ibid. Hollar and his counterpart in Paris, "Mason, the famous Burinator," both kept track of their time with an hourglass.

54. Campbell, *The London Tradesman,* quoted in Timothy Clayton, *The English print: 1688–1802* (New Haven: Yale University Press, 1997), 13.

55. Pond, "Journal of Receipts and Expenses," fols. 158, 159, 160.

56. Clayton, *English print,* 21; Pond, "Journal of Receipts and Expenses," fol. 135.

57. Clayton, *English print,* 21. Pond to Macardell for portrait of David Garrick; to Charles Grignion for Anson.

58. Welsh, *A bookseller,* 361.

59. Tyacke, "Map-sellers," 71.

60. Pond, "Journal of Receipts and Expenses," fol. 10.

61. Ibid., fol. 20.

62. Robinson, *Marine cartography,* 81.

63. Pond, "Journal of Receipts and Expenses," fol. 125.

64. Ibid., fols. 146, 154.

65. Harley, "Society of Arts," 271. For 500 copies of a nine-sheet map; 4,500 sheets cost £27.

66. Welsh, *A bookseller,* 361.

67. Tyacke, "Map-sellers," 73.

68. Ibid., 73 n. 50, citing D. F. MacKenzie, *Cambridge University Press* (1966), 1:93.

69. Ibid., 73 n. 51, citing Salter, ed., *Remarks and Collections of Thomas Hearne* (Oxford, 1914), 9:75.

70. For printing 18,000 engraved drawings. Ibid., 73, citing George Vertue, *Notebooks,* 2:155.

71. Clayton, *English print,* 22. Pond to the printer Wyatt.

72. Robinson, *Marine cartography,* 82.

73. Pond, "Journal of Receipts and Expenses," fol. 152.

74. Harley, "Society of Arts," 271. Harley illustrates Andrew Armstrong's "Estimate of Expense . . . for the Map of the County of Northumberland." Armstrong's map was engraved on nine copper plates. The paper and printing for 500 maps would therefore incorporate 4,500 sheets. It is not clear if his estimate of expenses for paper and printing is based on 4,500 sheets or 500. Printing for 4,500 sheets would average about 11 shillings per 100.

75. Welsh, *A bookseller,* app. 10.

76. Tyacke, "Map-sellers," 71.

77. Pond, "Journal of Receipts and Expenses," fol. 113.

78. BL, Add. Ms. 9767 (53). "Establishment & Accompts of the Committee for Trade and Plantations."

79. Tyacke, "Map-sellers," 71.

80. Ibid.

81. Harley, Petchenik, and Towner, *Mapping the American Revolution,* 27; D. C. Harvey, *Holland's Description of Cape Breton Island and other documents* (Halifax: Public Archives of Nova Scotia, 1935), 37–38.

82. Evans, *Uncommon obdurate,* 63 n. 11.

83. Tyacke, *London map-sellers,* xvi.

84. Nancy Valpy, "Plagiarism in prints: the *Musical Entertainer* affair," *Print Quarterly* 6, no. 1 (1989): 54.

85. BL, Add. Ms. 38729, Palairet to Nourse. My thanks to Jill Shefrin for alerting me to this reference.

APPENDIX 3

1. Bernard Romans, *A concise natural history of east and west Florida* (New York, 1775), 175.

2. Albert H. Heusser, *George Washington's mapmaker: a biography of Robert Erskine* (1928; new ed., New Brunswick: Rutgers University Press, 1966), 208. These are the evaluations of Robert Erskine, who complained that two continental dollars was not enough for a surveyor. He was receiving $35 per day for his survey team, out of which he paid $2–4 per day to the surveyors and 50 cents per day to the chain bearers.

3. Massachusetts Historical Society, Boston, David S. Greenough Papers, William Erving to Joseph Callender, 10 November 1786. My thanks to David Bosse for this reference.

4. Ibid.

5. Massachusetts Historical Society, Boston, Belknap Papers 161.D.141, 15 August 1791, Rev. Jeremy Belknap to Samuel Hill, for printing maps. My thanks to David Bosse for this information.

6. Susan Danforth, "The first official maps of Maine and Massachusetts," *Imago Mundi* 35 (1983): 56–57 nn. 55, 60.

APPENDIX 4

1. De Dainville, *Cartes anciennes de l'église,* 259.

2. Ibid., 258.

3. Pastoureau, *Atlas français,* 230. AN, MC, XV (229), 11 December 1670, contract between A. H. Jaillot and G. Sanson.

4. AN, MAR 2JJ/60 (12), Delisle to Renard, 11 December 1706. Delisle tells his Dutch partner that he usually sells his maps at 10 to 12 sous uncolored, 15 sous with the divisions colored.

5. BNF, Ge. FF. 13732 (II). Buache to Maurepas, 10 February 1737, proposing a lower price of 15 sous to undercut sales of Dutch maps.

6. Mary Sponberg Pedley, "The subscription list of the 1757 *Atlas universel:* a study in cartographic dissemination," *Imago Mundi* 31 (1979).

7. *Mémoires pour l'Histoire des Sciences et des Beaux-Arts,* November 1759.

8. François de Dainville, S.J., "Cartes de Bourgogne du XVIIIe siècle," in *La cartographie reflet de l'histoire,* ed. Michel Mollat du Jourdin (Geneva: Editions Slatkine, 1986), 75.

9. Roch-Joseph Julien, *Nouveau catalogue de cartes géographiques et topographiques . . . divisé en deux parties* (Paris: chez R.-J. Julien, 1763). In Julien's catalogue, prices for maps vary by author; e.g., maps by d'Anville tend to be more expensive than those by Robert de Vaugondy, Julien, or Homann, but d'Anville also tends to publish maps of more than one sheet. Maps considered to be antiquarian and rare were also more expensive, e.g., Roussel's eight-sheet map of the Pyrenees, priced at 36 livres, p. 13.

10. Didier Robert de Vaugondy, "Ouvrages de Mr Robert de Vaugondy," in *Institutions géographiques* (Paris, 1766).

11. De Dainville, *Cartes anciennes de l'église,* 262 n. 131. Advertised by Le Rouge in his *Nouveau catalogue des cartes géographiques et topographiques* (Paris, 1763), 18–19.

12. Robert de Vaugondy, "Catalogue des Ouvrages," 1777. BNF, Ge. DD. 7808.

13. AN, MAR 1JJ/16/3, January–March 1779. From sales by Buache de la Neuville. Chapuis, *À la mer,* 192–93.

14. AN, MAR 1JJ/17/10.

15. BNF, Ge. EE. 3119 (46): Dezauche.

16. Katie Scott, *The rococo interior* (New Haven: Yale University Press, 1995), 247.

17. Pierre Casselle, "Le commerce des estampes à Paris dans la seconde moitié du 18ème siècle," unpublished thesis, Ecole nationale des Chartes, Paris, 1976, 67. A price of a print could depend on format, type of engraving, color, fame of engraver and artist, the importance of the pull. Small portraits usually cost from 1 livre 4 sous to 2 livres. Very large complex prints were as much as 16 livres.

18. Jean Lattré, "Catalogue du fonds du Sr Lattré Graveur rue St Jacques près St Severin, à la Ville de Bordeaux," in *Atlas moderne* (Paris, [1777]).

APPENDIX 5

1. Tyacke, *London map-sellers,* 118. These valuations were placed on Thomas Jenner's stock-in-trade for probate purposes.

2. BL, Add. Ms. 9767, fols. 14–16.

3. John Ogilby and William Morgan, *Pocket Book of the Roads of England* (London: William Morgan and Christopher Wilkinson, 1689), advertisement.

4. Tyacke, *London map-sellers,* xvii.

5. Ibid., 138. This price was advertised "for a quick sale"; they were usually priced between 1s and 1s 6d.

6. John Senex, *A treatise of the description and use of both globes to which is annexed a geographical description of our Earth* (London: John Senex, 1718).

7. Herman Moll, *The World Described: or, a New and Correct Sett of Maps . . .* (Dublin:

George Grierson, 1732). Noted in Dennis Reinhartz, *The cartographer and the literati: Herman Moll and his intellectual circle* (Lewiston: Edwin Mellen Press, 1997), 34.

8. William Faden and Thomas Jefferys, *A Catalogue of Modern and Correct Maps, Plans, and Charts, chiefly engraved by the late T. Jefferys* (London: Faden and Jefferys, 1774); Robert Sayer and John Bennett, *Sayer and Bennett's enlarged Catalogue of new and valuable Prints, in sets, or single; also useful and correct Maps and Charts; likewise Books of architecture, views of antiquity, drawing and copy books, etc etc in great variety at No. 53, in Fleet-Street, London; . . .* (London: Sayer and Bennett, 1775), passim.

9. William Faden, *Catalogue of the geographical works, maps, plans, &c.* (London: William Faden, 1822).

10. The most comprehensive survey of prices of English prints is Clayton, *English print*; see esp. 22–23, 118–19.

APPENDIX 6

1. Scott, *Rococo*, 36, citing AN, F12 1456B.

2. AN, MAR 1JJ/2/1–6.

3. Etienne Martin Saint-Léon, *Histoire des Corporations de Métiers depuis leurs origines jusqu'à leur suppression en 1791* (Paris, 1922), 538. The author's estimate of a typical working-class wage, midcentury.

4. Berthaut, *Ingénieurs-géographes*, 1:24.

5. Michael Sonenscher, *Work and wages: natural law, politics, and the eighteenth century French trades* (Cambridge: Cambridge University Press, 1989), 203, citing AN, Y 12768, 27 December 1768. Sonenscher notes a worker's *journée* was divided into four, five, or six two-hour periods.

6. Philippe Minard, "Agitation in the work force," in *Revolution in print: the press in France 1775–1800*, ed. Robert Darnton and Daniel Roche (Berkeley: University of California Press, 1989), 119 n. 46.

7. Berthier, "Tableau."

8. Sonenscher, *Work*, 204.

9. Chapuis, *À la mer*, 297.

10. WLCL, manuscripts, Mildmay papers, letter to the Earl Fitzwalter, 19 December 1750. Sir William Mildmay was sent to Paris to negotiate on behalf of the British the return of prisoners and the boundary of Nova Scotia according to the terms of the Treaty of Aix-la-Chapelle (1748). A good early example of diplomatic inflation of cost-of-living allowances.

11. AN, MAR 1JJ/2/1–6, passim.

12. Martin Saint-Léon, *Corporations*, 538. This is the author's estimate of what a working-class family of four required to live.

13. Emmanuel duc de Croÿ, *Journal inédit du duc de Croÿ, 1718–1784* (Paris: Flammarion, 1907), 4:262.

14. Sonenscher, *Work*, 204, citing BNF, Ms. Joly de Fleury 557, fol. 27. The figures are quoted from documents concerning legal disputes and so may be inflated on the

part of the plaintiffs, usually a workman or group of workmen who were complaining about nonpayment of wages. Note that Mildmay's allowance for his rent is roughly 16 times the workman's rent (see "English diplomat" above in this table).

APPENDIX 7

1. BL, Add. Ms. 9767, accounts for the Lords of Trade and Plantations.

2. BL, Add. Ms. 28075 (238), grant of office of cosmographer and geographer to the king for annual fee paid by William Morgan and John Ogilby.

3. Stuart Mason and Barber, "Capt. Thomas," 280.

4. BL, Egerton 3807, Prince George of Denmark: letters, accounts and papers relating to the household and estate of Prince George of Denmark, consort of Queen Anne.

5. *The New State of England under Queen Anne* (London, 1707), 174–80.

6. A. Stuart Mason and John Bensusan-Butt, "P. B. Scalé, surveyor in Ireland, gentleman of Essex," *Proceedings of the Huguenot Society* 24, no. 6 (1988): 509. Peter Scalé Sr. had served in the Regiment of Foot under Theodore Vezey in Portugal, Spain, and Ireland, probably from the War of Spanish Succession until 1717.

7. Edward E. Curtis, *The organization of the British army in the American Revolution* (New Haven: Yale University Press, 1926), 158–60.

8. Tyacke, "Map-sellers," 69.

9. Clayton, *English print,* 13.

10. Heusser, *George Washington's mapmaker,* 10.

11. Pond, "Journal of Receipts and Expenses," passim.

12. My thanks to Laurence Worms for sharing detailed information he has accumulated on the apprenticeship fees of map engravers. He has studied the records of thirty-eight master engravers over a period from 1710 to 1806. Fees varied widely, from £5 to £125.

13. James Raven, *Judging new wealth: popular publishing and responses to commerce in England, 1750–1800* (Oxford: Clarendon Press, Oxford University Press, 1992), 51.

14. Carington Bowles, *New London Guide and Hackney Coach Directory* (London, 1786).

15. Louis-Sébastien Mercier, *Parallèle de Paris et de Londres: un inédit de Louis-Sébastien Mercier* (Paris: Didier Erudition, [1789?]; reprint, 1982), 204 n. 103.

BIBLIOGRAPHY

PRIMARY SOURCES

Anonymous. "A British navy yard contemplated in Newport, R.I. in 1764." *Rhode Island Historical Magazine* 6, no. 1 (1885): 42–47.

Anville, Jean-Baptiste Bourguignon d'. *Analyse géographique de l'Italie.* Paris: la Veuve Estienne et fils, 1744.

———. Trans. from French. *Compendium of Ancient Geography by Mons. D'Anville.* London: R. Faulder, 1791.

———. *Considérations générales sur l'étude et les Connoissances que demande la composition des Ouvrages de géographie.* Paris: Lambert, 1777.

Arrowsmith, Aaron. "Map of England and Wales: the result of fifteen years labour/ dedicated by permission to his Royal Highness the Prince Regent by H. R. Highness's dutiful servant and hydrographer, A. Arrowsmith." London: A. Arrowsmith, 1815.

Bacler-d'Albe, Louis Albert Ghislain. "Notice sur la gravure topographique et géographique." In *Mémorial du Dépôt de la Guerre* (Paris, 1803), 125–40.

Barnard, Frederick Augusta. *Bibliothecae Regis Catalogus.* London, 1820.

Bellin, Jacques-Nicolas. *Observations sur la carte de la Manche, dressée au Dépôt des cartes, plans & journaux de la marine, pour le service des vaisseaux du roi . . . en 1749.* Paris: l'imprimerie de Didot, 1750.

———. *Petit atlas maritime.* Paris, 1764.

———. "Remarques de M. Bellin . . . sur les Cartes et les Plans, qu'il a été chargé de dresser . . ." In le révérend père de Charlevoix, *Journal d'un voyage fait par ordre du Roi dans l'Amérique Septentrionale.* Paris: chez Didot, libraire, 1744.

Berthier, J. B. "Tableau de conviction, 16 Octobre 1771." In *Observations de M. Berthier, gouverneur de l'Hotel de la Guerre, en réponse à celles de M. DuPeron, directeur de l'Imprimerie Royale.* [Paris], [1771].

Bolton, Charles Knowles, ed. *Letters of Hugh Earl Percy from Boston and New York, 1774–1776.* Boston: Charles E. Goodspeed, 1902.

Bolton, S. *A complete system of geography.* London, 1747.
Boswell, James. *Life of Johnson.* Oxford: Oxford University Press, 1980. Originally published 1791.
Bowles, Carington. *New London Guide and Hackney Coach Directory.* London, 1786.
Brion de la Tour, Louis. Errata de l'Atlas Moderne ou Appel au Public de l'accusation de Plagiat intentée par le Sieur *** [Bonne] contre M. Brion, ingénieur-géographe du roi. [Paris], 1766. [In AN, MAR 1JJ/3/11.]
Buchotte. *Les regles du dessein et du lavis pour les plans particuliers des ouvrages et des bâtimens et pour leurs coupes, profils, elevations, & façades, tant de l'Architecture militaire que civile. . . .* 2nd ed. Paris: C. Jombert, 1755.
Campbell, R. *The London Tradesman. Being a compendious View of all the Trades, Professions, Arts, both Liberal and Mechanic, now practised in the Cities of London and Westminster. Calculated for the information of Parents, and instruction of youth in their Choice of Business.* London: T. Gardner, 1747.
Croÿ, Emmanuel duc de. *Journal inédit du duc de Croÿ, 1718–1784.* 4 vols. Paris: Flammarion, 1907.
Des Barres, Joseph Frederick Wallet. *A statement submitted by Colonel Desbarres for Consideration.* [London], [1796].
DeVorsey, Louis. *De Brahm's report of the general survey in the southern district of North America.* Columbia: University of South Carolina Press, 1971.
Donn, Benjamin. *An epitome of Natural and Experimental Philosophy including Geography and the Uses.* London: Law & Kearsley; London: Heath and Wing; Bristol: Donn, [1769].
Examen Impartial de la critique, sans nom d'auteur, des cartes de la mer Baltique et du golfe de Finlande, présentées à M. le Maréchal de Castries, ministre et secrétaire d'État de la marine, par Le Clerc, écuyer, chevalier de l'Ordre du Roi et membre de plusieurs Académies. Paris: l'Imprimerie de Clousier, imprimeur de Roi, 1786.
Faden, William. *A Catalogue of Maps, Charts, and Plans.* London: William Faden, 1778.
———. *Catalogue of the geographical works, maps, plans, &c.* London: William Faden, 1822.
———. *The Province of New Jersey Divided into East and West, commonly called The Jerseys.* London: Faden, 1777.
Faden, William, and Thomas Jefferys. *A Catalogue of Modern and Correct Maps, Plans, and Charts, chiefly engraved by the late T. Jefferys.* London: Faden and Jefferys, 1774.
Fortescue, J., ed. *The Correspondence of King George the Third from 1760 to December 1783.* Vol. 1. London: Macmillan, 1927–28.
Green, John [Bradock Mead]. *Remarks in support of the New Chart of North America in six Sheets.* London: Thomas Jefferys, 1753.
Harvey, D. C. *Holland's description of Cape Breton Island and other documents.* Halifax: Public Archives of Nova Scotia, 1935.
An inventory of the contents of the Governor's Palace taken after the death of Lord Botetourt: an inventory of the personal estate of his Excellency, Lord Botetourt, Royal Governor of Virginia, 1768–1770. Williamsburg, VA: Colonial Williamsburg Foundation, 1981.
Jefferys, Thomas. *Atlas des Indes Occidentales:* London: Robert Sayer and John Bennett; Paris: Roch-Joseph Julien, 1777.

Julien, Roch-Joseph. *Nouveau catalogue de cartes géographiques et topographiques . . . divisé en deux parties.* Paris: chez R.-J. Julien, 1763.

La Lande, Joseph-Jérôme Le Français de. *Journal d'un voyage en Angleterre, 1763.* Oxford: Voltaire Foundation, 1980.

La Tour du Pin, la Marquise de. *Journal d'une femme de cinquante ans, 1778–1815.* Paris: Librairie Chapelot, 1920.

Langford, Auctioneers. "Sale of Library of James West, Esq. deceased, late President of the Royal Society, 19 January 1773 and 12 following days." In *Sales Catalogues.* London: Langford, 1773. [In British Museum, Prints, ScA. 1.10]

Lattré, Jean. "Catalogue du fonds du Sr Lattré Graveur rue St Jacques près St Severin, à la Ville de Bordeaux." In *Atlas moderne.* Paris: [1777].

Le Rouge, Georges-Louis. *Atlas Amériquain.* Paris: Le Rouge, 1778.

———. *Catalogue des cartes et plans de Le Rouge, Ingénieur-géographe du Roi, ancien Lieutenant au Regiment de Saxe.* Paris: Le Rouge, 1773.

Le Sage. *Le géographe parisien ou le conducteur chronologique et historique des rues de Paris. . . .* Paris: Valleyre, Duchesne, Prault, Desaint, Delalain, 1769.

Lenglet Du Fresnoy, l'abbé Nicolas. *Méthode pour étudier la géographie . . . et un Catalogue des Cartes géographiques, Relations, Voyages et Descriptions nécessaires pour la Géographie.* 3rd ed. Paris: chez Rollin, fils, 1741.

Mead, Bradock. *The Construction of Maps and Globes.* London: Horne, Knapton, et al., 1717.

Mercier, Louis-Sébastien. *Parallèle de Paris et de Londres: un inédit de Louis-Sébastien Mercier.* [1789?]; reprint, Paris: Didier Erudition, 1982.

Moll, Herman. *[Atlas of Europe].* [London]: [c. 1715].

———. *The World Described: or, a New and Correct Sett of Maps . . .* Dublin: George Grierson, 1732.

Niceron, Jean-Pierre. *Mémoires pour servir à l'histoire des Hommes illustres dans la république des lettres . . .* Paris: Briasson, 1727.

Ogilby, John, and William Morgan. *Pocket Book of the Roads of England.* London: William Morgan and Christopher Wilkinson, 1689.

Plot, Robert. *The Natural History of Oxfordshire.* 2nd ed. Oxford and London: Leon Lichfield, Charles Brome, and John Nicholson, 1705.

Pownall, Thomas. *A topographical description of the dominions of the United States of America [being a revised and enlarged edition of] a Topographical description of such parts of North America as are contained in the (annexed) map of the Middle British Colonies etc. in North America.* Ed. Lois Mulkearn. 1775; reprint, Pittsburgh: University of Pittsburgh, 1949.

Robert de Vaugondy, Didier and Gilles. *Atlas universel.* Paris: Gilles and Didier Robert de Vaugondy and Antoine Boudet, 1757.

Robert de Vaugondy, Gilles. *Introduction à la géographie des Srs. Sanson, Géographes du roi, Quatrième édition, revûe, corrigée, et augmentée.* Paris: Durand, 1743.

Romans, Bernard. *A concise natural history of east and west Florida.* New York, 1775.

Sanson, Guillaume. *Introduction à la Géographie.* Paris, [1681].

Sayer, Robert, and John Bennett. *Sayer and Bennett's enlarged Catalogue of new and valuable Prints, in sets, or single; also useful and correct Maps and Charts; likewise Books of architecture,*

views of antiquity, drawing and copy books, etc etc in great variety at No. 53, in Fleet-Street, London; . . . London: Sayer and Bennett, 1775.
Scull, G. D., ed. *The Montresor Journals.* New York: New York Historical Society, 1881.
Senex, John. *A treatise of the description and use of both globes to which is annexed a geographical description of our Earth.* London: John Senex, 1718.
Smith, John. *The art of Painting in Oyl, to which is added The whole art and mystery of coloring maps and other prints with Water colors.* London: Hawes, Clarke & Collis, 1769.
Sterne, Laurence. *The Life and Opinions of Tristram Shandy, Gentleman.* Ed. James Aiken Work. 9 vols. London: R. and J. Dodsley, 1760. Reprint, New York: Odyssey, 1940.
Stiles, Ezra. *The literary diary of Ezra Stiles.* 3 vols. Ed. Frank Dexter. New York: Scribner's Sons, 1901.

SECONDARY SOURCES

Andrews, J. H. "The French school of Dublin land surveyors." *Irish Geography* 5, no. 4 (1967): 275 – 92.
———. *Plantation Acres: an historical study of the Irish Land Surveyor and his maps.* Belfast: Ulster Historical Foundation, 1985.
Barber, Peter. "Maps and monarchs in Europe, 1550 – 1800." In *Royal and republican sovereignty in early modern Europe,* ed. Robert Oresko, G. C. Gibbs, and H. M. Scott, 75 – 124. Cambridge: Cambridge University Press, 1996.
Bendall, Sarah. *Dictionary of land surveyors and local map-makers of Great Britain and Ireland 1530 – 1850.* 2nd ed. London: British Library, 1997.
———. "Estate maps of an English county." In *Rural images: the estate plan in the Old and New Worlds,* ed. David Buisseret, 63 – 90. Chicago: University of Chicago Press, 1996.
Berthaut, Henri. *La Carte de France.* 2 vols. Paris, 1898.
———. *Les ingénieurs-géographes militaires 1624 – 1831.* 2 vols. Paris: Imprimerie du Service géographique, 1902.
Bosse, David. "The Boston map trade of the eighteenth century." In *Mapping Boston,* ed. Alex Krieger and David Cobb, 36 – 55. Boston: Muriel G. and Norman B. Leventhal Family Foundation, 1999.
———. "'To promote useful knowledge': An *Accurate Map of the Four New England States* by John Norman and John Coles." *Imago Mundi* 52 (2000): 143 – 57.
Bradley, Margaret. "The financial basis of French scientific education and scientific institutions in Paris, 1790 – 1815." *Annals of Science* 36 (1979): 451 – 91.
Bridenbaugh, Carl. *Peter Harrison: first American architect.* Chapel Hill: University of North Carolina Press, 1949.
Broc, Numa. "Une affaire de plagiat cartographique sous Louis XIV: le procès Delisle-Nolin." *Revue d'Histoire des Sciences et de leurs applications* 23 (1970): 141 – 47.
———. "Un géographe dans son siècle: Philippe Buache (1700 – 1773)." *Dix-huitième siècle* 3 (1971): 223 – 35.
———. *Les montagnes vues par le géographes et les naturalistes de langue française au XVIIIe siècle.* Paris: Bibliothèque nationale, 1969.

Brooke, John. "The Library of King George III." *Yale University Library Gazette* 52, no. 1 (1977): 33–45.
Brown, Lloyd A. *The story of maps.* New York: Bonanza Books, 1949.
Buisseret, David, ed. *Rural images: estate maps in the Old and New Worlds.* Chicago: University of Chicago Press, 1996.
Bullion, John L. *A great and necessary measure: George Grenville and the genesis of the Stamp Act 1763–1765.* Columbia: University of Missouri Press, 1982.
Carhart, George. "The significance of craft practices for early modern map production." *Imago Mundi* 56 (2004): 194–97.
Casselle, Pierre. "Le commerce des estampes à Paris dans la seconde moitié du 18ème siècle." Unpublished thesis. Ecole nationale des Chartes, Paris, 1976.
Chapin, Howard Millar. *Cartography of Rhode Island.* Providence, 1915.
Chapuis, Olivier. *À la mer comme au ciel: Beautemps-Beaupré et la naissance de l'hydrographie moderne (1700–1850).* Paris: Presses de l'Université de Paris-Sorbonne, 1999.
Chatelus, Jean. "Thèmes picturaux dans les appartements de marchands et artisans Parisiens au XVIIIe siècle." *Dix-huitième siècle* 6 (1974): 302–24.
Chaussinand-Nogaret, Guy. *La noblesse au XVIIIème siècle: de la féodalité aux lumières.* 1976; reprint, Brussels: Editions Complexe, 1984.
Chauvet, Paul. *Les ouvriers du livre en France des origines à la Révolution.* Paris: Presses Universitaires de France, 1959.
Chipman, Willis. "The life and times of Major Samuel Holland, surveyor-general, 1764–1801." *Ontario History* 21 (1924): 11–90.
Clark, Jane. "Metcalf Bowler as a British spy." *Rhode Island Historical Magazine* 23, no. 4 (1930): 101–17.
Clark, Murtie June. *Loyalists in the southern campaign in the Revolutionary War.* 3 vols. Baltimore: Genealogical Publishing, 1981.
Clarke, G. N. G. "Taking possession: the cartouche as cultural text in eighteenth-century maps." *Word and Image* 4, no. 2 (1988): 455–74.
Clayton, Timothy. *The English print: 1688–1802.* New Haven: Yale University Press, 1997.
Clouzot, E. "Une enquête séismologique au XVIIIe siècle." *La Géographie* 24 (1914): 1–22.
Cohn, Ellen. "Benjamin Franklin, Georges-Louis Le Rouge, and the Franklin/Folger chart of the Gulf Stream." *Imago Mundi* 52 (2000): 124–42.
Coldham, Peter Wilson. *American loyalist claims . . . abstracted from the PRO, Audit Office series 13, bundles 1–35 & 37.* Washington: National Genealogical Society, 1980.
Cook, Andrew. "Alexander Dalrymple's *A Collection of Plans of Ports in the East Indies* (1774–1775): a preliminary examination." *Imago Mundi* 33 (1981): 46–64.
Cordier, Henri. "Du Halde et d'Anville (Cartes de la Chine)." In *Recueil de mémoires orientaux par les Professeurs de l'École des Langues orientales,* 389–400. Paris: Leroux, 1905.
Courboin, François. *L'estampe Française.* Brussels: Librairie d'Art et d'Histoire, G. Van Oest & Co., 1914.
Crane, Elaine Forman. *A dependent people: Newport, Rhode Island, in the revolutionary era.* New York: Fordham University Press, 1985.

Crone, G. R. "Further notes on Bradock Mead, alias John Green, an eighteenth century cartographer." *Imago Mundi* 8 (1951): 69–70.
———. "John Green: notes on a neglected eighteenth century geographer and cartographer." *Imago Mundi* 6 (1949): 85–91.
Cumming, William P. *British maps of colonial America.* Chicago: University of Chicago Press, 1974.
Curtis, Edward E. *The organization of the British army in the American Revolution.* New Haven: Yale University Press, 1926.
de Dainville, François, S.J. *La carte de la Guyenne par Belleyme 1761–1840.* Bordeaux: Delmas, 1957.
———. *Cartes anciennes de l'église de France.* Paris: Vrin, 1956.
———. "Cartes de Bourgogne du XVIIIe siècle." In *La cartographie reflet de l'histoire,* ed. Michel Mollat du Jourdin, 59–83. Geneva: Editions Slatkine, 1986.
———. *L'éducation des Jésuites (XVIe–XVIIIe siècles).* Paris: Les Editions de Minuit, 1978.
———. "Enseignement des 'géographes' et des 'géomètres.'" In *Enseignement et diffusion des sciences en France au XVIIIe siècle,* ed. René Taton, 481–91. Paris: Hermann, 1986.
———. *La géographie des humanistes: les Jésuites et l'éducation de la société française.* Paris: Beauchesne, 1940.
———. *Le langage des géographes: termes, signes, couleurs des cartes anciennes 1500–1800.* Paris: Editions A. et J. Picard & Co., 1964.
———. "La levée d'une carte en Languedoc à l'entour de 1730." In *La cartographie reflet de l'histoire,* 361–73. Geneva: Editions Slatkine, 1986.
Danforth, Susan. "The first official maps of Maine and Massachusetts." *Imago Mundi* 35 (1983): 37–57.
Darnton, Robert. *The business of the Enlightenment: A publishing history of the Encyclopédie 1775–1800.* Cambridge: Harvard University Press, 1979.
———. *The literary underground of the old regime.* Cambridge: Harvard University Press, 1982.
Daumas, Maurice. *Les instruments scientifiques au XVIIe et XVIIIe siècles.* Paris: Presses Universitaires de France, 1953.
Davis-Allen, Lisa. "The national palette: painting and map coloring in the seventeenth-century Dutch republic." *Portolan,* spring 1999, 23–36.
Dawson, Nelson-Martin. *L'atelier Delisle: l'Amérique du Nord sur la table à dessin.* Sillery, Quebec: Editions du Septentrion, 2000.
Delano-Smith, Catherine. "The map as commodity." In *Plantejaments I Objectius d'una Història universal de la Cartografia,* ed. David Woodward, Catherine Delano-Smith and Cordell D. K. Yee, 91–109. Barcelona: Institut Cartogràfic de Catalunya, 2001.
Delano-Smith, Catherine, and Roger J. P. Kain. *English maps: a history.* London: British Library, 1999.
Desreumaux, Roger. "Relations entre les arpenteurs et leurs employeurs au XVIIIe siècle dans la région lilloise." *Revue du Nord, histoire et archéologie: Nord de la France, Belgique et Pays Bas,* no. 66: April–September (1984): 531–41.
DeVorsey, Louis. "William Gerard De Brahm." In *Geographers biobibliographical studies,* ed. T. W. Freeman, 41–47. London: Mansell, 1986.

Dion, Marie-Pierre. *Emmanuel de Croÿ (1718–1784): Itinéraire intellectuel et réussite nobiliaire au siècle des Lumières.* Brussels: Editions de l'Université de Bruxelles, 1987.

Dörflinger, Johannes. "Time and cost of copperplate engraving: illustrated by early nineteenth century maps from the Viennese firm Artaria & Co." *Imago Mundi* 35 (1983): 58–66.

Drayperon, Ludovic. "Enquête à instituer sur l'éxécution de la grande carte topographique de France de Cassini de Thury." *Revue de géographie* 38 (1896):1–16.

Dull, Jonathan. *The French navy and American independence.* Princeton: Princeton University Press, 1975.

Edney, Matthew. *John Mitchell's map: an irony of empire* (Web site). Osher Map Library, University of Southern Maine, 1997. http://www.usm.maine.edu/~maps/mitchell/.

———. "Reconsidering Enlightenment geography and mapmaking: reconnaissance, mapping, archive." In *Geography and Enlightenment,* ed. Charles Withers, 165–98. Chicago: University of Chicago Press, 1999.

Egmond, Marco van. "The secrets of long life: the Dutch firm of Covens & Mortier (1685–1866) and their copper plates." *Imago Mundi* 54 (2002): 67–85.

Evans, G. N. D. *Uncommon obdurate: the several public careers of J. F. W. Des Barres.* Salem: Peabody Museum of Salem; Toronto: University of Toronto Press, 1969.

Evans, Ifor M., and Heather Lawrence. *Christopher Saxton: Elizabethan map-maker.* Wakefield: Wakefield Historical Publications and the Holland Press, 1979.

Fuhring, Peter. "The print privilege in eighteenth century France, pt. 1." *Print Quarterly* 2 (1985): 175–93.

———. "The print privilege in eighteenth century France, pt. 2." *Print Quarterly* 3 (1986): 19–33.

Garrioch, David. *Neighbourhood and community in Paris: 1740–1790.* Cambridge: Cambridge University Press, 1986.

Gasnault, Pierre. *Érudition Mauriste du XVIIIe siècle.* Paris: Institut d'études Augustiniennes, 1999.

Godlewska, Anne. *Geography unbound: French geographic science from Cassini to Humboldt.* Chicago: University of Chicago Press, 1999.

Griffiths, Antony. *Prints and printmaking.* London: British Museum, 1980.

Grivel, Marianne. "Le cabinet du roi." *Révue de la Bibliothèque nationale,* no. 18 (1985): 36–58.

Gruber, Ira D. *The Howe brothers and the American Revolution.* New York: Atheneum, 1972.

Guthorn, Peter J. *British maps of the American Revolution.* Monmouth Beach, N.J.: Freneau, 1972.

Hammett, Charles E. *A contribution to the Bibliography and Literature of Newport, R.I. comprising a list of books published or printed in Newport, with notes and additions.* Newport: Charles E. Hammett, Jr., 1887.

Hancock, David. *Citizens of the world: London merchants and the integration of the British Atlantic community, 1735–1785.* Cambridge: Cambridge University Press, 1995.

Harley, J. B. "The bankruptcy of Thomas Jefferys: an episode in the economic history of eighteenth-century map-making." *Imago Mundi* 20 (1966): 27–48.

———. Introduction. In *The County Maps from William Camden's Britannia by Robert Morden (1695): a facsimile.* [London]: David and Charles Reprints, 1972.
———. "John Strachey of Somerset: an antiquarian cartographer of the early eighteenth century." *Cartographic Journal* (June 1966): 2–7.
———. "Maps, knowledge, and power." In *The iconography of landscape: essays on the symbolic representation, design, and use of past environments,* ed. Denis Cosgrove and Stephen Daniels, 277–312. Cambridge: Cambridge University Press, 1988.
———. *The new nature of maps: essays in the history of cartography.* Ed. Paul Laxton. Baltimore: John Hopkins University Press, 2001.
———. "Origins of the Ordnance Survey." In *A history of the Ordnance Survey,* ed. W. A. Seymour, 1–20. Folkestone: Dawson, 1980.
———. "Power and legitimation in the English geographical atlases of the eighteenth century." In *Images of the world: the atlas through history,* ed. John A. Walters and Ronald E. Grim. Washington: Library of Congress, 1997.
———. "The re-mapping of England, 1750–1800." *Imago Mundi* 19 (1965): 56–67.
———. "Silences and secrecy: the hidden agenda of cartography in early modern Europe." *Imago Mundi* 40 (1988): 57–76.
———. "The Society of Arts and the Surveys of English Counties 1759–1809." *Journal of the Royal Society of Arts* 112 (1964): 43–46; 119–24; 269–75; 538–43.
Harley, J. B., and J. C. Harvey. Introduction. In *A survey of the county of Yorkshire by Thomas Jefferys, 1775,* ed. J. B. Harley and J. C. Harvey. Lympne Castle: Harry Margary, 1973.
Harley, J. B., and Donald Hodson. Introduction. In *The Royal English Atlas . . . by Emanuel Bowen and Thomas Kitchin,* 7–14. Newton Abbot: David and Charles Reprints, 1971.
Harley, J. B., and Paul Laxton. Introduction. In *A survey of the County Palatine of Chester by P. P. Burdett (1777).* Liverpool: Historical Society of Lancashire and Cheshire, 1974.
Harley, J. B., and Yolande O'Donoghue. Introduction. In *Old Series Ordnance Survey.* Lympne Castle: Harry Margary, 1975.
Harley, J. B., Barbara Petchenik, and Lawrence Towner. *Mapping the American Revolutionary War.* Chicago: University of Chicago Press, 1978.
Harley, J. B., and Gwyn Walters. "English map collecting 1790–1840: a pilot survey of the evidence in Sotheby sale catalogues." *Imago Mundi* 30 (1978): 31–55.
Harley, J. B., and David Woodward. *The history of cartography.* Vol. 1: *Cartography in prehistoric, ancient, and medieval Europe and the Mediterranean.* Chicago: University of Chicago Press, 1987.
Harris, Michael. "Moses Pitt and insolvency in the London booktrade in the late seventeenth century." In *Economics of the British booktrade 1605–1939,* ed. Michael Harris and Robin Myers. Cambridge: Chadwick-Healy, 1985.
Harvey, Miles. *The island of lost maps.* New York: Random House, 2000.
Harvey, P. D. A. *The history of topographical maps: symbols, pictures, and surveys.* London: Thames & Hudson, 1980.
Heawood, E. A. *Watermarks.* Hilversum, Holland: Paper Publication Society, 1950.
Heinz, Markus. "A research paper on the copperplates of the maps of J. B. Homann's First World Atlas (1707) and a method for identifying different copperplates of identical looking maps." *Imago Mundi* 45 (1993): 45–58.

Heinz, Markus, Michael Diefenbacher, and Ruth Bach-Damaskinos. *"Auserlesene und allerneueste Landkarten": Der Verlag Homann in Nürnberg 1702–1848.* Nuremberg: W. Tümmels, 2002.
Heusser, Albert H. *George Washington's mapmaker: a biography of Robert Erskine.* 1966 ed. New Brunswick: Rutgers University Press, 1928.
Hodson, Donald. *County atlases of the British Isles published after 1703.* 3 vols. Welwyn: Tewin Press, 1984–97.
Hofmann, Catherine. "L'enlumineure des cartes et des atlas imprimés, XVI–XVIIIe siècles." *Bulletin du comité français de cartographie* 159 (March 1999): 35–47.
Holland, T. R. "Yeakell & Gardner maps of Sussex 1778–83." *Sussex Archaeological Collections* 95 (1957): 94–104.
Howgego, James. *Printed maps of London circa 1553–1850.* 2nd ed. London: Dawson, 1978.
Hunter, David. "Copyright protection for engravings and maps in eighteenth century Britain." *Library* 9, no. 2 (1987): 128–47.
Hyde, Ralph. "The making of John Rocque's map." In *A-Z of Georgian London,* v–viii. London: London Topographical Society, 1982.
Jacob, Christian. *L'empire des cartes: approche théorique de la cartographie à travers l'histoire.* Paris: Albin Michel, 1992.
Johns, Adrian. *The nature of the book: print and knowledge in the making.* Chicago: University of Chicago Press, 1998.
Johnston, William Dawson. *History of the Library of Congress.* Vol. 1: 1800–1864. Washington: Government Printing Office, 1904.
Jolly, David C. *Maps in British periodicals, pt. 1: Major monthlies before 1800.* Brookline, Mass.: David C. Jolly, 1990.
Jones, Colin. *Madame de Pompadour: images of a mistress.* London: National Gallery, 2002.
Karrow, Robert. "Lake Superior's mythic isles: a cautionary tale for users of old maps." *Michigan History* 69, no. 1 (1985): 24–31.
Kish, George. "Early thematic mapping: the work of Philippe Buache." *Imago Mundi* 28 (1976): 129–36.
Koeman, Cornelis. *Atlantes Neerlandici.* 6 vols. Amsterdam: Theatrum Orbis Terrarum, 1967–1985. Currently being revised by Peter van der Krogt as *Koeman's Atlantes Neerlandici,* 3 vols. ('t Goy-Houten: HES Publishers, 1997–).
Konvitz, Josef. *Cartography in France, 1660–1848: science, engineering, and statecraft.* Chicago: University of Chicago Press, 1987.
Krieger, Alex, David Cobb, and Amy Turner, eds. *Mapping Boston.* Boston: Muriel G. and Norman B. Leventhal Family Foundation, 1999.
Krogt, Peter van der. *Globi Neerlandici.* Utrecht: HES, 1993.
Laxton, Paul. Introduction. In *Two hundred and fifty years of map-making in the County of Hampshire.* Lympne Castle, Kent: Harry Margary, 1976.
Le Bitouzé, Corinne. "Le commerce de l'estampe à Paris dans la première moitié du XVIIIe siécle." 2 vols. Unpublished thesis. Ecole nationale des Chartes, Paris, 1986.
Le Guisquet, Bernard. "Le Dépôt des cartes, plans et journaux de la Marine sous l'ancien régime (1720–1789)." *Annales Hydrographiques* 18, no. 765 (1992): 1–27.

Lemoine-Isabeau, Claire. "Note sur la gravure de la carte de Ferraris." *Gazette des Beaux-Arts* 80 (1972): 371–74.

Lippincott, Louise. "Arthur Pond's journal of receipts and expenses, 1734–1750." *Walpole Society* 54 (1988): 220–333.

———. *Selling art in Georgian London: the rise of Arthur Pond.* New Haven: Yale University Press, 1983.

Machemer, Grace S. "Headquartered at Piscataqua: Samuel Holland's coastal and inland surveys, 1770–1774." *Historical New Hampshire* 57, nos. 1 & 2 (2003): 4–25.

Marcel, Gabriel. "Correspondance de Michel Hennin et de d'Anville." *Bulletin de la Section de Géographie Historique et Descriptive* 3 (1907): 441ff.

———. "À propos de la Carte des Chasses." *Revue de géographie* (1897): 3–15.

Marion, Michel. *Bibliothèques privées à Paris au milieu du XVIIIe siècle.* Paris: Bibliothèque nationale, 1978.

Marshall, Douglas W. "Military maps of the eighteenth century and the Tower of London drawing room." *Imago Mundi* 32 (1980): 21–44.

Marshall, Douglas W., and Howard H. Peckham. *Campaigns of the American Revolution: an atlas of manuscript maps.* Ann Arbor: University of Michigan Press, 1976.

Martin Saint-Léon, Etienne. *Histoire des Corporations de Métiers depuis leurs origines jusqu'à leur suppression en 1791.* Paris, 1922.

McCusker, John J. *Money and exchange in Europe and America, 1600–1775.* Chapel Hill: University of North Carolina Press, 1978.

McKendrick, Neil, John Brewer, and J. N. Plumb. *The birth of the consumer society: the commercialization of eighteenth-century England.* Bloomington: Indiana University Press, 1982.

Meinig, D. W. *The shaping of America: a geographical perspective on 500 years of history.* Vol. 1: *Atlantic America, 1492–1800.* New Haven: Yale University Press, 1986.

Meyer, Jean. *La Noblesse bretonne au xviiie siècle.* Paris: S.E.V.P.E.N., 1966.

Michel, Christian. *Charles-Nicolas Cochin et l'art des Lumières.* Rome: École Française de Rome, 1993.

———. *Charles-Nicolas Cochin et le livre illustré au XVIIIe siècle.* Geneva: Librairie Droz, 1987.

Minard, Philippe. "Agitation in the work force." In *Revolution in print: the press in France 1775–1800,* ed. Robert Darnton and Daniel Roche, 107–23. Berkeley: University of California Press, 1989.

Monmonier, Mark. *How to lie with maps.* Chicago: University of Chicago Press, 1991.

Nebenzahl, Kenneth, and Don Higginbotham. *The atlas of the American Revolution.* Chicago: Rand McNally, 1974.

Oxford universal English dictionary. Ed. William Little, H. W. Fowler and J. Coulson. Oxford: Oxford University Press, 1937.

Pastoureau, Mireille. *Les atlas français XVIe–XVIIe siècles.* Paris: Bibliothèque nationale, 1984.

———. "Confection et commerce des cartes à Paris aux XVIe et XVIIe siècles." In *La carte manuscrite et imprimée du XVIe au XIXe siècle: Journée d'étude sur l'histoire du livre et des documents graphiques,* ed. Frédéric Barbier, 9–17. Munich: Saur, 1983.

———. *Nicolas Sanson d'Abbeville: atlas du monde 1665.* Paris: Sand & Conti, 1988.

Pedley, Mary Sponberg. *Bel et utile: the work of the Robert de Vaugondy family of mapmakers.* Tring: Map Collector Publications, 1992.

———. "'Commode, complet, uniforme, et suivi': problems in atlas editing in Enlightenment France." In *Editing early and historical atlases,* ed. Joan Winearls, 83–108. Toronto: University of Toronto Press, 1995.

———. "Gentlemen abroad: Jefferys and Sayer in Paris." *Map Collector* 37 (1986): 20–23.

———. "'I due valentuomini indefessi': Christopher Maire and Roger Boscovich and the mapping of the Papal States (1750–1755)." *Imago Mundi* 45 (1993): 59–76.

———. "The map trade in Paris, 1650–1825." *Imago Mundi* 33 (1981): 33–45.

———. "Map wars: the role of maps in the Nova Scotia/Acadia boundary disputes of 1750." *Imago Mundi* 50 (1998): 96–104.

———. "New light on an old atlas: documents concerning the publication of the *Atlas universel* (1757)." *Imago Mundi* 36 (1984): 48–63.

———. "The subscription list of the 1757 *Atlas universel:* a study in cartographic dissemination." *Imago Mundi* 31 (1979): 66–77.

———, ed. *The map trade in the late eighteenth century: letters to the London map sellers Jefferys and Faden.* Oxford: Voltaire Foundation, 2000.

Pelletier, Monique. *La carte de Cassini: l'extraordinaire aventure de la Carte de France.* Paris: Presses de l'école nationale des Ponts et Chaussées, 1990.

———. *Les cartes des Cassini: la science au service de l'État et des régions.* Paris: Comité des travaux historiques et scientifiques, 2002.

Petto, Christine. "Mapping the body politic in early modern France." Ph.D. thesis, Indiana University, 1966.

Phillips, A. D. M. "Introduction." In *A map of the county of Stafford from an actual survey begun in the year 1769 and finished in 1775 [by William Yates].* Lympne Castle: Harry Margary, 1984.

Phillips, Philip Lee. *Notes on the life and works of Bernard Romans.* Deland: Florida State Historical Society, 1924.

Potter, Jonathan. *Antique maps from Jonathan Potter [catalogue 22].* London: Jonathan Potter, 2003.

Prest, Wilfrid. *Albion ascendant: English history, 1660–1815.* Oxford: Oxford University Press, 1998.

Raven, James. *Judging new wealth: popular publishing and responses to commerce in England, 1750–1800.* Oxford: Clarendon Press, 1992.

Ravenhill, William. "Joel Gascoyne, a pioneer of large scale county mapping." *Imago Mundi* 26 (1972): 60–70.

———. "Joel Gascoyne's Stepney: his last years in pastures old yet new." *Guildhall studies in London history* 2, no. 4 (1977): 200–15.

———. *A map of the County of Devon . . . by Benjamin Donn.* Exeter: Devon and Cornwall Record Society and the University of Exeter, 1965.

Reinhartz, Dennis. *The cartographer and the literati: Herman Moll and his intellectual circle.* Lewiston: Edwin Mellen Press, 1997.

———. "An exploration of the 'silences' on various late eighteenth-century maps of northern New Spain." *Terrae Incognitae* 35 (2003): 43–53.

Reynaud, Jean. "Les Cassini et la carte de Provence (1776–1790)." *Bulletin de la Section de Géographie* 37 (1922): 113–26.
Ristow, Walter W. *American maps and mapmakers: commercial cartography in the nineteenth century.* Detroit: Wayne State University Press, 1985.
Robinson, A. H. W. *Marine cartography in Britain.* Leicester: Leicester University Press, 1972.
Robinson, F. J. G., and P. J. Wallis. *Book subscription lists: a revised guide.* Newcastle upon Tyne: Harold Hill & Son Ltd for the Project for Historical Bibliography, 1975.
Roche, Daniel. *A history of everyday things: the birth of consumption in France, 1600–1800.* Translated by Brian Pearce. Cambridge: Cambridge University Press, 2000.
———. *Les républicains des lettres: gens de culture et lumières au XVIIIe siècle.* Paris: Librairie Arthème Fayard, 1988.
Rodger, Elizabeth. "An eighteenth century collection of maps connected with Philippe Buache." *Bodleian Library Record* 7, no. 2 (1963): 96–106.
———. *Large scale county maps of the British Isles 1596–1850: A union list.* Oxford: Bodleian Library, 1972.
Rostenberg, Leona. "Moses Pitt, Robert Hooke, and the English atlas." *Map Collector* 12 (1980): 2–8.
Sandler, Christian. *Die Reformation der Kartographie zum 1700.* Munich: R. Oldenbourg, 1905.
Schilder, Günter. *Dutch folio-sized single sheet maps with decorative borders, 1604–60.* Alphen aan den Rijn: Uitgeverij Canaletto/Repro-Holland, 2000.
———. *Monumenta cartographica Neerlandica.* 7 vols. Alphen aan den Rijn: Uitgeverij Canaletto, 1986–2003.
Scott, Katie. *The rococo interior.* New Haven: Yale University Press, 1995.
Seymour, W. A. *A history of the Ordnance Survey.* Folkestone: Dawson, 1980.
Shennan, J. H. *Philippe, duke of Orléans: regent of France 1715–1723.* London: Thames and Hudson, 1979.
Shipton, Nathaniel. "General James Murray's map of the St Lawrence." *Cartographer* 4, no. 2 (1967): 93–101.
Skelton, R. A. "Copyright and piracy in eighteenth century chart publication." *Mariners Mirror* 46 (1960): 207–12.
———. *County atlases of the British Isles, 1579–1850: a bibliography.* Folkestone: Dawson, 1978.
———. *James Cook: surveyor of Newfoundland.* San Francisco: David Magee, 1965.
———. "The origins of the Ordnance Survey of Great Britain." *Geographical Journal* 128 (1962): 415–26.
Sonenscher, Michael. *Work and wages: natural law, politics, and the eighteenth century French trades.* Cambridge: Cambridge University Press, 1989.
Stroup, Alice. "Le comté Venaissin (1696) of Jean Bonfa, S.J.: a paradoxical map by an accidental cartographer." *Imago Mundi* 47 (1995): 118–37.
Stuart Mason, A., and Peter Barber. "'Captain Thomas, the French engineer': and the teachings of Vauban to the English." *Proceedings of the Huguenot Society* 25, no. 3 (1991): 279–87.
Stuart Mason, A., and John Bensusan-Butt. "P. B. Scalé, surveyor in Ireland, gentleman of Essex." *Proceedings of the Huguenot Society* 24, no. 6 (1988): 508–18.

Tardieu, Alexandre. "Notice sur les Tardieu, les Cochin et les Belle, graveurs et peintres." *Archives de l'art Français: recueil de documents inedits* 4 (1876): 49–68.

Taton, René. *Enseignement et diffusion des sciences en France au XVIIIe siècle.* Paris: Hermann, 1986.

Taylor, E. G. R. "The 'English Atlas' of Moses Pitt 1680–1683." *Geographical Journal* 95 (1940): 292–99.

———. *Late Tudor and early Stuart geography, 1583–1650: a sequel to Tudor geography, 1485–1583.* 1934; reprint, New York: Octagon, 1968.

———. "Notes on John Adams and contemporary map makers." *Geographical Journal* 97 (1941): 182–84.

———. "Robert Hooke and the cartographical projects of the late seventeenth century (1666–1696)." *Geographical Journal* 90 (1937): 529–40.

Thompson, Edmund. *Maps of Connecticut before the year 1800.* Windham, Conn.: Hawthorn House, 1940.

Thorpe, F. J. "Holland, Samuel Johannes." In *Dictionary of Canadian biography,* 425–29. Toronto: University of Toronto Press, 1983.

Thrower, Norman J. W. "Edmond Halley and thematic geo-cartography." In *The compleat plattmaker: essays on chart, map, and globe making in England in the seventeenth and eighteenth centuries,* ed. Norman J. W. Thrower, 195–228. Berkeley: University of California Press, 1978.

———. *Maps and civilization.* 2nd ed. Chicago: University of Chicago Press, 1996.

———, ed. *The compleat plattmaker: essays on chart, map, and globe making in England in the seventeenth and eighteenth centuries.* Berkeley: University of California Press, 1978.

Tillyard, Stella. *The aristocrats: Caroline, Emily, Louisa, and Sarah Lennox 1740–1832.* New York: Farrar, Straus and Giroux, 1994.

Tooley, R. V. *The French mapping of the Americas: the Delisle, Buache, Dezauche succession (1700–1830).* London: Map Collectors' Circle, 1967.

Tyacke, Sarah. *London map-sellers 1660–1720.* Tring: Map Collector Publications, 1978.

———. "Map-sellers and the London map trade c. 1650–1710." In *My head is a map: essays and memoirs in honour of R. V. Tooley,* ed. Helen Wallis and Sarah Tyacke, 63–80. London: Francis Edwards and Carta Press, 1973.

Valerio, Vladimiro. *Società uomini e istituzioni cartografiche nel mezzogiorno d'Italia.* Florence: Istituto Geografico Militare, 1993.

Valpy, Nancy. "Plagiarism in prints: the *Musical Entertainer* affair." *Print Quarterly* 6, no. 1 (1989): 54–59.

Véchambre, Michèle. "Lire la carte topographique à 15 ans." In *L'oeil du cartographe et la représentation géographique du Moyen Age à nos jours,* ed. Catherine Bousquet-Bressolier, 269–77. Paris: Comité des travaux historiques et scientifiques, 1995.

Verner, Coolie. "Copperplate printing." In *Five centuries of map printing,* ed. David Woodward, 51–75. Chicago: University of Chicago Press, 1975.

———. "Mr. Jefferson makes a map." *Imago Mundi* 14 (1959): 96–108.

Vuacheux, Ferdinand. "Un mémoire et une lettre du géographe Philippe Buache (1763)." *Bulletin de la Section de Géographie* 16 (1901): 269–74.

Welsh, Charles. *A bookseller of the last century: being some account of the life of John Newbery and of the books he published.* London, 1885.

Withers, Charles W. J. *Geography, science, and national identity: Scotland since 1520.* Cambridge: Cambridge University Press, 2001.

———. "The social nature of map making in the Scottish Enlightenment, c. 1682–c. 1832." *Imago Mundi* 54 (2002): 46–66.

Wood, Denis. *The power of maps.* New York: Guilford Press, 1992.

Woodward, David. "English cartography, 1650–1750." In *The compleat plattmaker: essays on chart, map, and globe making in England in the seventeenth and eighteenth centuries,* ed. Norman J. W. Thrower, 159–93. Berkeley: University of California Press, 1978.

———. *Maps as prints in the Italian Renaissance: makers, distributors, and consumers.* London: British Library, 1996.

Worms, Laurence. "Location in the London map trade: a talk given to members of IMCOS on 3rd June 2000." *International Map Collectors' Society Journal* 82 (autumn 2000): 33–42.

———. "Thomas Kitchin's 'journey of life': hydrographer to George III, mapmaker, and engraver." *Map Collector* 62 (1993): 2–8, 14–20.

———. "William Faden." In *Dictionary of National Biography: Missing Persons,* 218. Oxford: Oxford University Press, 1993.

Zupko, Ronald Edward. *French weights and measures before the Revolution.* Bloomington: Indiana University Press, 1978.

INDEX

DISCARD